Rhetoric's Earthly Realm

LAUER SERIES IN RHETORIC AND COMPOSITION
Series Editors: Catherine Hobbs, Patricia Sullivan, Thomas Rickert, and Jennifer Bay

The Lauer Series in Rhetoric and Composition honors the contributions Janice Lauer Hutton has made to the emergence of Rhetoric and Composition as a disciplinary study. It publishes scholarship that carries on Professor Lauer's varied work in the history of written rhetoric, disciplinarity in composition studies, contemporary pedagogical theory, and written literacy theory and research.

OTHER BOOKS IN THE SERIES

Techne, from Neoclassicism to Postmodernism: Understanding Writing as a Useful, Teachable Art, by Kelly Pender (2011)

Greek Rhetoric Before Aristotle, , Revised and Expanded Edition, by Richard Leo Enos (2011)

Walking and Talking Feminist Rhetorics: Landmark Essays and Controversies, edited by Lindal Buchanan and Kathleen J. Ryan (2010)

Transforming English Studies: New Voices in an Emerging Genre, edited by Lori Ostergaard, Jeff Ludwig, and Jim Nugent (2009)

Ancient Non-Greek Rhetorics, edited by Carol S. Lipson and Roberta A. Binkley (2009)

Roman Rhetoric: Revolution and the Greek Influence, Revised and Expanded Edition, by Richard Leo Enos (2008)

Stories of Mentoring: Theory and Praxis, edited by Michelle F. Eble and Lynée Lewis Gaillet (2008)

Writers Without Borders: Writing and Teaching in Troubled Times by Lynn Z. Bloom (2008)

1977: A Cultural Moment in Composition, by Brent Henze, Jack Selzer, and Wendy Sharer (2008)

The Promise and Perils of Writing Program Administration, edited by Theresa Enos and Shane Borrowman (2008)

Untenured Faculty as Writing Program Administrators: Institutional Practices and Politics, edited by Debra Frank Dew and Alice Horning (2007)

Networked Process: Dissolving Boundaries of Process and Post-Process, by Helen Foster (2007)

Composing a Community: A History of Writing Across the Curriculum, edited by Susan H. McLeod and Margot Iris Soven (2006)

Historical Studies of Writing Program Administration: Individuals, Communities, and the Formation of a Discipline, edited by Barbara L'Eplattenier and Lisa Mastrangelo (2004). Winner of the WPA Best Book Award for 2004–2005.

Rhetorics, Poetics, and Cultures: Refiguring College English Studies (Expanded Edition) by James A. Berlin (2003

RHETORIC'S EARTHLY REALM

Heidegger, Sophistry, and the Gorgian Kairos

Bernard Alan Miller

Parlor Press
Anderson, South Carolina
www.parlorpress.com

Parlor Press LLC, Anderson, South Carolina, USA

© 2011 by Parlor Press
All rights reserved.
Printed in the United States of America

SAN: 254-8879

Library of Congress Cataloging-in-Publication Data

Miller, Bernard, 1946-
  Rhetoric's earthly realm : Heidegger, sophistry, and the Gorgian kairos / Bernard Miller.
    p. cm.
  Includes bibliographical references (p.   ) and index.
  ISBN 978-1-60235-147-9 (pbk. : alk. paper) -- ISBN 978-1-60235-148-6 (hardcover : alk. paper) -- ISBN 978-1-60235-149-3 (adobe ebook : alk. paper) -- ISBN 978-1-60235-211-7 (epub : alk. paper)
  1. Heidegger, Martin, 1889-1976. 2. Sophists (Greek philosophy) 3. Time--Philosophy. I. Title.
  B3279.H49M494 2011
  111'.1--dc22
                            2011010080

The initial cover design was the work of Jean Kearns Miller.
Printed on acid-free paper.

Parlor Press, LLC is an independent publisher of scholarly and trade titles in print and multimedia formats. This book is available in paper, cloth and Adobe eBook formats from Parlor Press on the World Wide Web at http://www.parlorpress.com or through online and brick-and-mortar bookstores. For submission information or to find out about Parlor Press publications, write to Parlor Press, 3015 Brackenberry Drive, Anderson, South Carolina, 29621, or email editor@parlorpress.com.

Contents

Acknowledgments *vii*

1 Introduction: Earthly Realms and the Pre-Socratic Mystery *3*

2 The Platonic *Kairos* *34*

3 The Gorgian *Kairos* *74*

4 *Das Sein, Dasein,* and *Doxa*: Attending to the Way of Heidegger's Thought *117*

5 Heidegger and the Gorgian *Kairos* *166*

6 Paradox and the Power of the Possible: *Kairos* as the Mark of the Trickster *237*

Notes *321*
Works Cited *361*
Index *379*
About the Author *387*

Acknowledgments

This book is the result of the efforts of a number of people. Whatever is worthwhile in it I owe to them; all that is otherwise is my responsibility alone. Therefore, I express my appreciation for the suggestions and insights I have gathered on this topic of *kairos* and Sophistic rhetoric in discussions with Susan Evenson, Virgil Lokke, Bud Weiser, Rob Carr, Linda Mercer Learman, Maryam Barrie, Gwen Fleming, Larry Juchartz, and Jack Barden. And to Calvin Schrag, who without fail brought to all our discussion a fabulous mix of enthusiasm, kindly temperament, and boundless intelligence. In this regard I also express my gratitude to the Hunkpapa Lakota of the Standing Rock Indian Reservation of North Dakota. I came to them a tempest-tossed refugee from the US Army and the regimented enlightenment of academia, and years later I left re-imagining all I think and know through what I learned from them of the American Indian's experience and literature.

I also express my thanks to David Blakesley, the editor of Parlor Press. Given all my rewrites, revisions, and delays, I came to realize if patience is a virtue, as surely it is, he is a very virtuous person indeed. He is also a very capable and helpful editor. This manuscript has also benefited considerably from the critical reading given by Thomas Richert, Purdue University and Editor of the Lauer Series of Parlor Press. His encouragement convinced me the manuscript was worth completing, and his suggestions have made it a better document from top to bottom. This book has also benefited from the copy editing of Ethan Sproat, whose knowledge of the MLA style sheet and diligence in applying it to this script went far beyond what I would ever be able to muster. Finally, I wish to thank my wife, Jean, and my kids, Adrian and Cassandra, all of whom have taught me there are far more important things in the world than writing a book, even one about the Greeks of antiquity. Lest I forget, Jean, Adrian, and Cassie are here and now, and those Greeks are long gone, with neither wife nor children of mine among them.

Inasmuch as the support of an academic institution is essential to complete a project of this sort, I expressed my appreciation to Purdue University, in particular its splendid program in rhetoric and composition. And here I close with my acknowledgement of the debt I owe my teacher, mentor, friend, and Godmother to my son, Janice Lauer. This manuscript grew out of a presentation I gave many years ago at a RSA conference in Arlington, Texas, and from that presentation to my PhD dissertation, and now, at long last, to this book. The distances between each, in time and content, seem best measured in light years, and for all these years Janice has been with me and this book every step of the way. I express my deepest appreciation to her.

Rhetoric's Earthly Realm

God is day and night, winter and summer, war and peace, surfeit and hunger, changing as fire consuming incense, each being called according to its scent.
—Heraclitus

Tree and grass, eagle and bull, snake and cricket first enter into their distinctive shapes and thus come to appear as what they are. The Greeks early called this emerging and rising in itself and in all things physis . . . We call this ground earth.
—Martin Heidegger, Poetry, Language, and Thought

Where language touches the earth, there is the holy, there is the sacred. In our deeper intelligence we know this: that names and being are indivisible.
—N. Scott Momaday, Man Made of Words

1 Introduction: Earthly Realms and the Pre-Socratic Mystery

The philosophy of the Greeks conquered the Western world not on the strength of what it was in the original beginning but through what it had become towards the end of the beginning, in its incipient end.

—Martin Heidegger, *Introduction to Metaphysics*[1]

Nietzsche says the will to truth is the essential prejudice of philosophers. If the remark begs the question of its own truth, it is nevertheless the case that the practice of discerning truth, cleaving it from mere appearances, has been the purpose of philosophers since Plato. And often, in more prosaic ways, the matter can seem as compelling for the rest of us. When I drive to campus looking for a place to park, I am often beset by the recklessness of pedestrians who gather in hordes at intersections, edging their way into traffic, and others who line the streets and then singly or in groups of two or three scurry across, dodging cars of rushed and harried drivers such as myself. Yet, once I am parked and become a pedestrian myself, I step into a different world. I am then no less beset by the recklessness of drivers as I join the hordes, rushed and harried once more, waiting for my chance to dart out in front of traffic to cross streets that seem to require such risks to be crossed at all. And only afterwards, in the relative quiet of the workday, I reflect upon dangers and near-tragedies, presuming to discern a truth of my own as I come to terms with the prime ingredients of deconstruction, hermeneutics, and the entire poststructuralist enterprise: that is to say, philosophers to the contrary, truth and enduring realities are impossible to come by, that all things are necessarily matters of

3

perspective, just as they so evidently appear to be in the course of the morning commute.

To be sure, if ever I suffer some misadventure and am run down bolting off to class, it would seem that I—or someone on my behalf—will likely be able to speak of this particular issue with more certainty than mere perspective would allow. But just as likely not. If the trip to work is never so harrowing, the point remains if there is some determinate truth of this or any other matter in the world, I am not sure that we can ever come to know it. As far as I am aware, there have never been any serious injuries associated with campus traffic at this place, and perhaps never will be. And even if tragedy should one day strike, as it surely has elsewhere, the nature of the danger would continue to remain open to perspective, to say nothing of what brought it about, who might be at fault for it, what might be done to correct it. So it would seem on the basis of this example, prosaic as it may be, that the world is as it appears to be, and whatever possibly exists beyond appearances is, in particular, open to perspective. Of course, things are far more complicated than my narrative could ever indicate, but to pursue the poststructuralist ethic to its bitter end is to conclude that there is no place of pure and perfect truth to which we have access, even though, under any given set of circumstances, some perspectives can be judged more viable or probable than others.

The ambiguities involved in such matters have been traditionally a concern of rhetoric, leading some to suggest that insofar as they are now embedded in philosophy by virtue of poststructuralist thought, the usefulness of the distinction between rhetoric and philosophy is diminished or undone entirely. Yet these ambiguities have been traditionally a concern of philosophy as well. One of the most celebrated and consequential challenges to the idea of a realm of ultimate reality underlying and explaining surface features given in appearances is evident in Immanuel Kant's contention that "things-in-themselves" constitute reality, that they reside in a transcendental noumenal world and are, therefore, not only unknown but intrinsically unknowable to us. We can become aware of them only through appearances in this world or how they are given in *phenomena*, as they are, thereby, "things-for-us." Hence, lacking any traditional approach to stable reality, we can know things only from particular and therefore inevitably limited perspectives. That Kant should postulate the existence of a noumenal world at all was part of his peculiar response to rationalist metaphysics of the seventeenth and eighteenth centuries, yet one wholly consistent also with his unshaken belief that appearances must themselves be grounded in a reality that is not itself comprised of mere appearances.

In any event, Kant's distinction between "things-in-themselves" and "things-for-us" is the basis of what is called the "Copernican Revolution." That idea is now, peculiarly, the philosopher's stone of the poststructuralists, though it was originally Kant's as well. In the Second Preface to *The Critique of Pure Reason*, he says that we learn from the Copernican hypothesis that nature is constrained "to give answer to questions [only] of reason's own determining," and despite the "happy thought" that reason seeks and achieves untainted truth in nature, what reason actually finds in nature is first put there by reason itself (20). All along we have assumed on the basis of the classical idea of *logos* that our ways of knowing conform expressly to the nature of what is known, but here Kant insists the meaning of any experience is the result of our manner of perceiving it, which is, in turn, a product of predispositions, or perspectives, that shape our experience and, in effect, implicate the knower within the known. Thus, what we know is circumscribed, limited by our ways of knowing, by perspectives that come prior to any particular experience or encounter, and these ways of knowing include not only social, cultural, and political imperatives, but those of the natural sciences as well—all of which create and channel our capacity for experience.

For the philosopher, and surely for us all, these are implications that go beyond issues of whether we see the world from the curb or from behind the steering column of a car. Indeed, the Copernican Revolution puts in doubt the warrant of the Western philosophical tradition, and Kant's *Critique* was an effort to deal precisely with that problem, to quite literally reconstruct metaphysics from the ground up. Why not let the observer revolve, he asks, and leave the stars at rest? As we deal only with the essentials of his argument, it is as follows: whether the sun revolves around the earth, as Ptolemy said, or the earth revolves around the sun, as Copernicus said, is *wholly* a matter of perspective. Barbara Johnson puts it more concisely, saying that "the idea that the earth goes around the sun is not an *improvement* of the idea that the sun goes around the earth. It is a shift in perspective which literally makes the ground move."[2]

Indeed, this take on the Copernican Revolution is centuries old, if not in some ways more ancient yet, and, as I say, has become central to the analyses of the poststructuralists. But no thinker, ancient or modern, has realized its implications as thoroughly and ably as Martin Heidegger. To Heidegger the issue of perspective utterly shifts the focus of philosophic inquiry from specific things-in-themselves to the grounds of their being things-for-us in the phenomenal world. The nature of Being, traditionally conceived as some transcendental idea or essence, is for Heidegger active alone in the world of appearances, in the here and now of the *Lebenswelt*. As such, his philosophy has been described as an "ontology of immanence," where he claims that "appearing is the very essence of Being" (*Introduction* 101), as appearing and Being coincide, thereby being the same rather than the former a mere rudiment or reflection of the other. The traditional concept of time as an indistinguishable sequence of "nows" is discarded as well, so that truth and reality, insofar as we can know them, are confined to an "horizon of temporality" where not only the appearance of things has precedence over whatever is considered to stand behind appearances, but the latter is effaced altogether, to include all principles, ideas, and eternal verities associated with it. Accordingly, the matter of perspective is reconfigured in Heidegger, as it is here rooted in a peculiar phenomenology, in a process of letting things become manifest themselves in their revealing and concealing, where objects "show themselves" and thereby encounter us as much as we encounter them. Indeed, to speak of the various imperatives or even preeminence of perspectives is all too often to affirm the priority of the self, of consciousness, ego, or soul. But in this view the sense of self is necessarily reconfigured as well, made exclusively dependent upon prior perspectives or historically based predispositions. Here, all fundamental issues of our thinking are construed ontologically as much as they are epistemologically, which for Heidegger gives rise to questions concerning the very grounds and nature of human knowledge and reason, and, ultimately, for Heidegger, to the question concerning the meaning of Being—*Die Frage nach dem Sinn von Sein*—that he announces at the very outset of his stellar *Being and Time*. The question is basic to Heidegger's thought, and persistent throughout his corpus. It takes its measure altogether from the mystery of Being, and is at once prosaic and profound: "Why is there something rather than nothing?" he asks.

This is, at any rate, the gloss I presently put on Heidegger's ontology. What is of more immediate importance is to stress its relation to poststructuralism, the movement Heidegger did much to inspire.[3]

In the most general sense, poststructuralism is an activity undermining the presumptions, propositions, and ideas that we regularly and often implicitly accept about the world. What is more, it challenges the intent of traditional philosophy, its very *raison d'etre*, condemning as futile the quest for certainty or pure knowledge based on inviolate first principles that would explain or themselves embody the truth or reality of things; and the history of philosophy is replete with such principles, postulated and discarded in turn, to be replaced with yet others that are discarded as well, by principles of essence, existence, substance, consciousness, will to power, and so on. And if, throughout these shifts and changes, the idea of first principles at all times endures in the perspective of traditional philosophy, it is never clear with Heidegger whether poststructuralism is indeed a way of thinking that comes after traditional philosophy or is instead a return to something prior to it, far more traditional yet.

Realms of Life and Growth

Hesiod tells us that back in the Golden Age all humankind ever desired was to live on friendly terms with the blessed immortals. Indeed, to be numbered among them was Plato's noble aspiration for us all, or at least those such as himself, the philosopher kings among us. In this quest he is the archetypal structuralist, providing in his dialogues a series of themes the poststructuralists subject to censure. In the *Philebus* he says the philosopher's calling is to achieve realization of *what is*, to gain pure knowledge of things fixed in perpetuity without flux or mixture, the essence or orderliness that in his vision of the ideal world would be rendered in perfect clarity by the ideal philosopher (58a–59c). Moreover, the philosopher's pure art of truth can neither be achieved nor even adequately expressed in words, and thereby stands in opposition to the work of the Sophists, what Plato denounces as the rhetoricians' secondary and inferior art of discourse. Philosophy fixes our gaze upward to the heavens, as well it should, for the earth harbors in contrast the nether regions of malevolence and decay, at worst the evils and agonies of the flesh that poets in myths and tales of harrowing have depicted as graphically as any theologian or inquisitor. As language is distinct from the ideal reality of what is enduring and true, it partakes of the earth in particular, a world of mere appearances condemned by Plato as corporeal and corrupt. Therefore rhetoric, surely Sophistic rhetoric, shares in this legacy, at times reviled as akin to death itself. It is as cosmetics in dealing only with appearances, being as the art of the mortician in this respect. It is "a mischievous, deceitful, mean, and ignoble activity," Plato says, where by "shapes and colors" and "smoothing and draping" things take on "an alien charm," obscuring or concealing reality altogether (*Gorgias* 465b–c).

As would any good structuralist, Plato establishes a *privileged* perspective, a point of absolute meaning outside and above the world of discourse that gives meaning to the whole. As I will attempt to establish, it is to such a rarefied region that Plato's dialectic leads. Despite contentions of some rhetoricians to the effect that the dialectic is manifest in open, benevolent disputation, in the mutual and unconstrained verbal give and take of questions and answers, the way Plato himself describes the dialectic in the *Phaedrus* and *Republic*, and here more cogently in the *Seventh Letter*, the process carries us beyond mere sense perception, specifically beyond language, and ends in a "flash of understanding," with the mind "flooded with light" (344b). Plato's attack on the Sophists is itself a mere implication of this goal, a wayside skirmish brought on by the philosopher's essential responsibility to do battle against the bewitching powers of language. Invariably, he seeks to serve the philosopher's classical ideal of "*aneu logou*," the truth that resides beyond words.[4]

Heidegger identifies the philosopher's faith in certainty as the "metaphysics of presence," that at this point can be defined as the acceptance of some basic truth or reality that is itself outside the play of language but to which language must ultimately refer. But his poststructuralism, evident in so many ways by a peculiar paradigm he calls "earth," crushes the philosopher's stone outright, and from the dust of the settling debris it should come as little surprise that language looms in the ascendant, distinctly triumphant. As Anthony Cascardi says, poststructuralism, specifically in its variant of modern hermeneutics, focuses on language "both as an instrument which produces understanding in us and as a form of life in which we make ourselves understood" (225). Heidegger says this much, and far more. In his phenomenology particular emphasis is necessarily given to the priority of language, of it functioning by its very nature as a "form of life." Not only does language bring things into appearance and thereby reveal the world to us, but to Heidegger it discloses us to the world, to the degree that human nature itself "takes place out of the speaking of language." Indeed, we are "given over" or "appropriated" utterly by language (*Poetry* 207–208). In this perspective language is not a tool, and only in some highly qualified sense can it be said to lead to truth. Here we must lay aside ideas that we are the source of language, or that language merely "re-presents" thoughts, concepts, or things. Rather, language *presents*. As Heidegger says, language is the "house of Being," such that Being transforms itself and its essence into speech, into an "original dimension" through which we are able to respond to Being.

Here, we would be hard pressed to say there is anything beyond discourse or, if there is, to say what it might be, and Heidegger would claim it was always thus. Nevertheless, he gathers and presents these themes of language, Being, and time under the umbrella or paradigm he calls "earth." The term appears only in his later works, but the idea itself is present from the outset. It seems to have a number of meanings, though I see them sharing an essential congruity as Being in its immanence. Earth is most simply the good earth under our feet, the abode in which our being is grounded. This is the wisdom of Zarathustra, entreating us to remain true to the earth and renounce extraterrestrial hopes, to exult the body and spurn the soul (Nietzsche, *Zarathustra* 42). Or it is the wisdom Eric Havelock sees in the fate of Oedipus, where our nature is as his, springing from the "putrefaction of worms" yet reaching down "with twisted roots not to realms of death but to life and growth, formless, vital, without prejudice and prejudgment" (*Liberal* 28). Still, we are concerned with earth in its Heideggerian aspect, complex and multifaceted. And among its meanings is earth as the "concealing" event of Being, at the root of a phenomenology placing the focus on the power of Being and relegating the perceiving subject to scarcely more than yet another manifestation of this power. And earth means "native soil," a place of the "rootedness" of a people, where the forces of nature and culture are joined to constitute a community. The earth is then conceived in ways congruent with *doxa*, leading to its most comprehensive meaning. Earth is *physis*, the word designating both the power and process of nature, the power of emergence that is at once the process of being rooted in native soil. Here, in full measure, our knowledge conforms to things rather than things conforming to our knowledge of them, such that, in our conformity, our proper concern can only be Being itself. This power is never our own, but through it we come into our own, being authentically who and what we are only as it is expressed though us. In Heidegger, this power of *physis* is apparent most conspicuously as language.

Thus, we are concerned with ontological matters. We are concerned with the imperatives of Being in its immanence, of life lived flushed with the earth, as language itself is conceived as "rising" from the earth, providing our dwelling, our "house of Being." Our bond with the earth is thereby achieved neither through our souls nor bodies but by virtue of the event unfolding in our journey from birth to death, in the resulting narrative that comprises both body and soul as one. Here, to be aware of language's power is to be stirred by the power of *physis*, by this mystery of life and death. At the same time it is to be aware that language is this mystery, that in its alliance with earth it bears the mark of our deliverance as surely as it does of our mortality. In Heidegger, then, we might begin to appreciate this idea of earth, particularly in its connection to the early Greek experience, by means of the concept of autochthony. The sentiment, if scarcely the full substance of the idea, appears in Heidegger as *Bodenständigkeit*, a *mythos* that is the sum of native soil and culture.

Autochthony means literally "born of the earth," and in fifth century Greece was a commonplace trope used in Attic funeral orations, referring to the spontaneous creation of heroes rising from native soil, the womb of the motherland. The myth finds its origin and widest application in funeral orations by the likes of Pericles, Demosthenes, and Gorgias. In Athens these orations were part of periodic rituals mandated by law, so fixed and momentous that they were a genre unto themselves, the *epitaphios logos*. The earth bore these warriors and nourished them, these orators say, and then received them and in her bosom they now repose. In fact, archaeologists of late have unearthed the remains of warriors in Athens that evidence indicates are the very ones over whom Pericles delivered his funeral oration ("Fallen"). No greater sense of authentication is possible. These orations were given over earth containing the warriors, their flesh being one with it, in a bond both immediate and perfectly explicit. In compelling ways there is reflected in these orations an ancient, even primeval awareness of the wholeness of language and corporality, of the word being one with flesh in a most fitting expression of our being, indeed of the Heideggerian expression of earth itself.

We might then see the myth of autochthony as the effort to complete a circle, a movement not from birth to death but birth to rebirth, investing the warrior with spiritual being, where his glory, his peculiar *doxa*, is achieve by way of confirming in the funeral oration a spirituality that was there always, from his very birth and forever of the flesh. What the oration offers is this alchemy, where the sacrifice of death itself transcends death. Some might see this alliance of earth and language as unwholesome in any event, little more than an effort to distract us with soothing words from the burden of our mortality, which in itself could be the principal purpose of encomiums, eulogies, and funeral orations. However, what I have in mind is not a particular type of discourse but the power of language as such—as, in fact, is sometimes seen in funeral orations—where the need for utterance is not based on an effort to elude death or our dread concerning it but fully to realize them both. Such is the measure of "authentic" language, of the word becoming one with the flesh, determining our being. In this manner we are "released" into our own nature, as Heidegger says, allowing language to secure for us our being, "to grant an abode for mortals" (*Way* 129; *Poetry* 192).

This perspective not only contests our timeworn conceits of Being as a fixed and indubitable existent but evokes in its place a belief in Being as quickened by language, animate and pliable, necessarily eliminating recourse or access to a "higher" reality beyond language, whether within this world or not. If the philosophic perspective is directed upward, this perspective is of the earth, disclosing a very different world indeed, and in the process disclosing much about the nature of perspectives as such. Heidegger says that Kant brings ontology to the fore, and therefore the proper sense of the Copernican Revolution is that knowledge is possible only by virtue of the prior "ontological comprehension" of the knower (*Kant* 22). Comprehension of this sort is implicit in our very nature by virtue of our participation in the process of Being. "Perspective" is thereby not primarily a subjective activity but the result of the revealing and concealing power of Being, in that manner obliging the union of knower and known.

These dynamics are the basis of Heidegger's ontological phenomenology, and in his later works he equates this view of Being with *physis*, with earth itself, and that with language. Thus, he rejects from the start the "correspondence" theory of truth, that effort of matching statements to realities fixed and prior to any statements about them. Indeed, the scope is broad, even all-embracing, and what we thereby gather in the synoptic view of Western intellectual development is a series of voices in the "conversation of mankind," as Richard Rorty says, none inherently more benign nor correct than another. Perspectives abound. And claims to reality in any larger sense are perspectives still. That is what is fundamentally at issue, and it should be seen as neither a fault nor impediment but, insofar as we can know it, the way of the nature of things. Thus, "reality" finds its most complete and honest expression only in a myriad of perspectives.

This is a book that deals in perspectives. Perspectives on the Sophist Gorgias in particular and Sophistic rhetoric in general. And on *kairos,* the concept I consider to provide one of the fullest and clearest measures of each. *Kairos* means "time." More specifically it means the "right" time, and in this context has been associated with insight, discovery, and revelation. It is most notably a religious term, in Christian history and theology consummately the event of the word made flesh. Yet the term is more ancient still and rooted no less securely in rhetoric, though here its meaning seems staid and settled, fixed in the Platonic perspective. So, by way of a departure from the conventional view of *kairos* in rhetorical theory, I attempt to fashion in this book a viable interpretation of Gorgias and Sophistic rhetoric that is, at the same time, congenial to a more favorable understanding of each. For any number of reasons, such an approach has its share of dangers. The first is apparent to anyone who has attempted any research on the Sophists. Very little of their work survives complete and intact, and much of what remains is given in scattered, disjointed fragments presented by others in contexts often inimical to the Sophistic purpose. As Mario Untersteiner says, the Sophists, unlike the philosophers, never perceived it as their purpose to impart wisdom for the ages but were concerned primarily with what they took as the more pressing and practical affairs of their age alone. As a result, much of what we do know of the Sophists comes in antitheses to their views expressed by Greek philosophers engaged in a process of repudiation "bound gradually to result in the *damnatio memoriae* of their immediate predecessors" (9)—namely, the Sophists themselves.

And Gorgias, more than any of the Sophists, has suffered the ill-fortune of having his place in history marked out by an adversary. Indeed, Plato has more to say about Gorgias and his rhetoric than what is to be found in the fragments and surviving works of Gorgias himself. And what Plato has to say is seldom good news for Gorgias, and more seldom still is it good news for Sophistic rhetoric. Yet, the curse that Plato casts upon the Sophists is mixed. If we did not understand them through Plato's eyes, I am not sure we would understand them in any coherent sense at all, and could conceivably be reduced to reading their works and fragments as we do ink blots, assigning meaning at our whim or, more likely yet, on the basis of some other set of propositions that surely would be more inchoate than Plato's.[5] On the more malevolent side of the curse, our intellectual heritage has been forged to such an extent by Plato that even those accounts otherwise favorable to Gorgias and Sophistic rhetoric seem haunted by his presence lurking still at the very heart of these accounts. Thus, any discussion of the Sophists must necessarily deal with Plato, but to whatever extent a less jaundiced view of Sophistic rhetoric is grounded in "condemnations of Plato's condemnations" of the Sophists, we are also well advised, I believe, to move beyond Plato as best we might, to follow the path of Heidegger's thinking and thus "to bring the silence of ancient thinking to expression" (*Early* 85).

The purpose of this book, then, is to offer an account that more fully accommodates the idea of Sophistic rhetoric, allowing us to appreciate it as nearly as possible on its own merits. If the goal is not that novel, I believe the route to the goal is both untried and propitious. At any rate, this book offers an analysis that by virtue of a Heideggerian perspective is vested in the pre-Socratic mentality through which Sophistic rhetoric developed and flourished. Central to this analysis is the concept of *kairos*. I use it as a touchstone throughout, as the locus where Heidegger's poststructuralism and Gorgias's "prestructuralism" might be joined, connected in their essentials by Heidegger's purpose "to think what the Greeks have thought in an even more Greek manner" (*Way* 39). Here, in the thick of this peculiar purpose, the idea of earth provides the overall ambiance, the pattern through which this connection to the pre-Socratic mentality might be accomplished. In this context it therefore makes sense to speak of "rhetoric's earthly realm," not only in terms of the contrasts to Plato's celestial realities but more conspicuously as Sophistic rhetoric bears the expression of *physis* and native soil.

Kairos and Its Place in Sophistic Rhetoric

The idea of *kairos* dates back at least as far as the rhapsodes of ancient Greece, and aphorisms that advise prudence or justice identified with the right time abound in Pindar, Solon, and the Seven Sages. In the Christian tradition, particularly in the Pietistic strains of Protestant theology, *kairos* has its most critical function, persisting as part of a more embracing scheme of "salvation history," of the myth of the incarnation wherein the eternal becomes manifest in the temporal affairs of humankind. As salvation history designates the process of the eternal becoming accessible and known within the purview of human history, *kairos* marks the special moment, that opportune time in history of the coming of the Christ when the word is made flesh.

But it is in theories of rhetoric, both classical and modern, that the idea of *kairos* is just as persistent, where *kairos* has an equally detailed development and, for the rhetorician, an equally critical function. Here, *kairos* is interpreted as designating the power of the opportune moment in gaining advantage in speech, in obtaining a persuasive edge in delivering one's message that would otherwise be lost at a time less opportune. Here, also, *kairos* is associated with propriety and decorum, and more specifically concerns the rhetor's skill in the adaptation of language to a given situation. In any event, that is the rhetorical dimension of *kairos* as it is found in Plato's dialogues, and similar notions of *kairos* persist in Aristotle, Quintilian, and Cicero, as they do in modern theory. I refer to the substance of these notions as the "Platonic *kairos*" and contend that the Platonic *kairos* constitutes the accepted meaning of *kairos* in rhetorical theory. Indeed, as I shall argue in the next chapter, if there is a Platonic endorsement of rhetoric, it lies not in the subtleties of the dialectic but most clearly in *kairos*, the singular and unique recognition that Plato seems to concede to Gorgias in regard to the practical activity of discourse. And even here he gives the devil his due most grudgingly. In effect, Plato devalues *kairos* to the significant extent he devalues rhetoric itself.

Plato's understanding of *kairos* would seem to be grounded in his exclusion of Sophistic rhetoric from theoretical justification. Indeed, to the extent that rhetoric is justified at all in Plato's system of thought, it is subordinate to philosophy, resulting in a peculiar hybrid that is the wont of many to call "philosophical rhetoric."[6] Here, *logos*—which we can understand as reason, language, or some fortuitous combination of the two—is reason alone in Plato's presentation, to the exclusion of language. More specifically, the "genuine rhetorician," he says in the *Phaedrus,* is the one who knows the "truth" of matters on which he speaks (260d); he is therefore first and foremost a philosopher skilled in the method of dialectic. Indeed, insofar as there is an "art" of rhetoric, it is vested in philosophy alone.[7] Ornamentation and stylistic displays of speech are generally spurned as a result, because in Plato's scheme truth not only exists independently of its verbal embodiment in language, but the effect of language is very often to subvert or utterly destroy it.

Thus, a further implication of Plato's philosophical rhetoric is that the rhetorical act is relegated to the subsequent expression of truth gained by means of the dialectic, philosophy's method. This is, of course, an arguable proposition, and will be argued presently. But the point being emphasized for the moment is that on the theoretical level philosophical rhetoric is not about rhetoric at all, for in the expression of Plato's idea of truth, language—as a sensuous medium—can only be used analogically, and the phrase that thus resonates throughout Plato's dialogues to depict this extralinguistic truth of *aneu logou* (literally "without speech") is "to give an account of," the process of translating truth into language and thereby necessarily defiling truth, removing it from reality once or twice over. Hence, philosophical rhetoric is conceived only in terms of the structuralist's project of proof by reason. As Jane Sutton points out, the sanctions of "reasoned argument" in Plato's philosophical rhetoric constitute the "essential" nature of rhetoric, where *logos*, as reason, is "the structure (bottom, cause, foundation) of the rhetorical act" (207).

Sutton provides as well a series of interesting extensions to George Kennedy's distinction between "primary" and "secondary" rhetoric, arguing that philosophical rhetoric is not only "linked in a powerful logic to adumbrate any other aspect of rhetoric such as character and emotional appeals" (214), but in its "hyper-affirmation of reason" is *primary* in contrast to the inferior status of the *secondary* rhetoric. As the former comprises the "living substance" (223) of rhetoric, secondary rhetoric is concerned basically with matters of style, focusing upon "linguistic invention as alogical" and having "a highly poetical, mystical, and even superstitious bent" that seduces and overpowers reason and moral judgment (214–15). Though Sutton does not address the issue of *kairos* directly, given her analysis it would seem that *kairos* takes it place in rhetorical theory as secondary, subordinate to the use of reasoned argument inherent in "primary" rhetoric. And Kennedy, though he would surely object to some features of Sutton's analysis, gives us no reason to see the issue of *kairos* any differently.

Ironically, it is with the Sophists, Gorgias in particular, that *kairos* is most closely associated. And here *kairos* is not just primary but pervasive, essential in all respects. As Untersteiner says, and many others agree, Gorgias's entire system of thought—his ethics, aesthetics, epistemology, and rhetoric—is based on *kairos*.[8] And there are further ironies still. Given the number of scholarly works that have appeared in recent years offering a re-examination of the Sophists, we have surely reached the point where we can no longer claim that Sophistic rhetoric is ignored, neglected, or underappreciated. Indeed, a movement within rhetorical theory has gained such salience it even has a name: "neo-Sophistic." Among its purposes is to "rehabilitate" Gorgias, to reassess his world view placed under such vigorous attack by Plato, to bring about his transformation from a minor and much maligned figure concerned only with inflated, ornate style. In this reappraisal, Gorgias has gained recognition as the source or chief proponent of a number of features vital to rhetor's art. A series of journal articles concerned specifically with Gorgias, and books by Eric White, Susan Jarratt, Victor Vitanza, and John Poulakos—and more recently those by Michelle Ballif, Bruce McComisky, and Scott Consigny—dealing with Gorgias and Sophistic rhetoric in general, have pointed to the value of the rhetorical implications manifest in Gorgias's world view: specifically to the concept of *doxa*, the medium in which the rhetor works; of *apate*, the means by which the rhetor "deceives" his audience; and, above all, *peitho*, or persuasion itself, and how it functions in the very practical, day to day affairs of men and women.[9]

However, to the degree that this re-examination has been concerned specifically with *kairos*, the problem of a prejudicial attitude persists, and is particularly acute in the case of Gorgias where, ironically, the effort has been specifically to place his work in a perspective apart from the prodigious influence of Plato. Despite the elements of *doxa*, *apate*, and *peitho* that have been duly recovered from the Gorgian world view, some fuller significance they might bring to bear on the nature of Sophistic rhetoric in general is neglected, curiously left to languish. Once again, even the works most enterprising in their appreciation of the Sophists very often proceed on the basis of premises laid down by Plato. There seems to be, in other words, certain terministic screens in place that are part of our intellectual inheritance, those shades of Plato I spoke of earlier, and for lack of a more congenial perspective, we necessarily condemn, often against our own best intentions, the complete inventory of Sophistic rhetoric as subordinate to philosophical rhetoric. More specifically, we tend to form our conception of Sophistic rhetoric from the perspective of *aneu logou*, and in the cluster of binary oppositions that ensue, Sophistic rhetoric is relegated to a less favored status, serving mainly to illustrate in the contrast the superior principles of philosophical rhetoric. Thus, *doxa* is "mere" opinion, the fabric of unfounded beliefs allowing language to distort and dissemble, standing in contrast to "real" truth that can exist only beyond words; *apate* is "deception" precisely in contrast to this truth, with the result that style is, at best, elegant variation and, at worst, gratuitous display, sham and pretense; and *peitho* is inescapably secondary to philosophy and left to designate the rhetor's grab bag of verbal ploys and tricks in contrast to the dialectic as the way leading to this truth beyond words.

This list of oppositions could be extended, or reduced to those more fundamental, such as the distinction between appearance and essence or even that between body and soul. These are matters to be addressed in some detail later on, but for the purpose of this introduction they can be put in perspective by another opposition more fundamental yet. Here I focus on *res* and *verba*—the distinction between "things that are" and "words" respectively—as what is more crucial in understanding *kairos* and its place in both philosophy and Sophistic rhetoric. Particularly here, even as the current movement to re-examine Sophistic rhetoric is animated by an attack on Platonic metaphysics and epistemology, the basic idea of Plato's conception of rhetorical *kairos* continues to be confirmed, resulting in definitions of *kairos* as precisely the Platonic *kairos*, depicting it solely as a means to govern choice of style, as a matter of decorum or appropriateness, patently as the adaptation of language to a given situation. Therefore, the problem is that the Gorgian *kairos* remains "unrehabilitated" and, for that reason, so does Sophistic rhetoric because of the pivotal role that the idea of *kairos* could play in getting beyond the Platonic perspective and providing what I would consider a more complete and productive understanding of Sophistic rhetoric.

Thus, in this account I intend to isolate and explore Gorgias's notion of *kairos*, for in the nature of the Gorgian world view there is a dimension of *kairos* that has yet to be fully exploited, one allowing a series of useful distinctions to be made between philosophical and Sophistic rhetoric that leads, I believe, to a fuller appreciation of the latter. It is an enigmatic dimension based on the primacy of language rather than reason, of *verba* rather than *res*, investing *kairos* with a much more mystical imprint than otherwise sanctioned in theories of rhetoric. In this interpretation the implications of the "right moment" are carried to lengths beyond the pale of the traditional, Platonic interpretation of the term. In the evidence available to us, especially in his "Encomium of Helen" and the "Funeral Oration," Gorgias stresses the role of language as fundamental, free from any necessary correspondence to a higher order of reality designating "things that are." In these discourses and elsewhere he features his celebrated contention that language is "Lord," the great "*dynastes*" or power. The clear implication, of course, is that in the Sophistic attitude no art of rhetoric is possible without deference to language as an independent, external power, exerting an active force impinging on the *psyche* from without. For Gorgias, as Thomas Rosenmeyer says, "speech is not a reflection of things, not a mere tool or slave of description, but . . . it is its own master" (231–32). This idea that "*logos* is free," entailing the "complete 'suspension' of rational belief" (Segal 112), has been associated by both Charles Segal and Jacqueline de Romilly—and most critically by Gorgias himself—with the enchantment of magic or the influence of drugs. In such a context, the Gorgian *kairos* itself designates a "spontaneity" in the generation of language that, at first glance, seems to have obvious connections to the darker side of the "pre-Socratic mystery," to processes of psychic intervention, poetic frenzy, and divine madness.

I am speaking more specifically of that peculiar Gorgian manifestation of *kairos* that Untersteiner says "releases" the persuasive force of an "irrational *logos*" in the instant of decision (177). And it is a decision, Untersteiner adds, that comes not from the speaker involved but from *kairos* alone, which, of itself, "wills" or "imposes" the decision (178, 181). Though differing interpretations can be placed on this idea of *kairos*—and even, indeed, on Untersteiner's explanation of the phenomenon—what is clear is that the *logos* in question is "irrational," an affirmation of language displacing reason, and that it is, as well, associated with a "mysterious, divine gift" that is "incomprehensible to ordinary mortals" (Segal 120, 126). In contrast to its place in Platonic rhetoric, *kairos* in this Gorgian sense, as C. J. de Vogel says, has no foundation in a rational *taxis,* that instead a "complete irrationality" reigns (118).

Though an elaboration of these ideas and their significance will be provided in due course, it is apparent that Gorgias's notion of *kairos* follows an inexorable logic of its own, far removed from the regularity and precision that governs Plato's conception of the term. It is here, in the discord of the two ideas of *kairos*, that the problem I am attempting to address is most apparent. There is a dissonance in these claims that *kairos* is, on the one hand, simply the adaptation of language to a given situation or that it is, on the other, a mysterious force that prompts an "irrational *logos*." The problem is manifest particularly in a peculiar impasse we face whenever *kairos* is discussed. Whether it is treated in Gorgian or Platonic terms, very little is said, or can be said, about the place of *kairos* in rhetorical theory. To the extent *kairos* is understood at all as the mysterious and irrational force Untersteiner suggests it to be, it seems to be too mysterious and irrational to be amenable to the strictures of rhetorical theory. That we should therefore acknowledge only Plato's conception of *kairos* is clear, but it is a perspective that not only preempts the promise of worthwhile insights into the nature of Sophistic rhetoric but often results, outright, in sweeping condemnations of the latter. But with Gorgias, as well, the impasse remains, as is evident in the comment of Richard Engell, one of the architects and most perceptive scholars of the movement to rehabilitate Gorgias. After discussing Gorgias's "Funeral Oration," Engell defines *kairos* as "the appropriateness of a conclusion to a given situation" that governs belief and acceptance, "rather than a correctness according to strict logic." He then offers the abrupt rejoinder that "there is little more that can be said about the concept" (178). Engell does indeed say more, and his further analysis suggests that *kairos* has its place in rhetorical theory as an epistemological feature that submits demonstrative logic to a "new standard of rational persuasion" (179). Thus, even in this account, as *kairos* might imply a "new standard," it remains "rational" persuasion just the same.

Engell's account may be one of the more obvious but is typical of interpretations of *kairos* among rhetoricians attempting to "rehabilitate" Gorgias. Whatever role *verba*, as the primacy of the language, might play in an understanding of *kairos* is ignored. As a result, our approach to the concept is restricted by a fixation on *res*, excluding *verba*, and all tenets remain "rigorously non-verbal," as Cascardi says in making a similar point concerning Plato's art of rhetoric (225). The strictures of primary rhetoric are kept in place, and *kairos* is constrained to yield only the above version of the Platonic *kairos*, of which "little can be said." Perhaps it is because of its failure to provide anything of deeper substance in rhetorical theory than that of a technique or skill circumscribed by secondary rhetoric—as much as any lack of scholarship of the issue—that once provoked James Kinneavy to say *kairos* has been a "neglected concept" in studies of rhetoric and "almost completely overlooked in rhetorical scholarship" ("Relation" 13).[10]

Thus, at a glance, the basic problem prompting this study is that the concept of *kairos* seems to have been interpreted by rhetoricians only in terms subordinate to primary rhetoric, where its significance is sharply diminished, specifically by way of the structuralist warrant of *aneu logou*. A major consequence of this interpretation has been not only to misrepresent the Gorgian *kairos* but, as a result, to misrepresent Sophistic rhetoric as well, such that both take their place in rhetorical theory exclusively in terms of a Platonic frame of reference. But what persists as the most evident and serious effect of this problem is that an interpretation of *kairos* as a vital, dynamic force in rhetoric is foreclosed, as the concept takes its place among other features of secondary rhetoric, those that are the mere skills and techniques of rhetoric rather than part of its "living substance."

Heidegger and a Path to the "Original Beginning"

My purpose in this book is to offer a reassessment of the Gorgian *kairos* and thereby assert its place precisely as part of the "living substance" of rhetoric. I will attempt to gain access to this Gorgian *kairos* in light of what Heidegger has to say about Being and time, phenomena and reality, and, most definitely, language itself—all of which I gather under his rubric of "earth." Not only do the more mysterious features of the Gorgian *kairos* seem to have their apparent development in these themes from Heidegger but, by way of this idea of earth, language is the express manifestation of *physis*, being prior to us—ultimately *being us*—informing my interpretation of the Gorgian *kairos* and Sophistic rhetoric as a whole. It is hardly coincidental, then, that the idea of *kairos* is present throughout Heidegger's corpus, and present not only in his discussions of the nature of time but more prominently in his theory of language. What is also germane in this context is that Heidegger offers one of the most cogent analyses of Plato to be found in Western philosophy, or at least one of a type very appropriate to my purposes here. It is an analysis based upon "destruction" of the metaphysical tradition, and, therefore, a "destruction" of Platonism in particular. Indeed, Heidegger quite roundly equates metaphysics with the latter, as he claims, presumably not to wane too subtly on the issue, that "Metaphysics is Platonism" (*Time* 57). Yet "destruction" in this Heideggerian sense is neither an attack nor even a refutation. Heidegger presents it as an effort to "loosen up" the "hardened" philosophical tradition, to remove the "screens" imposed by the tradition, so that the "birth certificate" of basic ontological concepts can be uncovered. Destruction's aim, then, is positive. Heidegger says it is an effort to "reawaken" or evoke "those primordial experiences in which we achieve our first ways of determining the nature of Being" (*Being* 44). Thus, by way of "destruction," Heidegger's purpose is to recover "original insights" uncorrupted by metaphysics, and here again my purpose for using Heidegger in this study is apparent.

Heidegger's thought can be traced to Husserl, Kierkegaard, Kant, or even Aquinas, but as a "thought which seeks origins," Heidegger's lies with the pre-Socratics. William Richardson identifies it as a "species of re-trieve" of thought prior to Plato (89, 545). William Kluback and Jean Wilde make essentially the same point. Indeed, George Seidel claims that Heidegger's interpretation of the pre-Socratics is itself one of the more direct and dependable approaches to Heidegger's thought, which can, in turn, be understood at its "root and basis" only as pre-Socratic (3). John Poulakos, in purposes similar to the one engaging me here, says that "by overcoming the metaphysical tradition, Heidegger opens the path to the sophistical, whose glory was overshadowed by the imposing dynamics of systematization." More pointedly, he concludes, "Heidegger's post-metaphysical thought discloses the premetaphysical tradition by removing the veil of metaphysics" ("Rhetoric" 216).[11] The testimony from yet others establishing the connection between Heidegger and the pre-Socratics is substantial, but in due course that connection will be made clearly enough through Heidegger himself, as he moves by the means most integral to his poststructuralism, his "destruction" of the metaphysical tradition.[12] It is a movement specifically to uncover early Greek thought, to what it was prior to Plato, "in the original beginning," as he is quoted in the epigraph to this chapter. The distance separating this beginning from its "incipient end" in Plato is evident, pitting the old gods of Gorgias against the new ones conjured by Plato. Indeed, the victory of the latter has been so complete that the distance separating Gorgias and Plato, who lived as contemporaries, is in many ways greater than that between Plato and us, though he lived twenty-five centuries ago.

In any case, my aims in this book are modest enough, to reaffirm the importance of the Gorgian *kairos* and reinstate it within the bounds of rhetorical theory by means of a perspective that provides insights otherwise excluded by traditional interpretations of the rhetorical *kairos*. In this effort I am guided by that paradigm previously mentioned and one to be invoked repeatedly—namely, the Heideggerian paradigm of earth, that power of *physis* manifest as language and its way of revealing, concealing, and ultimately creating phenomenal reality. With this paradigm the principal concerns of my analysis are put into clearer relief, and I narrow my approach to three basic themes in Heidegger's thought that are either present or implied in Gorgias's world view. All three are critical to an understanding of the Gorgian *kairos* and, by means of the Gorgian *kairos*, critical to an enhanced understanding of the pre-Socratic world view in which Sophistic rhetoric had its source: First of all, in Heidegger, an ontological dimension of language is stressed, in which human being does not possess language but is itself possessed by language, and where, as a result, the *logos* of language is conceived as an "indigenous field," as *physis* precisely, evoking a relationship with the experiential, physical world that is primarily linguistic.[13] Secondly, in Heidegger, human being is never manifest as an isolated ego or soul, a free-floating consciousness that perceives, investigates, and manipulates the things of the world, but is *dasein*—a "being-there," a "being-in-the-world," such that the nature of the human being cannot be considered separate from the world and can be considered at all only insofar as the world is of its essence. And lastly, in Heidegger, analysis moves on the basis of a "primordial temporality" encompassing in human being an integration of past, present, and future that fashions the web of occurrence. Hence, the past exists presently as a past preserved, the future exists presently as a future anticipated, and time itself exists only by virtue of our implication in it. In essence, the temporality of *dasein*, of "being-in-the-world," precludes not just the eternal but time taken only in terms of mere chronology or homogeneity, the "vulgar" conception of time, as Heidegger says. He instead puts the emphasis always on temporality, and therewith on situational context.

It is on the basis of these categories that I isolate the more obscure features of the Gorgian *kairos*. Such an analysis, I believe, implies a more provocative and embracing consequence than the Platonic *kairos*, of *kairos* as it is understood conceptually as an element of secondary rhetoric, as a rhetorical skill in the adaptation of language to a situation. In contrast, *kairos* in this Gorgian sense seems to have its roots in notions of time and language expressed above, in the emphasis on *verba* and the resulting emphasis on the phenomenological dimension of reality that, if ever such a reduction is possible, is the essence of Heidegger's thought. He seeks to retrieve the original conception of *logos* as language, a notion in the thick of *Sehenlassen*, of "letting things show themselves," where, as Ernesto Grassi says, "words permit the meaning of things to appear" (*Heidegger* 19).

Sophistic Rhetoric as Mystery and Praxis

These ideas concerning the ascendancy of language move in areas rife with age-old antagonisms between philosophy and rhetoric. I broach the issue here only to introduce Heidegger's now commonplace thesis of the "end of philosophy" and resulting contentions from various quarters of a "rebirth of rhetoric" apparent in Heidegger's thought. Indeed, Heidegger's radical reflection on the nature of language not only strips philosophy of its metaphysical and epistemological functions as necessary conditions but prompts such critics as Henry Johnstone to argue that Heidegger conceives of philosophy as inherently rhetorical, in that its antecedents, as those in rhetoric, are based on "evocative-elucitory" functions of language rather than fundamentally "rational" principles.[14]

Rorty makes a similar point, claiming that ultimately there is no exclusive "philosophic domain" of pure fact or reason beyond discourse, that instead all is vibrant with the hermeneutical imperative. Thus, of its own momentum philosophy undermines the very premises that would make it the essential source of authentic knowledge and means of inquiry; it "deconstructs itself," to the effect that it can no longer exclusively provide legitimate epistemological and metaphysical foundations but must assume the role of a contributing voice among many others in the conversation of mankind (*Mirror* 390–91). To Calvin Schrag, the deconstruction is necessarily mutual, affecting both philosophy and rhetoric, so that in rhetoric the uses of argumentation and persuasion are transformed or, at least, in transition from techniques of control and manipulation to means of expression of the self and the world within the "hermeneutical texture of communicative praxis." In any event, the bearing of the argument appearing in the literature of such critics is that to whatever extent we have lost the traditional function of philosophy, we have gained, or regained, a revitalized rhetoric in its place. In summing up the transformation in an argument that I suspect he only conditionally endorses, Schrag says, "We may have lost a father . . . but in the end we have gained a son. Philosophy dies so that rhetoric can be born" ("Resituated" 166).

For reasons such as these, especially those giving prominence to language in place of "fundamentally rational principles," Heidegger's thought mystifies. Kluback and Wilde say it partakes of the "pre-Socratic mystery," being "strange and difficult to comprehend" (12). Assuredly one effect of this analysis will be to affirm that this is true. Indeed, it is impossible to imagine a more pressing question than that concerning the meaning of Being, yet even as Heidegger has dealt with it as thoroughly as any thinker in the Western tradition, it remains inevitably a question in his analysis. That is to say the mystery of Being remains a mystery, shaped as much by "non-Being" as "Being" itself, given the latter's pervasive negativity, its aptness to conceal as much as reveal. In this vein Gadamer points to the "peculiar paradox" of critics attempting "to order" Heidegger's thought, "to write a systematically developed summary" of it, when, given Heidegger's ways, the effort is "not only futile but pernicious" (*Heidegger's* 20). Indeed, his work mystifies as much by means and method as content. So Heidegger himself encourages us to pay attention to the path of his thinking, as that is equally its content (*Identity* 23), and then invites us to take the journey, as he offers a series of hints and intimations on the way to the "happening" of truth. And here I take the journey as best I can for what is in the offing, a remarkably well imagined world where *physis* is affirmed expressly as language, to that place I identify as Sophistic rhetoric. If the haze of mystery clouds it, Heidegger offers a cohesive and relatively complete means of inquiry nevertheless, a channel through which this examination of *kairos* can proceed unencumbered by mystery's perceived flaws, providing instead a way to understand *kairos* as mystery resolved in terms of the very practical work of language, thereby never allowing its significance to dwindle to the inchoate whisper of a muse.

Indeed, insofar as we perceive it only in terms of the "essential prejudice" of philosophers, the idea of mystery is itself flawed. It stands opposed to the truth that reason yields, obtaining its definition in contrast to what Plato and the entire Western philosophic tradition are disposed to take as "what is," as "things that are." Be that as it may, what I suggest in the following chapters is that the Gorgian *kairos*, perceived in terms of *verba*, is a way of insight achieved not only in defiance of reason and *res* but as purely linguistic in origin and manifestation, and in that sense it *is* mysterious, inevitably and properly so. Thus, from this perspective we come to know Sophistic rhetoric not as an activity of mystery alone nor praxis alone, nor as some fortuitous combination of the two, but as essentially prior to divisions of this sort, prior to certain philosophic premises that work distinctions between rational and irrational, thinking and being, mystery and praxis. This perspective thereby allows us to overcome that barrier to our understanding of the Sophists, namely that capricious attitude we often assume concerning their rhetoric, that it is either representative of an illogical and mysterious process or is exemplary of bedrock practicality. The issue is apparent in what Charles Segal calls the "double aspect" of the Sophists' *techne* of rhetoric, of the rhetor's art coming together in the peculiar nexus of his role as "linguistic technician" and "magicianlike charmer" (117). Moreover, by means of a Heideggerian perspective, there is no need to resort to arcane mysteries or magic to explicate Sophistic rhetoric. Nor, as significantly, is there a need to avoid situating Sophistic rhetoric exclusively in the practical realm. In this perspective the claims I make regarding *kairos* are hardly outlandish nor even all that remarkable, but constrained by the understanding that *because* Sophistic rhetoric partakes of the "mysterious" it is a very practical discipline indeed. Here, not only can the Gorgian *kairos* be seen as distinct from the traditional rhetorical *kairos* by means of the respective emphases of *verba* and *res*, but through the emphasis of the former on *verba*, on that creative presence of language as an "indigenous field," emerge the elements of the timely, the possible, and the imperatives of situational context that underscore rhetoric at its source.

The demise of philosophy notwithstanding, it is important to realize that if we are indeed witnessing a rebirth of rhetoric, on that basis alone the role of *kairos* in the Gorgian sense can be seen as nothing less than that of midwife—for it is, in a Heideggerian perspective, that moment of "de-cision" (*Ent-scheidung*) coming in the nexus of earth and *dasein* that is "authentic" language. Moreover, in the context provided by a Heideggerian conception of philosophy as inherently rhetorical, the Gorgian *kairos* has not only a congenial ambiance but, within this ambiance, an explicit connection to Heidegger's notion of the *Augenblick*, of the "phenomenon of the instant"—the *kairos*—as prompting in the right situation the creation of language itself. *Kairos*, then, is the instant which to Gorgias triggered the persuasive force of an "irrational *logos*" free from privileged representations of the "things that are"; and, to Heidegger, in the context of *Being and Time*, it is that which prompts "authentic" language as a response to an "aboriginal *logos*," whose call comes in our openness to it—in the *kairos* of a "primordial hearing," a "hearkening" to "the voice of a friend whom every Dasein carries with it" (*Being* 206).[15]

Thus, to say simply that Heidegger's thought mystifies is to put the matter far too simply. But if "mystery" is the way the world works, then appreciating it as such is a matter of utmost practicality. The legacy of the neo-Sophists is to define Sophistic rhetoric in patterns of humanism, relativism, pragmatism and praxis, based very often on the *homo mensura* doctrine, the belief that man is the measure of all things. Surely Sophistic rhetoric is all of this, the epitome of practicality, though never for that reason in opposition to mystery but inevitably complementary to it. Language is the measure here, Being's mystery immanent as earth, and in the revealing and concealing power of Being language points always to the mystery. So my focus remains the Gorgian *kairos* as the consummate expression of this mystery, and thereby a most viable means to understand Sophistic rhetoric. Besides, as I suggest, the distinction between mystery and practicality is itself a result of the priority given reason, as language itself brooks no such opposition. Though never so apparently as we speak it, it is so most essentially as we hear it, as it comes to pass, Heidegger says, as the "aboriginal" or "irrational" *logos*. Indeed, as we have no adequate language to speak of the mystery, it speaks to us, or so Heidegger claims, though always in hints, riddles, paradoxes, and various asides. Here, "truth," as the essential prejudice of philosophers, yields us nothing. At least, no answers we can trust, and answers are all that it has to offer. And so the question remains, as all referents are enclosed within language, within the enigma of earth as the source of this aboriginal *logos*, and by that measure the source of *our* being. Ultimately, as Michael Haar tells us, the question of Being concerns how we can enter into this ontological relationship with earth, even as we are, simultaneously, of the earth; and through this relationship, as Heidegger himself tells us, the thinker's essential responsibility can only be to "poetize on this riddle of Being" (*Early* 58).

A Preview

This question concerning the meaning of Being was the source of Heidegger's thought and remained his underlying theme throughout. This was Heidegger's own position on the matter, and the idea of *kairos* can be seen as one of the more obvious implications in this quest. I am, however, aware of the distinctions many critics make between the "early" Heidegger and the "later" Heidegger, and in this analysis I will treat in context any considerations that may arise from this possible shift in Heidegger's thought.[16] Whatever the case, I reiterate what I feel will result from this application of the Heideggerian *Augenblick* is an understanding of *kairos* more in the spirit of Gorgias's sense of the term. Generally what is gained through an analysis of this sort are conceptions of human being and the nature of reality distinct from those imposed by traditional metaphysics, providing in their place a perspective more kindly disposed to Sophistic rhetoric than those that have been hitherto available. Specifically what is gained is an understanding of the Gorgian *kairos* as the aperture through which the world of the speaker opens, of language itself, through human being, exerting a claim to become manifest.

In the following chapter I examine the rhetorical *kairos*, exploring its basis in terms of the peculiar Platonic take on the three modes of inquiry mentioned earlier: namely, the themes concerning the nature of language, human being, and time. Special emphasis is thus given to the subordinate role that language plays in Plato's scheme of things, in his metaphysics and epistemology that result in his doctrine of the human soul and conception of time as homogeneous, as "tainted" and ultimately "unreal." The chapter begins with a brief overview of *kairos*, of it meanings in various contexts and disciplines—such as religion, psychology, and literature—to further familiarize the reader with the concept and to provide a context for what follows. I move then to a discussion of the dynamics of language and myth in oral culture and, from there, to a more detailed development of the critical components of Plato's philosophy pertinent to his attack on these dynamics. Of specific relevance in this context is Heidegger's concept of the "metaphysics of presence" that he characterizes as the foundation of Plato's philosophy and as pervasive in Western philosophy as a whole. I use the concept as a means to locate the source of the rhetorical *kairos* in Plato's thought and then to discuss the persistence of this idea of *kairos* in modern theories of rhetoric. My purpose in this chapter is to make explicit the connection of the conventional notion of the rhetorical *kairos* and the Platonic *kairos*. Two significant implications of this argument concern Plato's dialectic, developed here as the essential *techne* underscoring philosophical rhetoric, and, what is collateral to the dialectic, Plato's doctrine of the soul, his idea of an "interiorized" spirituality that is the essence of the self and foundation of human being as subject.

In the third chapter I examine the Gorgian *kairos*, providing a comparison and contrast to the Platonic *kairos*, with an emphasis on the contrast in terms of the three modes of inquiry used in the second chapter. Particular attention is given to Gorgias's "Funeral Oration," where the clearest and most sustained presentation of *kairos* is to be found among the fragments and surviving works of Gorgias. Indeed, nowhere are the motifs of native soil and the power of *physis* more in evidence than in the Greek funeral oration and some others of more recent vintage, where the homage given earth is extensive and very explicit. Of crucial concern here is the Gorgian idea of an "irrational *logos*," which I examine in terms of a contention critical to an understanding of the Gorgian *kairos*, the idea that "*logos* is free." Closely related to this notion of earth and *logos*, and equally crucial in this context, is the pre-Socratic conception of the "*psyche*." In contrast to the Platonic idea of the soul, the *psyche* is associated purely with physical capacities or functions, and far from being an autonomous entity or "interiorized" self, the *psyche* is nothing in itself until given content and shaped by language. Indeed, in this context the *psyche's* relationship to language is reciprocal, where the meaning of each is based on the "co-belonging" of each. In terms of this vague borderline between inside and outside, subject and object, *psyche* and language, the idea of a Sophistic *techne* of rhetoric is introduced and discussed in this chapter, especially as it is manifest as a matter of style.

In the fourth chapter I clear the way for a full discussion of Heidegger and the Gorgian *kairos* by explicating the "ontological" conception of language apparent in a retrieval of a Sophistic sense of *doxa*.

Again I follow what I interpret to be Heidegger's lead, applying his explanation of *doxa* as "glory" and contending that it can profitably be applied in the context of Sophistic rhetoric. Specifically, I argue that in its essential sense *doxa* is not mere opinion but necessarily comes prior to all notions of *subjectum*, of the soul or self as isolated or pre-given, and therefore comes prior to subjective conviction itself. In that *doxa* is here considered as the creative emergence of beliefs rather than just the beliefs themselves, this view involves a shift in perspective drastic indeed, from *doxa* as product to *doxa* as process. It thus occupies a mysterious region that is the essence of earth—of *physis*—where the distinction between "to be" and "to be said" is extremely problematic, difficult to be made if it can justly be made at all. In this conception "reality" is thereby focused as much within the domain of rhetoric and language as within Plato's or any other philosophical or metaphysical system.

In the fifth chapter, "Heidegger and the Gorgian *Kairos*," I offer an interpretation of *kairos* that not only makes more explicit the Gorgian sense of the phenomenon but, on the basis of the ontological conception of language derived from Heidegger's idea of *doxa*, one that differs significantly from the Platonic *kairos*. Though a Heideggerian perspective informs the entire study, in this chapter alone I apply directly the ideas and speculations of Heidegger to the subject at hand. Not only are Heideggerian modes of inquiry employed as an organizing principle, but a pattern of exegesis based on Heidegger's thought emerges throughout as a means to recover, isolate, and specify the otherwise obscure notions that underscore the nature of the Gorgian *kairos*. What will result in the end, once again, is not a "correct" meaning of *kairos* proposed in the place of the Platonic *kairos*, but a contrast that I hope will make the distinction between the two, preserve the integrity of the Gorgian *kairos*, and plot its rhetorical implications and significance in terms of Sophistic rhetoric as such.

The last chapter of the book serves what I have tried to foster throughout, to provide as many familiar, accessible illustrations of the nature of the Gorgian *kairos* that I can gainfully offer. In this chapter the themes of language, the self, and time are present, but are here employed with the purpose of linking the Gorgian *kairos* with Heideggerian ontology through the ancient mythos of the trickster. My approach is broad and inclusive, ringing in whatever attributes of the trickster seem to serve this purpose, but for that reason the effort is hardly misplaced, especially as we are attentive to the fact that *kairos* was first of all a trickster figure in Greek mythology whose name only belatedly, for very compelling reasons, was appropriated and applied to a rhetorical strategy. More essentially, however, this paradigm of the trickster allows an understanding of *kairos* in terms of the "co-belonging" of *dasein* and Being, placing *kairos* itself squarely in realm of what Heidegger calls the "uncanny." The latter is essentially the pre-Socratic mystery made ubiquitous, earth as the revealing and concealing power reflecting in all its wonders and bewilderments the boundlessness of Being. A rather peculiar, even ironic sense of *techne* results, yet one that not only accounts for the surface features or effects of Gorgias's emphasis on style but one that is ultimately a function of our participation in Being—and thus our being *as* language—in that peculiar mix of ontology and language that is so typical of Heidegger. Here, of course, *kairos* is not a rationally fixed or qualified knowledge that we have of the right time to speak but is a force in its own right, invoking *us* in its own good time, using *us* to speak. This paradigm of the trickster therefore shapes an understanding of *kairos* as principally a force of the uncanny; in effect, a fortuitous or opportune opening by way of chance or fate to insight and innovation. By grace of the trickster we thus become aware of the unfamiliar in the familiar, of "para-*doxa*" and *doxa* as complementary forces vital to one another. *Kairos* is then apparent by dint of its mythological roots, in the Yoruba invocation of the trickster Eshu, for instance, of he "who translates yesterday's words into novel utterances."

Moreover, this last chapter is also a summation, bringing to bay ideas introduced, developed, and sometimes left at loose ends in previous chapters. The problem of their significance does not simply lie in recovering them from a time unimaginably remote, but more decidedly it comes by way of Heidegger's peculiar purpose "to think what the Greeks have thought in an even more Greek manner." Here, perspective is merely another measure of the mystery that stands in place of the pretense that it is possible to know what the world is really like, specifically the pretense in this case that we can then extract from it the ideas of the Sophists in their full integrity. Such perfect knowledge is not for us and, insofar as we presume to possess it, serves no purpose other than to make us docile and indolent. Heidegger says that at best we can remain open to the questioning of the early Greeks as our only sure passage to the primordial experiences we ultimately share with them, but "all later thinking which seeks dialogue with ancient thinking should listen continually within its own standpoint," to thereby "bring the silence of ancient thinking into expression." It is a process that inevitably accommodates the earlier thought to the later, "into whose frame of reference and ways of hearing it is transposed" (*Early* 85). Thus, there is no "thing-in-itself" that is Sophistic rhetoric. What we discover in the end is contingent as much on ourselves and our relation to Being—to earth—as it is on something we call Sophistic rhetoric.[17]

It is here, in the co-belonging of who we are and what we would otherwise conceive as being apart from us, that the pre-Socratic mystery is most in evidence, that we are most obliged to think as the Greeks in an even more Greek manner. So I persist with the issue of perspective that I broached at the outset, as I attempt as best I can to express the full effect of the mystery, though the variations and extensions I make on the motif of perspective move well beyond those innocuous musings on campus traffic. To see things from a different perspective can mean not only to inhabit a different world but to be fundamentally a different person or, at this point in our discussion, to be something other than a person in any traditional sense whatsoever. What is ultimately at issue, as I will attempt to explain and apply in due course, is not essentially a matter of perspective at all, but a view to the contrary, one abating the influence of subjective ego to the extent that the proper question becomes, as Heidegger projects Being as language, not how we see things but how things reveal themselves to us. As the sense of self is diminished or compromised, the issue becomes less a matter of our perspective but one of how the world encounters us, of how, as a matter of language, we are then "disclosed" to the world.

Thus, through language we participate in Being rather than the latter being what we make of it through any particular perspective, a shift in perspective truly "to make the ground move." We must therefore remain "open" to Being according to Heidegger. It is a problematic move, explicit in his curious phrase to "let Being be." Though herein, perhaps, the essential wisdom of Sophistic rhetoric lay all along. Certainly Gorgias devises a rhetoric that aims to annul or dilute the sense of self-will in the listener. But as the variables of audience, situation, place and purpose are brought into play, the speaker himself becomes less and less a possibility as the controlling subject, and at best is merely another feature of the equation, in ways no more consequential than the rest. Language then assumes its role as the "Great Lord" or "*dynastes*," just as Gorgias claims it to be in his "Encomium of Helen." Thus, things have less to do with perspectives unique to the individual subject but are, instead, given in phenomena, ultimately the result of the revealing and concealing power of *physis* and thereby the result of the splendidly creative power of the word.

And now there is a final matter concerning Heidegger's life and thought that some would say bear upon issues to be raised in this book. In 1933 Heidegger joined the Nazi Party and in that capacity was a telling presence as Rector of Freiburg University for nearly a year. The degree of his commitment to Nazi purposes and principles remains in dispute, though in later years he never adequately explained nor apologized for his involvement. Richard Rorty says Heidegger joined the party because he was a "ruthless opportunist" and "political ignoramus" who saw in party membership the chance to advance not just his career but his philosophy (*Essays* 19). I see little in Heidegger's personal life before or after his participation in Nazi politics to gainsay Rorty's assessment, and this book should then be interpreted neither as an apology for Heidegger's politics nor, least of all, for his thought. The real question, however, is whether Heidegger's affiliation with Nazi politics was in some sense an extension of his philosophy. There are those who claim many German soldiers died in Russia and North Africa with copies of *Being and Time* in their backpacks, though I can scarcely imagine a book less likely to incite patriotic fervor, to say nothing of heroic or manly airs.[18] And certain themes of National Socialism, certainly those of autochthony and native soil, are present in Heidegger's thought, as they are in Nietzsche's, definitely Fichte's, even Goethe's; or, indeed, in Continental philosophy as a whole, to include Jean Paul Sartre's. There are, as well, aspects of Eastern thought in Heidegger's philosophy, especially in his discussion of the mystery of Being. And, as will be discussed, aspects of American Indian beliefs are present in Heidegger's views on *doxa* and language, especially as each is implied in the wholeness of earth.

Heidegger spent his entire life in the region of the Black Forest and surrounding environs, seldom venturing beyond Freiburg, and it could be that his provincialism, however self-inflected and arrogant, caused him to be deceived, even blinded by his own philosophy, though I see nothing intrinsically fascist in the latter and believe it to be as profound and original as any in modern times. In addition to all the sources of

Heidegger's thought mentioned above, surely the one most compelling is ancient Greek thinking, specifically as it is apparent in the "pre-Socratic mystery," where language in connection with autochthony is given particular emphasis in the paradigm of earth. Charles Bambach describes autochthony "not simply as rootedness in the soil, in the past, or in the tradition from out of which one views the world," but as signifying "something concealed, mysterious, and chthonic whose meaning lies hidden beneath the surface of the earth, or rather whose meaning needs to be worked out in confrontation (*Aus-einander-setzung*) with this concealment in order to grant one an authentic identity" (19). It is on this premise that Heidegger proceeds, to frame the purpose of philosophic discovery in terms of this confrontation, to retrieve what is concealed; and, as I say, I follow his way of thinking as best I can to retrieve and in that sense reveal the nature of the Gorgian *kairos*. The success or failure of that effort speaks for itself in this book, and I will speak no more of Heidegger's politics.

2 The Platonic *Kairos*

Of that region beyond no one of our earthly poets has ever yet sung, nor will ever sing worthily. . . . It is there that Reality lives, without shape or color, intangible, visible only to reason, the soul's pilot; and all true knowledge is knowledge of her.

—Plato, *Phaedrus*

Plato's thinking is rooted in Parmenides and other pre-Socratics, most conspicuously in Socrates himself, and to a degree even in such Sophists as Isocrates. But none of these measure up to Plato in all that he had to offer in the magnificent *tour de force* that is his philosophy. Nor, it seems, have any since, as we take to heart Whitehead's observation that the whole of Western philosophy is little more than a series of footnotes to Plato. It is not so unusual, then, that I suggest the rhetorical *kairos* had its origin and pattern of subsequent development in Platonic epistemology and metaphysics, that it is, to all intents, the Platonic *kairos*. In this chapter I will attempt to identify and detail some of the critical ingredients of Plato's philosophy that shape this understanding of *kairos*, presenting them as they pertain, in counterpoint, to the Heideggerian themes introduced in the last chapter, those pertaining to the nature of the self, language, and time.

The first of these themes involves, in Plato, a dichotomy of subject and object based upon the Socratic doctrine of the soul, which meant to Plato, if not expressly to Socrates, the identification of the soul with the self and the self with reason (Guthrie, *Socrates* 150). The second theme is implied in the first, in that particular theories of language are necessarily a function of the manner in which the interaction of subject and object is perceived. In Plato's philosophy, *logos* is a property of the isolated subject, a rational faculty (*ratio*) entailing our jurisdiction over language, through which we represent a non-verbal, pre-existent reality by means of words. The third of these themes, concerning the nature of time, involves Plato's conception of reality as a "detemporalized presence" lodged in the eternal Forms (*eidos*), resulting in an interpretation of time as a series of abstract and homogeneous "now points" detached from specific, concrete situations.

All three themes bear directly on Plato's theory of language that, in substance, rests upon logical rather than strictly rhetorical considerations. In this connection all three themes can be gathered under the rubric of the "metaphysics of presence." The phrase is most often associated with Derrida, but he takes it from Heidegger, and I use it here to provide focus to my contention that the current, accepted notion of *kairos* in rhetorical theory is not only the inevitable result of Plato's epistemology and metaphysics but can be traced to specific passages in his work where the rhetorical *kairos* is given its particular meaning and significance.

I begin this chapter with an overview of the idea of *kairos* as it appears in contexts other than rhetoric, a brief and very selective survey that I offer to provide a general introduction to *kairos* so that the main business of isolating its nature within rhetorical theory might proceed more agreeably. I then move on to a discussion of the oral tradition and the "poetized state of mind" that Plato consistently and often vehemently attacked in his dialogues. I use this discussion not only as a backdrop against which Plato's own theory of language can be put into clearer relief but to show that Plato's attack on poetic experience may, in fact, be seen as an attack on those particular elements of the oral tradition that constitute the basis of what he found most disconcerting in Sophistic rhetoric. Here I broach the idea of the "oneness of language and life" through which I contend Sophistic rhetoric has its origin—thereby providing, in the process, points of reference for an examination of the Gorgian *kairos* to follow in the next chapter. In this context I deal specifically with Plato's interpretation of *kairos* as it appears in the *Phaedrus*, where it very evidently means "propriety of time" but in terms of his philosophical rhetoric is associated generally with the adaptation of language to a given situation. I conclude by discussing the meaning of *kairos* in some of its later uses in rhetorical theory, establishing and briefly examining its presence in Aristotle, Cicero, Quintilian, and modern rhetorical theory; a process that entails tracing the changing appearances of *kairos*, especially as it moves under guises of decorum, *to prepon*, and even *stasis*.

Kairos, A Digressive Synopsis

The etymological roots of the word *kairos* are uncertain, and translations from the Greek are indefinite and diffuse, such that *kairos* has "no single or precise equivalent in any language."[1] Nevertheless, a "basis sense" of the words persists, and given its development in myth and religious texts, *kairos* means "time." More specifically, it means the "right time," the "opportune" or "decisive" moment.[2] This is the basic sense of the word in Greek mythology, where *kairos* was personified as the god of opportunity. He—Kairos—was the youngest son of Zeus and is depicted in engraved carnelians and stone reliefs as wing-footed and limber, sometimes carrying a razor and balance scales, but always as tonsured, with a long, thick forelock. Fleeting and elusive, he was not easily apprehended and would only come one's way but once. To let Kairos pass was to be left grasping in vain at the slick pate in back. But to those with an eye for opportunity, Kairos could be confronted and seized by the tresses in front, where the grip was firm, so that time was thereby "taken by the forelock."

Arthur Bernard Cook explains the purpose of the balance scales as reflecting the vicissitudes of chance and fate, and it is therefore significant that in some of the carnelians and reliefs Kairos can be seen tipping the scales to one side or the other with his finger. The purpose of the razor, however, is more enigmatic. It is open to scholarly dispute whether its presence is linked to a role played by Kairos in some arcane puberty rite or whether it served merely as a pivot for the scales or was, in fact, wielded by Kairos "offering naught but a knife to his followers" (Cook 867). But a carnelian in the Berlin Collection of Classical Art shows Kairos himself poised on the razor, treading its edge. To Cook, it is a posture that captures the essence of the old Greek proverb, "It stands on the razor's edge" (862). Here, indeed, it would seem that Kairos assumes the attitude of the decisive place or point, personifying that sense of *moira*—or fate—coming through the agency of the specific, critical situation so graphically displayed in the carnelian.

Such an interpretation is in accord with A. MacC. Armstrong's contention that *kairos* represented to the ancients an "uncanny power" that placed them in the grip of two opposing and overwhelming forces. In its "weird inescapability" *kairos* was "fate"; in its "baffling incalculability" it was "luck." In either event *kairos* was what "made things turn out well or badly for the agent irrespective of his forethought and abilities" (210). Thus, it could be that the very reason *kairos* was personified as a god at all was not because of the power inherent in the idea, but because of the inscrutable nature of what that power might portend. The enigma is not only deeply rooted in the idea of *kairos* but is integral to it. In one sense *kairos* manifests the decisive power of human will in seizing the moment, as we indeed "take time by the forelock." In another sense the power of *kairos* never comes directly at our behest, being instead precisely as Armstrong says, baffling and uncanny in its reflection of fate's utter capriciousness, beyond our ability to control or even comprehend.

However, insofar as we do not insist upon a complete separation of human will and human fate, the contradictions are assuaged. In this sense Kairos is not so clearly a god but a trickster, assuming there is a difference, as he simultaneously embodies elements of good and evil, creator and destroyer, the giver and taker. As the trickster is understood as a force for proportion and harmony between the community and the individual, and between the conscious and unconscious within the individual, it would seem in this context that Kairos is quite aptly associated with the balance scales. But equally pertinent is the role the trickster assumes as antagonist to the *status quo*, scorning the customs, rules, and laws of society. The idea awaits development in a later chapter, but for now a useful example is the *koyemshi*, the sacred fools of the Pueblo Indians, whose purpose is to smash taboos, mock the sacred chants and stories, and thereby embody, as Douglas Hill says, every individual's need to break free of the cumbersome and burdensome obligations owed to society.[3] Their antics provide a fitting instance of the trickster's purpose to present the "para-*doxa*," that secondary or contrary view to the prevailing body of cultural beliefs, the *doxa*. In this way the "para-*doxa*" is, at the same time, complementary to the prevailing *doxa*—indeed, the negation intrinsic to it—so that through the interface balance and harmony might be maintained or restored. Nonetheless, however germane these extrapolations might become subsequently in this study, at this point the basic sense of *kairos* is simply "decisive" or "crucial" time. It is "an occasion, whether favorable or unfavorable," as Armstrong says (2, 10). Otherwise, as far as its place in mythology is concerned, it would seem that a more specific meaning of the term remains lost in its complex and ill-seen inception.

In modern religion, however, the idea of *kairos* has a more complete and compelling presence. In Christian thought it is the basis of the hermeneutical effort, the essential means of explicating the mystery of the incarnation, insofar as that mystery can be explicated at all. *Kairos* thereby embraces fully the scheme of salvation history, constituting in source and essence the process of the materialization of the spirit, of the latter assuming bodily form. Here, specifically in the incarnation myth, *kairos* designates the idea of the transcendent God becoming flesh as He Himself partakes in His revelation to humankind. Thus, in an even more dynamic fashion than the way the idea is expressed in Greek mythology, *kairos* in Christian thought is quite explicitly the manifestation of a god. Though specific scriptural passages making use of the term are extensive, one of those most pertinent to the myth of the incarnation is Mark 1.15: "The time is fulfilled, and the kingdom of God is at hand." Another is First Thessalonians 5.1–5, counseling preparedness, for "the day of the Lord cometh as a thief in the night."[4] Indeed, it is this passage that enticed the early Heidegger and, according to Hans Kung, had considerable influence on *Being and Time* (499).

But the more particular theological implications of *kairos*, and its fullest conceptual basis in Christian thought, are given by Paul Tillich. In a sense he demythologizes *kairos*, submitting it to a meticulous, hermeneutic interpretation, such that his theology is established largely by means of the exegesis of the term. He speaks of *kairos* in terms of the "times in which the eternal breaks into the temporal between an 'already' and a 'not yet'," subjecting the Christian "to the infinite tensions of this situation in personal and in historical existence" (*Systematic*, 2: 164).

He isolates the concept further, contrasting it with *chronos*. The latter is clock time, time that is abstract, homogeneous, and divisible into identical units. *Kairos*, on the other hand, is not quantitative and measurable, the dull cadence of what Heidegger calls "everydayness," but is organic to human experience. It expresses the way our lives are meted out in critical events that disrupt the normal sequence of *chronos*—a measuring, say, in opposition to the way Newton measured the universe. It yields a dynamic view of the world understood in terms of "creation, conflict, and fate" in contrast not only to *chronos* but to the more general concept of *logos* that Tillich understands as being profoundly evident in today's world by way of its derivative functions of "methodical reflection" and scientific interpretation. *Kairos*, then, is precisely as stated in Mark and First Thessalonians. "Not empty time, pure expiration," and not "mere duration" either, Tillich says, "but rather qualitatively fulfilled time, the moment that is creation and fate" (*Interpretation* 129). In this perspective theology is thereby enlarged to embrace a historiography based upon a polarity of *logos* and *kairos* that is resolved "in history" through a peculiar Christian dialectic of the rational, methodical, and static in opposition to the irrational and dynamic forces inherent in *kairos*.[5] Thus, as Tillich sees it, this dialectic constitutes the nature of *Heilsgeschichte*, of "salvation history."

With a markedly similar emphasis to that in Christian thought, a concept of *kairos* is also present in psychology, especially in the work of psychotherapists who base their methods of diagnosis and treatment of certain psychological disorders on insights taken from the existential philosophy of Tillich and others.[6] Henri Ellenberger, for example, identifies the event of *kairos* as the last and most critical of three phases of existential psychotherapy. In the first phase, diagnosis is structured upon the concept of "existential neurosis," an affliction based not in repressed traumas or specific instances of emotional stress *within* life but in the patient's inability to find meaning in life in *general*. In the second phase, treatment is conducted by means of "encounter" therapy to bring "sudden liberation from ignorance and illusion." The "arrival" of *kairos* marks the third phase, seen by Ellenberger as the decisive moment when the patient is, abruptly, "inwardly ready" to respond to therapy. At this juncture the therapist is most likely to successfully intervene and assist in bringing about a cure that would have been impossible at some point either sooner or later in time (119–20).

Harold Kelman, one of the few American psychologists to endorse either the precepts or practices of existential psychotherapy, expresses the specific relevance of *kairos* in dealing with patients suffering from clinically observed symptoms of "neurotic egocentricity." He says that once a "bottoming out" in despair and anxiety is experienced by a patient, a major shift in "personality manifestation" is prompted by a *kairos* bearing the qualities of an illumination that result in a "new self" and "unitive" world view, involving all dimensions of human values: factual, aesthetic, moral, and spiritual (67). In such a case, *kairos* "implies a right time in the course of events to do certain things that will favor a crucial happening" (80). This transformation is assisted by the therapist working with his client, but, to Kelman, the transformation is a fundamentally mysterious process whose outcome he attributes neither to the therapist nor the client, nor even to their association.

He says both "experience themselves as agents and as participants in a process moved by forces greater than either or both" (67). In the end, Kelman justifies the practice of existential psychotherapy on the basis of its practical results and explains the transformation brought about by *kairos* only through analogies to Christian revelation (76) and divine grace (80).

In the course of his argument Kelman cites instances of *kairos* apparent in literary works, giving particular attention to Tolstoy's *The Death of Ivan Illych*. Indeed, modern literature is rich in illustrations of the theme of *kairos*, and surely one of equal importance to *Ivan Illych* is Marcel Proust's *Remembrance of Things Past*. Here, in the opening pages of the novel, the Proustian hero wakes from sleep disoriented and confused, "more destitute than the cave dweller" (5). In that first moment of consciousness, he experiences a "primal simplicity," an *ennui* perhaps similar in nature to the "existential neurosis" discussed by the psychologists above, though Proust sees the tribulation as pervasive, afflicting life *in toto* with the burden of what is "mediocre, accidental, and mortal" (34). The source of the malady is that time has lost its human vector; it is "disincarnate." In the fourth book of the novel, *Cities of the Plain*, Proust relates the nature of the phenomenon, its essential manifestation:

> Then from those profound slumbers we awake in a dawn, not knowing who we are, being nobody, newly born, ready for anything, our brain being emptied of that past which was previously our life. . . . Then, from the black tempest through which we seem to have passed . . . we emerge, prostrate, without a thought, a "we" that is without content. (2: 271)

Proust is describing here a peculiar instance of *chronos* that proceeds from time as we understand it in the "gnostic" sense—of time in its expressly malignant emphasis, marking only the meaninglessness and transitory nature of life, as we would thereby live it imprisoned in the world of the flesh, closed off from any higher province, human or divine.[7] Coincidental to gnostic time is not only this deprivation of Being but a consequent sense of the strangeness of one's present situation in the world, a listlessness and boredom in an "immense patch of oblivion," as Proust describes it. In his account there is salvation or deliverance only by way of a deep and abiding mindfulness of the past. The past alone puts us in touch with our origins, overcomes *ennui*, transforms pure non-Being into the intricate texture of life's wondrous substance and significance. In Proust, then, *kairos* is the moment of remembrance, a "deep remembrance" when certain impressions are veritably full and flourish outside the current of gnostic time.

Though the entire text of Proust's seven volume novel displays this phenomenon persistently and in various ways, perhaps the passage that exemplifies the process of deep remembrance most concisely is that of the *petites madeleines* in *Swann's Way*. The hero, "only by chance," takes the cakes with his lime tea, yet their taste "infinitely transcends their savors," inspiring a moment of remembrance and "all-powerful joy" (1: 34). In that moment, the past is regained by the hero, even relived. Remembrance of things past comes first in a "whirling medley of radiant hues" and moves more distinctly to a time he had eaten them before, to a Sunday morning he spent as a child in Cambrey. He is then enveloped in the specific sounds, fragrances, and lush colors of the village. Proust writes, "the good folk of the village and their little dwellings and the parish church and the whole of Cambrey and its surroundings, taking their proper shape and growing solid, sprang into being, town and garden alike, from my cup of tea" (36).

Among the things apparent in this episode is the sense of the past that inspires Proust is not mere recollection, for the "deep remembrance" occasioned by the moment moves well beyond any reconstruction of the past achieved through acts of the will or intellect. Proust likens the recovery instead to the Celtic belief of souls held captive in material objects. Like the lost souls, the past lies hidden in objects, and is released in a rediscovery of the self by means of the sensations and feelings of the present that the objects give us (34). Thus, deep remembrance is not only linked to physical sensations but otherwise is wholly gratuitous. And there are interesting variations on the theme, not all of them Celtic. In *Swan's Way* the hero remains resigned to his malaise, to all that is mediocre and mortal, and then is delivered not by his efforts but by a process that Georges Poulet, in his analysis of Proust, says can be understood as the equivalent to divine grace in Christian thought (305). It is thereby an analysis that parallels Proust's in yet another significant respect. In deep remembrance memory functions as a mysterious solicitation that one does not bring about of his own accord but can only remain open to, in the form of desire (300). Poulet says that from a feeling of existence detached from time and place, deep remembrance acts to close the breach of subject and object, expressing "between the being who feels and the object felt a spontaneous accord in which the desire of the one meets with the solidity of the other; as if the external world were now precisely what we would wish it to be" (300). Thus, not only are the stifling effects of *chronos* overcome in Proust's particular formulation of *kairos*, but the past confers upon the present its authenticity. Indeed, in Proust, through the power of the moment the past in its authenticity *is* presence.

The theme of the integration of the past and present, one of many such themes to be found in the literature of Proust, is seen by Frank Kermode as inherent in modern literature as such. The focus of his book, *The Sense of an Ending*, concerns the discrepancy between the work of the imagination and the chaos of life, the distance between a sense of human justice and the "inhuman" reality of things as they are. In a radical extension of the Proustian idea of desire, Kermode says, "We have lost a literature which assumed that it was imitating an order for one that assumes it has to create an order, unique and self-dependent" (167). In essence, we move from a theory of art ultimately based on a *logos* or lawfulness inherent in nature or reality, then mimicked in our art, to one established on the premise that in the absence of such a *logos* we ourselves create a *logos* in the midst of the chaos, a human *logos* imposed by means of our literature. So we return to that atavistic sense of *logos*, seeing it as a function exclusively of language. "It is not that we are connoisseurs of chaos," Kermode says, "but that we are surrounded by it, and equipped for co-existence with it only by our fictive powers" (64).

Thus, in one sense, Kermode designates *kairos* as "a historical moment of intemporal significance" (47); but in this other, more embracing sense, *kairos* is temporal integration, accomplishing for Kermode "a bundling together of perception of the present, memory of the past, and expectation of the future" (46) that cancels the *angst* or *ennui* of our "intermediary position" between an unknown beginning and an unknown end, of our birth and death invariably *in medias res*. Given this interpretation, *both* the past and future are immanent in the present, in a "fullness of time" with beginning, middle, and end in concord. A "timeless literary order" is established as a result, and Kermode culls Aquinas' concept of *aevum*—the "time of angels" who stand poised as on a razor's edge between eternity and temporality—and interprets it as "the time-order of novels" (72). In this manner the "terrifying limitlessness of time" is overcome through *kairos*, through the "time redeeming" yet fundamental unreality—the deceit—of the "fictive power" of words. Though a particular difficulty with Kermode is his reluctance to conclude a number of arguments he barely begins, it is obvious that fiction, as he speaks of it in *The Sense of an Ending*, is a quest for order and meaning that ultimately becomes yet another form of desire, here having a linguistic disposition that finds its Christian equivalent in prayer and supplication.

These illustrations hardly present a clear and consistent definition of *kairos*, but they do offer some interesting extensions of the "basic sense" of *kairos* mentioned at the outset. First of all, it is apparent from these perspectives that there is no sense of *kairos* that is discipline specific to either psychology or literature. There is, for example, no specific "psychological" *kairos* or "literary" *kairos*. Moreover, Kelman's emphasis on the importance of *kairos* in psychotherapy can be contested, and Kermode's paradigm of *kairos* as the defining element of fiction is definitely suspect.[8] What is important, however, is that the sense of *kairos* used in each has its antecedents in Christian thought, especially as *kairos* is linked there with revelation and divine grace. Essentially what the above perspectives do provide are certain insights into the nature of *kairos*, of its attributes relating to language, situational context, and opportune time of decision that directly bear upon its place in rhetorical theory. Yet these attributes, as they are presented above, remain those exclusively of the Christian *kairos*. We might therefore conclude that one particularly vibrant "strain" of *kairos* lies in the Christian perspective, brought to fruition in contemporary thought in the theology of Tillich in particular, and clearly evident in the above illustrations and others that could be mentioned.[9]

A Preface on the Rhetorical *Kairos*

This Christian interpretation of *kairos* is in many respects at odds with that given in rhetoric. Though the origin of the rhetorical *kairos* seems to be the same as the Christian *kairos* and is likewise shrouded in myth, there is an independent development in each. Owing in large part to the good graces of Plato, rhetoric appears to represent a "demystified" lineage, offering a degree of clarity to the meaning of *kairos* seemingly absent in its Christian formulation, however appealing the various hermeneutical analyses might be. At the same time, however, it is obvious that rhetoric, as the most pragmatic of activities, would necessarily avail itself of any means as conducive as *kairos* in providing connections to the situational contexts of the practical world of human affairs. It is from this perspective, then, that we can give some focus to the definition of the rhetorical *kairos*.

I have said that in the rhetorical tradition *kairos* is the power of the opportune moment in gaining advantage in speech, that in Plato it is associated with propriety and decorum, specifically manifest in the rhetor's skill in adapting language to a given situation. This matter of "appropriateness"—the framing of discourse to the particular circumstances of time, place, speaker, and audience—prompts theorists such as James Kinneavy to see *kairos* as the "capstone" of Plato's conception of the rhetorical art ("Neglected" 86). Aristotle also affirmed the recognition of a set of circumstances involving situation, speaker, and audience as crucial to the purpose of rhetorical discourse. Indeed, in his definition of rhetoric as the "faculty of observing in any given case the available means of persuasion" (*Rhetoric* 1355b, 25–27), the presence of *kairos* is implicit, even integral to the rhetor's discipline, for rhetorical skill is here identified with the discovery of the persuasive elements "in any given case."[10] The place of *kairos* is given even greater emphasis in John Poulakos' definition of rhetoric as "the art which seeks to capture in opportune moments that which is appropriate and attempts to suggest that which is possible" ("Toward" 36). Here, *kairos* is pointedly expressed as rhetoric, being virtually equivalent to it.

However, it avails us little at this point to consider the role of *kairos* as so fundamental to rhetoric, even on Poulakos' terms, which he derives exclusively from what he identifies as Sophistic notions of rhetoric. By all accounts *kairos* has an important function in rhetoric, though it hardly possesses the vitality, to say nothing of the cosmic significance, that Tillich assigns to the term in Christian thought. On the contrary, the rhetorical *kairos* seems to affirm attributes Tillich would condemn as rigid and formulaic. Indeed, it is hard to see the matter otherwise, in that the imperatives implied by "philosophical rhetoric" comprise the defining ingredients of *kairos* in the rhetorical tradition, regardless of how rhetoric can be otherwise understood or defined apart from philosophical rhetoric.

Chief among these imperatives is *aneu logou*, the phrase literally meaning "without speech" and constituting truth's primary criterion to Plato, namely that it necessarily resides beyond discourse.[11] By the same measure it resides beyond rhetoric, and *kairos* thereby takes its place in Plato's rhetorical theory under this constraint. In this context philosophical rhetoric affirms *res* over *verba*, giving precedence in Plato's parlance to lovers of truth over lovers of language. Once again, Plato's "genuine rhetorician" knows the truth of the matters on which he speaks. In Plato, philosophic knowledge exists "independently of any verbal embodiment" and "precedes the correct application of language to a human situation," as Morris Partee says (116); it fosters a rhetoric that attempts to create "a formal, logical system of argument from universally affirmed and binding first principles," as Richard Enos says ("Epistemology" 51). By way of these limitations set in place by *aneu logou*, the rhetorical *kairos* can neither evoke nor participate in any appreciable way in the formulation of truth. Thus, unlike the meaning of the terms in Christian thought, especially as Tillich defines it in contrast to *logos*, the concept of *kairos* in Plato is subordinate to *logos*, relegated to the level of a rhetorical technique to be deployed once truth has been discovered through the use of other means, nonverbal and otherwise "rational" means that ultimately constitute the workings of the Platonic dialectic itself.

On this matter much the same can be said of Aristotle and others like him who accord rhetoric certain allowances denied by Plato, claiming that rhetoric is essentially persuasive rather than demonstrative, that it is, above all, an art governed by practical contingencies and does not strive for cognitive certainties, affirmations of logic, or articulations of universals. Nonetheless, consummate philosopher that he is, Aristotle frames a crucial issue of his rhetoric in much the same way as Plato, perceiving *verba* as secondary and inferior to *res*. As Friedrich Solmsen says in a note on the opening page of the *Rhetoric*, Aristotle's "modes of persuasion" (*Rhetoric* 13554a10)— *ethos*, *pathos*, and *logos*— are based on "the use of rational speech" (1355b5), being "attempts at *logical argument* (Solmsen's italics) on which Aristotle would himself like to see Rhetoric rely" (19, Roberts' translation). And even if we see the matter otherwise, or as necessarily more subtle, complicated, or problematic than Solmsen seems to indicate, on the general matter of *verba* and *res* Aristotle's methodical treatment of rhetoric is clearly a movement away from *verba* and of a piece with Plato's, one that de Romilly says departs from the Sophistic ideal of the "magic" of speech and moves toward "reason and austerity" (66). Altogether unceremoniously, Robert Pirsig makes the same point. His criticism of Aristotle is always heavy handed and in places, it seems, based on willful distortions, but given Aristotle's need for method and orderliness, and his deadly penchant for categorization, Pirsig's criticism of Aristotle, especially in terms of *verba* and *res*, is as worthwhile and imposing as that given above.[12]

Thus, however diverse definitions of rhetoric might be, it is significant that the meaning of *kairos* in rhetorical theory is resolved with remarkably consistency. Kinneavy, who has done as much as any other modern rhetorician in explicating the rhetorical *kairos*, says the term has at least three meanings relevant to rhetoric: 1) fitness, proportion, and due measure; 2) critical time or occurrence; 3) state of affairs or overall situation which structures rhetoric for both speaker and listener. The definition he ultimately endorses integrates all three components: "Kairos, when resolved into the rhetorical skill . . . can be defined as that which is fitting in time, place, and circumstance, which means the adaptation of the speech to the manifold variety of life, to the psychology of the speaker and hearer" ("Relation" 13).

Kinneavy takes his definition from Gino Funaioli, one that Untersteiner cites in full in his work, *The Sophists* (197). Using the same source as Kinneavy, George Kennedy offers the same definition of *kairos* in *The Art of Persuasion in Ancient Greece*, where, in context of his discussion of Gorgias, he adds that the rhetorical *kairos* is "the principle which governs the choice of organization, means of proof, and particularly the style" (67). Thus, in rhetorical theory the meaning of *kairos* is clear and consistent, most generally designating the rhetor's judicious adaptation of language to any particular situation he might encounter. It is a conception of *kairos* that is ably articulated by Plato, one that abides in Aristotle, Cicero and Quintilian, and extends to modern theorists other than Kinneavy and Kennedy. However, before the connections between the Platonic *kairos* and the rhetorical *kairos* can be drawn more explicitly, it is necessary to trace their source in philosophy. And that effort necessarily begins with Plato's attitude toward the poets, towards them and others he perceived as purveyors of deceitful words.

PLATO AND THE DECLINE OF POETIC EXPERIENCE

Philosophy is . . . only convention made consistent and deliberate.

—Santayana, *Animal Faith*

Scholars may disagree on whether Plato's attitude toward rhetoric was hostile in general or whether it was merely directed against certain

Sophistic notions of rhetoric.[13] However, there is less room for dispute concerning his attitude toward the poets. His position here seems quite settled and certain.[14] He condemned what he considered the malevolent influence of the Greek poetic tradition that stretched from Hesiod and Homer to Aristophanes, saying that poetic language projected a "dream state" yielding dilemmas and contradictions that eventually culminate in deceit and depravity. In the *Republic,* Plato says that the poets' work is destructive of truth and virtue (X.608b), that they engage in an imitation that corrupts the understanding of their listeners (X.595b). Though deferential references to specific poets can be found in Plato's work (II.363ff), he would censor them as a group, anoint them with myrrh, crown them with garlands, and then escort them to the borders and banish them from his Ideal State (III.398a —b). But far more than the poets alone, Plato's target was the prevailing *doxa* of the Hellenes, a domain of experience shaped by the "Homeric state of mind" prevalent in the religious and cultural institutions of his day and nourished by the whole of the Greek *paideia*. Precisely in this context poetry and literature represented in themselves a perverse morality to Plato, forces of darkness and decay of special danger to the young.

The nature of these ills and others are depicted most vividly in the parable of the cave. Indeed, the imagery could scarcely be more obvious, the sunlit brilliance of the day contrasted with the darkest regions of earth, subterranean places of concealment and mystery, eerie shadows and vague, muffled echoes. If the cosmology behind the parable is far more ancient than Plato, to him the distinction is preeminently philosophical and therefore moral, essentially the difference between a magnificent rapport with the divine and the evils of our earthly state (*Republic* VII.517e). The latter is the world of *doxa,* mere appearances and opinions dependent on the contingencies of time and place, of perceptions manipulated through the ugly talents of poets and Sophists, their clever and deceptive use of language in particular. So Plato laments what passes for wisdom in the cave, the ignorance that binds believer and belief as chains to the wall, to the point of the bondsmen being in the habit of conferring honors and prizes on those most adept in identifying the passing shadows and predicting which would follow which. He laments in particular the fate of those who have experienced truth in all its celestial radiance and then being compelled to testify in courts of law, as did Socrates, to meet the prejudices of those who know only appearances of truth, its quivering shadow on the wall of a cave (VII.516c –517d). So Plato's gaze is fixed upward, in the direction of the philosopher's ascent to the radiant vision that is the supra-sensual realm of the soul, a place of self-consciousness, an integrity or wholeness of the self that breaks the chains and perhaps as well the "spontaneous accord" of believer and belief that is desire, but we are thereby freed from the earth's pull in the realization that from these heights truth is eternal. And if that truth is abstract, detached from any particular earthly reality of time and place, it is universal and unsullied, the most exquisite expression of "what is."

Of course, it is the nature of ideals to exceed our reach, but the measure of Plato's success in his condemnation of poetic literature is manifest, and very ably presented in Eric Havelock's argument that Plato was the prophet of the transformation of Greek culture from one based on the dynamics of poetic expression to one maintained primarily on the abstract, conceptualizing thought of metaphysics. Walter Ong makes a similar point, seeing the transformation more comprehensively as a movement from orality to literacy, where "the philosophic thinking Plato fought for depended entirely on writing" (25). Moreover, both Havelock and Ong agree that the transformation was occasioned by the emphasis on visual experience over aural, by the eye supplanting the ear as the chief organ of gathering and storing knowledge (Havelock, *Preface* vii; Ong, *Orality* 31).

These are, by now, very familiar arguments, yet in this particular context they remain significant, still requiring our attention. Surely, two features of Ong's "psychodynamics of orality" help to provide an understanding of the nature of the oral culture criticized by Plato. Ong says that oral cultures "tend to use concepts in situational, operational frames of reference that are minimally abstract in the sense that they remain close to the living lifeworld" (49). Thus, "basic orality" marked a domain of experience that put the emphasis on what is immediate and practical as opposed to permanent rules and principles of morality and metaphysics. And secondly, Ong says, oral cultures not only "conceptualize and verbalize all their knowledge with more or less close reference to the human lifeworld," but they assimilate "the alien, objective world to the more immediate, familiar interaction of human beings" (42). That is, both Ong and Havelock agree, learning or knowing in oral cultures is equated with achieving empathetic and participatory identification of the knower and the known (Havelock, *Preface* 197–214; Ong, *Orality* 31–77).

In all of this there is a certain connectedness to the earth and curiously, in ways integral to oral cultures, to language itself. In cultures of basic orality there is no separation between "external reality" and a self-governing human consciousness of the nature to be found in cultures based on full literacy. Prior to the revolution wrought by Plato, the Greek ego was not the autonomous "soul" spoken of by Plato, but was transparent, the sheerest membrane or channel outright that opened to a "self-identification" with a "whole series of polymorphic vivid narrative situations" conveyed in poetic performance (Havelock, *Preface* 200). In response to the performance, the ego was "split up" into an endless series of moods—love, hate, fear, joy, despair—and in the absence of analytic categories of the "literate mind" was unable "to discover reasons for action in itself rather than in imitation of poetic experience" (200). To distinguish it from views of the ego or self as fixed, autonomous, and unchanging, the disposition of the soul spoken of by Havelock and Ong has the name *psyche,* that ancient sense of human being owing its makeup to *mythos* which, in turn, relies on a conception of *logos* shaped not by reason but utterly by language.

Again, the basic arguments are familiar. Not only are there fundamental distinctions to be made between oral and literate cultures, but these distinctions are based on alternate modes of consciousness that proceed, respectively, from mythic forms of conception and, in the absence of such forms, from abstract and universal principles removed from vivid narrative situations. Here, when Havelock and Ong speak of the communal identification of the knower and the known, they are speaking not only of an attribute of oral cultures but of cultures pervaded by a wholly different conception of language. In the oral tradition, thought "came into being" through mnemonic patterns of alliteration and assonance, epithetic and other formulaic expressions, where language was conceived "as a mode of action and not simply as a countersign of thought" (Ong, *Orality* 32). Given these considerations, far more lies in the balance than the surface features of language, mere matters of style. A radically different expression of human nature is at issue, one having its source and substance in language that comprises completely the *psyche.* Thus, we might see the latter as existing prior to the distinction between literal and metaphorical, subject and object, and a host of other distinctions brought about by the abstracting work of the literate mentality. Always, language is the determining factor. Previous to these distinctions, as Ernst Cassirer says, there was an "indissoluble unity" between word and referent, "a complete congruence between image and object, between the name and the thing," so that "the conscious experience is not merely wedded to the word but is consumed by it" (58).

In Cassirer's description, language in mythic thought, like the *logos* of the primordial first word, is creative, and creative in terms of a consciousness that, being sacred—or, as Havelock and Ong would have it, oral—provides a means of imagination that enables us to be attuned to the world, to hear it and realize it, thereby giving it new and spiritual existence in our relationship with it. There is here a notion of language—of *logos*—quite different from a conception of *logos* as tool or utility, or as simply the guidance of reason. On the contrary, *logos* conceived essentially as language pervades with a radical immanence, with a dynamism characteristic of living organisms themselves. Static descriptions of the universe are neither adequate nor complete. Being is but a becoming imbued with the flux and fluidity of the human lifeworld, devoid of any structure or internal organization established by dividing reality into categories of things or properties.

Logos retains its meaning as the lawfulness of the world, the basis of its intelligibility. But here its source lies in the prior imperatives of *verba*, in the web of cultural beliefs, social conventions, and shared meanings that comprise the *doxa* of a community as its hedge against chaos and turmoil. In this context *logos* puts one in the presence of *ousia*, that in its pre-Platonic manifestation meant the "process" of Being, as it thereby projects an openness in which both subject and object, life and language, are put in such proximity as to persist even as undivided. Here, *verba* does not so much eclipse *res* as comprise it, or is one with it in a wholeness prior to each, invoking in that sense the paradigm of "earth." In this view *logos* bears the signature of Heraclitus, of *logos* made manifest in language as a "collecting" or "gathering" (Minar 323). Hence, in one sense, a person collects or gathers in Being; yet, in another, he is similarly collected or gathered in *by* Being. "*Wenn ihr nicht mich, sondern das Wort vernehmt*" is the way Diels translates the fragment from Heraclitus (B50). "Heed not me, but the word"—the *logos*. Insofar as one does this—heeding not the speaker but the language that speaks through him—"it is wise to say that in *logos* all things are one," as the fragment concludes.[15]

In contrast to the definitions of our being in classical philosophy based upon more analytical frames of reference, it is difficult in Heraclitus to extract a principle of the separation of idea and thing, word and referent—subject and object—upon which these frames of reference would seem to rest. As George Thompson says, the Heraclitean *logos* is invested with a "mystical significance" based upon the "principle of the interpenetration of opposites" (79). There is no identity of a "consciousness-subject" or "ego-subject." Indeed, Bruno Snell has shown in his study of the language of Heraclitus that the syntax of the fragments themselves springs directly from personal experience and sympathetic involvement rather than the abstract and conceptualizing thought of logical understanding.[16] The Heraclitean *logos* therefore has a double sense, designating on one level the meaning which lies in human speech and, on another, the meaning which resides in things. In this manner *logos* means not only that which we say but that which "speaks" to us. ["*Logos (ist) nicht nur dies sinnvolle menschliche Rede, sondern auch der Sinn, der in den Dingen ruht, der zu uns spricht und uns die Dinge bedeutungsvoll macht*"] (Snell 365). The emphasis is on words revealing things, or of things revealing themselves through language, speaking to us, as it were. As such, Being and our existence within Being are contained in the wholeness of language, in the "mystical significance" of the Heraclitean *logos*.

From this perspective reality is not a ready-made realm of fact, manifest and indisputable. Nor can it be fully comprehended through reason or reflection but, such as it is, can be realized only in acts of language. It is an idea of language once articulated by von Humboldt, who claimed we are so utterly "situated" in language that it determines who we are, virtually that we are. We "spin" language out of our own being, he says, and ensnare ourselves in it (53, qtd. in Cassirer 9). In this strange dominion of language, what is primal and persistent is a reciprocal relationship between us and our environment, one reflecting the existence of a "presence" in nature, a reflection of our own humanity to be sure, but the presence nonetheless of a life force, a "thou" in all things that can not be ascertained in the least by analytical means. It is a presence, as Henri Frankfort says, claiming "recognition by the faithful" rather than "justification before the critical" (16). The expression given this encounter between us and this presence, the means through which it obtains realization in us, transpires as language and nothing else besides. Here, language represents—*re*-presents—absolutely nothing but is itself an "organ of reality" producing and positing "a world of its own" (Cassirer 8). We thereby live with objects, ideas, and events essentially as they are presented in language, so that language, rather than definite elements of external existence, constitutes the "forms" of mythological conception: "Man lives with *objects* only insofar as he lives with these *forms;* he reveals reality to himself, and himself to reality, in that he lets himself and the environment enter into this plastic medium, in which the two do not merely make contact, but fuse with each other" (Cassirer 10).

"Plastic medium" is indeed a fitting description of the élan of poetic experience, of the awareness through which the community rituals, storytellings, and oral performances of the Greek rhapsode took place, conveying that sense of the "illusion that comprehends reality." Manifest on the most tangible level as *doxa,* this was specifically the sort of mentality that came under attack by the Greek philosophers, Plato in particular. The metaphor he applies to this disposition of the poet and his audience is familiar and very apt: "It is a dream from which both of them, bewitched by the images which pass before them, have to be awakened before they can become aware of *what is.*"[17] This perspective of "what is," at least Plato's formulation of it, establishes a philosophic basis for determining truth as eternally fixed, abstracted from the contingencies of time and place, and therefore possessed of a viability that is bound neither to sensible phenomena nor to particular cases or situations. The "what is" of things resides in the Platonic Ideas, their Forms (*eidos*) existing above and unaffected by their appearances in the material world. In clear contrast to the "mythological forms" of which Cassirer speaks, these serve to mediate cognition, impose order on the disconnected flux, and guarantee the accessibility to the knower of the known. Plato serves this purpose of *aneu logou* through his dialogues, doing so most succinctly in the *Phaedrus,* as Socrates says that truth resides in that region beyond which "our earthly poets" have ever sung, where "reality lives, without shape or color, intangible, visible only to reason, the soul's pilot; and all true knowledge is knowledge of her" (247c).

Indeed, in the context given by Plato, to call them "Forms" at all places them beyond the manifold of sense perception, establishing their *a priori* status as intelligible realities independent of the mind but accessible only to the mind in the "unblemished vision" of the "mind's eye" rather than that of the body (249b–50d). Hence, the nature of the Forms, as Derrida says, lies in their "nonsensible visibility" and their "intelligible invisibility" (*Disseminations* 134). As opposed to the sympathetic self-identification inherent in poetic experience, a dichotomy of subject and object is established in place of the encompassing aural experience, an intellectual "seeing," if you will, implying that the one who sees is in all cases separate from that which is seen. Things are thus abstracted in the literal sense of the word, in a rational process that seeks reality in the "highest thing" (Grassi, *Four* 16), in the Forms themselves as the standard of stable and enduring truth. Heidegger calls this conception *Lichtmetaphysik,* Plato's "sight metaphysics," and it entails the ultimate separation of objective knowledge from the knowing subject. We are possessed by the light of reason, reified from our surroundings by the *lumin naturale* present within and comprising the integrity of the self.[18] Indeed, reason, conceived as the essence of the self, marks the epitome of the doctrine of the soul.

The idea of the soul, this "strange, new knowledge of the self" that Francis Cornford claims to be one of the most remarkable discoveries in Western philosophy (73–123), was incipient in the thought of the early Pythagoreans and was most profoundly evident in that of Socrates. Indeed, he was the first to identify the soul with "I"—with normal waking consciousness as distinct from mere dream states. But it was Plato who first elevated the concept to the level of a philosophical doctrine. The argument can surely be made that in the *Apology* Socrates dies for the sake of the soul. In fact, the phrase that resounds throughout the early dialogues is Socrates' injunction "to care for the soul."[19] The soul's predecessor in pre-Socratic thought, the *psyche,* existed only on the level of physical functions or, as we have seen, in moods of pride, courage, grief, and so on. It was a "feeble and witless thing," altogether dependent on the body (Burnet 247, 252). In Heraclitus, for instance, the *psyche* was associated with the circulation of the blood, the steam of one's breath (Cleve 58–59), or quite typically with the body's "double" which roamed at large in dreams as one slept. In this perspective "man regarded himself as a kind of mirror, in which the external world was seen through a glass darkly" (Gomperz 495).

When Plato speaks of the soul he uses the term "*psyche,*" as well he should, for that is the Greek word for "soul." However, as I have indicated, when Plato speaks of the *psyche* he is talking about something that differs in principle from the *psyche* spoken of by Homer, most of the pre-Socratics, and in crucial respects by the Sophist Gorgias. In the most general sense, Plato's idea of the soul is similar to our own, insofar as we take the soul to mean the essence of the self that is spiritual, immortal, and by those measures distinct from earthly things. However, relative to the pre-Socratic notion, the significance of the doctrine of the soul cannot be overstated. It is the single most important component of Plato's "rhetoric," as it is of his philosophy in its entirety. It is of such crucial significance that Plato's attack on Sophistic rhetoric is secondary to this more momentous effort to secure, by virtue of our souls, what is fixed and unchanging in our being and thereby in reality as such.

The soul's purpose is to rule the "inferior parts" of our nature associated with passions and physical appetites (*Timaeus* 69c–70a). Thus, to Plato, the soul is part impulse, but its rational essence is portrayed as the part that thinks and knows, governing physical and psychological "moods" in the well regulated man or woman. What is apparent in the myth of the soul so richly expressed in the *Phaedrus* is that the charioteer alone is identified with reason, with the soul's most noble aspect, while the steeds—surely the black one—represent the non-rational, animal facets of the human personality that must be directed or held in check by reason. This tripartite nature of the soul will be given more attention in due course, but here the point that needs emphasis is that the doctrine of the soul, specifically as the soul is identified with reason, establishes human being as ego-subject, whose integrity apart from all specific contexts or situations constitutes a necessary first premise governing statements and conclusions about any particular situation in which it might be involved. This attitude achieves its conceptual apotheosis in Descartes; it is the basis of Santayana's proposition that we are possessed of an "animal faith," the presumption as a matter of faith alone that there exists an absolute distinction between our consciousness and what surrounds it, that reality, as either substance or essence, is fixed and final and ultimately owes neither its being nor its nature to our awareness of it. In short, the doctrine of the soul not only preempts the identification of the knower with the known, replacing the influence of poetic expression with philosophy in the process, but more specifically it establishes the autonomy of the knower, calling him "to be awaken and converted away from becoming toward the abstract object which constitutes timeless and intelligible knowledge" (Havelock, *Preface* 245; Plato, *Republic* V.476c). In sum, Plato's doctrine of the soul establishes an eternal "essence" of the self that is distinct from the body and all that environs it, in the end obtaining its express warrant in the dominating influence of *aneu logou*.

Accordingly, the nature of philosophic knowledge is apparent in contrast to *doxa*, that condition of belief and conviction Plato defines in terms of the dream state preceding an awakened state of consciousness. *Doxa* is here associated with the poets, orators, and rhetoricians of fifth century Greece and is unfortunately, and sometimes ruinously, translated as "opinion." Essentially, *doxa* denotes the sheer presence of things believed, the uncritically accepted conventions, shared beliefs, and convictions that constitute the substance of the cultural cohesion of any community. Pierre Bourdieu says *doxa* rests on the "hither side of all inquiry," implying the "immediate fit" between subjective and objective structure (168–69). In the specific context of Plato's conception of *logos*, *doxa* thereby expresses the ontic status of the world of appearances, seen as the shadows or pale reflections of the objective essences of the Forms.

In the *Timaeus* Plato explores some of implications of this conception of *doxa*, and more doggedly the conception of language underscoring it. He opens the dialogue by contrasting myth and dialectic and forthwith develops the distinction between "lovers of language" and "lovers of truth." Socrates desires to know how the Ideal State depicted in the *Republic* would appear once it is "set in motion," how it might respond to some great struggle or challenge, and be "celebrated" in some fitting manner. He says poets are a "tribes of imitators" ill-equipped to perform the task, that the Sophists are also unsuited because they are neither philosophers nor statesmen and that, besides, their nomadic way of life precludes their allegiance to any state. Ironically, Socrates also says that he himself has not been given the proper language to celebrate his Ideal State, that he lacks the "gift of description." He therefore defers to Critias, who responds with a mythic tale that he claims was first related by Solon, concerning how Athens saved the Mediterranean world from Atlantis. In that the narrative rests on the authority of Solon, Critias claims further that it can therefore be accepted as actual fact rather than mere legend (19b–21b). He does not use the language of logic or reason but informs Socrates that the language he uses is, by virtue of its repetition of ancient lore, especially appropriate to reflect the "ingression of perfection," indeed of creation itself. After Critias completes his narration, Socrates responds with a number of philosophical speculations embracing such topics as the nature of God, the distinction of the sensible and intellectual, and peculiar conceptions of time and space that constitute the core of the *Timaeus*—all of them, it would seem, given in the spirit of Plato's belief that the mythic stage of explanation has been supplanted by a more sophisticated method of explanation reserved for dialectical reasoning.[20] Of specific concern are certain epistemological tenets of this method, those underscoring the nature of "what is." As they stand in contrast to those of *doxa*, Socrates tells Critias the following:

> If mind [*nous*] and true opinion [*doxa*] are two distinct classes, then I say that there certainly are these self-existent ideas [Forms] unperceived by sense, and apprehended only by the mind; if however, as some say, true opinion differs in no respect from mind, then everything that we perceive through the body is to be regarded as most real and certain. But we must affirm them to be distinct, for they have a distinct origin and are of a different nature; the one is implanted in us by instruction, the other by persuasion; the one is always accompanied by true reason, the other is without reason; the one cannot be overcome by persuasion, but the other can: and lastly, every man may be said to share in true opinion, but mind is the attribute of the gods and of very few men. (51e)

Plato's thought can be wonderfully compelling, as it is here especially. Insofar as mind is an attribute of the gods, thinking is one with Being, and as each is actualized by virtue of what lies in some heavenly sphere beyond this world of appearances, this union of thinking and Being marks indeed the apotheosis of the soul's excellence. But the passage also provides an apt extension of the ideal of *aneu logou*, for one obvious implication of Plato's separation of the world of mind from that of opinion is that true knowledge (*episteme*) is not a function of language. Another is that *ousia*, or Being in this Platonic sense, is fixed in the "self-existent ideas" of the Forms, consequently severing the relationship between language and Being as surely as it seals that between thinking and Being. Samuel Ijselling's point is well taken, as he claims Plato's Forms, his theory of Ideas, affirm an "*absolute* reality" detached and eternally removed from earth-bound influences of language, exempt "from any word or speech about it" (15).

The focus of Socrates' response to Critias is not only that mind—understood here as *nous* or reason, the soul's highest function—can be contrasted with the *doxa* of the poets and Sophists, but that truth as an exclusive function of the mind precedes language and therefore constitutes a necessary condition for anyone to achieve "genuine" speech. The idea is essential to Plato's notion of rhetoric as a *techne* or "art." Not only must we heed the dictum that the rhetor in all cases must know the truth of what he speaks, but, as Socrates says later in the *Phaedrus*, "Any man who does not know the truth but only has gone chasing after opinions, will produce an art of speech which will seem not only ridiculous, but no art at all" (262c). Thus, *logos* is neither language nor the essence of *ousia* in the Heraclitean sense of "becoming," but is one with reason, a faculty of the human soul, a property of the isolated subject who now, through Plato's conception of *logos*, is the "master" of language. As George Seidel observes, *logos* loses its moorings and drifts aimlessly in its transformation from Heraclitus to Plato, until it "becomes a derelict thrown up on the island called Reason" (98). And "derelict" is the likely word, for *logos* is here conceived not only as reason but therefore as untainted by the conventional beliefs through which the transformation was accomplished, indeed the beliefs that ultimately constitute its very nature; and "island" is the likely word, for *logos* is pure and isolated as envisioned here, the means of methodical reflection on all things to include itself. Thus, we become aware of *logos* as our way to bypass the treacherous routes of *doxa*, to escape the latter altogether. In fact, the transformation is all the more remarkable for its stealth, as if done by slight of hand to obscure elements of chance and fate, turning convention about to make it consistent, deliberate, and reasonable. It is a perspective called philosophy.

And it bodes ill for poets and Sophists. Armed with his philosophy, Plato directs his attack against the power of the word and its artificers. In that the dialectic proceeds by reason only, without the help of the senses, Plato claims the "sensual" influence of language is but deceit and indulgence in flattery. The poets bear the sweet influences of melody and rhythm that awaken and nourish feelings, but these affect the "rational part" of the soul only adversely, "impairing reason"; and once the poets' words and phrases are stripped of meter, harmony, and rhythm, it is apparent they are "far removed" from truth and reality (*Republic* X.601–05). It is little wonder, then, that in the Ideal State described in the *Republic* the simple style alone would be permitted (III.397–98b). "Linguistic ornaments are mere obfuscation," as Morris Partee says of Plato. "Utterances so brief as to be formless are best; the continuity of speech must not obscure the continuity of reason." In this manner Plato "consistently denies any value whatsoever to the particular embodiment of thought" that language provides. Plato thus "eviscerates" poetry of its "verbal organ," which would seemingly separate language utterly from thought. In any event, to Plato the stylistic powers of the poets, Sophists, and rhapsodes serve only to call attention to their discourse, obscuring truth and promoting deceit ("Plato" 385).

What we can conclude from this analysis is that Plato's theory of language is exemplary of what the poststructuralists call logocentric; that is, the Forms establish a basis or foundation that predetermines meaning. Ong calls the exclusively linguistic implication of Plato's logocentricism the "pipeline" model of communication (166), where content is distinct from *lexis*, the medium through which content is expressed. Concepts of the mind are seen as existing independently and prior to their expression in language, so that language is used as a tool to represent something outside itself. Under this presumption, then, knowledge not only exists independently of its verbal embodiment but due to that difference can never be truly expressed, and, as there are many particular embodiments of thought, there are many different ways of saying the same thing, that unavoidably remains the "same thing" however expressed.

As a result, truth in Plato is derived referentially, as a function of the correspondence between words and things. In place of *aletheia*—of truth apparent in the revealing and concealing power of Being manifest in those earthly recesses of the cave—truth is instead evident straight away in the sunlit world Plato shapes. Truth is then conceived as *veritas*, what Heidegger calls truth as a measure of the "correctness of intellectual judgment" achieved through the conformity an object bears to our expression of it (*Existence* 297). Here, not only are the Forms real entities, the ultimate constituents of reality to Plato, but they are meant to serve as objects of linguistic meaning or reference (Graeser 366–72; Kerford 76), constituting his "purely referential theory of meaning" (Kerford 77). Accordingly, there are two parallel realms in Plato: reality and discourse; and discourse is "genuine" only when it truthfully represents the reality of "what is." *Res* both precedes and supercedes *verba*, so that fixed and prior truth must always be the measure of what we say, insofar as we speak without deceit, without *apate*. And the proclivities of style, that are in all instances expressly matters of *verba*, embody the most sinister forces of deceit, for they bear the seductive powers of language at their greatest potency, often catching us unaware, bewitching us. In this manner the transformation is accomplished. We move from the immanence of poetic experience to Plato's epistemology of transcendence set out in the dialectic, and from there to his metaphysics putting us beyond the existential entanglements of the earthly realm of *doxa*.

Last of all, Plato's effort to overcome *doxa* can be seen in general as an effort to overcome the exigencies of time, the most persistent of *doxa*'s "existential entanglements." Events, to be "events" at all, transpire in time, bound to specific situations that determine our experience of them. In the place of *doxa*—of the "endless doings and events" of sheer "becoming," as Havelock says—Plato purpose was to substitute a "discourse of being," free from time-conditioning (*Preface* 180–82). If reality is interpreted in terms of essence, as a world of Forms that is abstract and timeless, then we, by participation in that essence by virtue of our immortal souls, escape time. This Platonic view, what Heidegger calls the "ontotheological," places us outside the rhythm of temporality, in a posture of disinterest and objectivity where the distinction between permanence and change, Being and becoming, is patent. Thus, temporality, understood here as involvement and participatory identification with time—of Being precisely *as* time—is effaced. An objectification comes to pass in its place, in the regularities of *chronos* and pure duration rather than in the events that mark the existential tensions of life lived out in the full sensibilities of our mortality in the thick of the here and now.

As Plato presents it in the *Timaeus*, time is resolved into a "moving image of eternity" that "abides in unity." When the Demiurge created the material universe, with his eyes fixed on the transcendent world of the Forms, temporality was just an imperfect image of eternity. As a result, the "past and the future are created species of time, which we unconsciously but wrongly transfer to eternal being." Thus, "we say that it 'was', or 'is', or 'will be', but the truth is that 'is' alone is properly attributed to it" (37d—e). As the "it was" and the "it will be" are banished from temporal organization, excluded as "not being," the "it is" alone is attached to being in the regularity of stability and presence. "True being is identical with what is *present* or with *presence* (Weiss 178), resulting in an exclusive emphasis on the *nunc stans*, a "standing" or "permanent now" that is alone the correct image of "eternal being."

This "metaphysics of presence," as Derrida says, shapes classical metaphysics on the basis of "intemporal kernel of time" (*Margins* 40), an "eternal now" isolated from duration. In that sense the metaphysics of presence is simply a reiteration of the doctrine of the Forms in that it "postulates a reality, or a truth," in Sharon Crowley's definition, "which exists outside the perceiving consciousness of man and which is unmoved, essentially unchanged, by his perception of it" (279). But in another sense, critical to the following discussion of the rhetorical *kairos*, the metaphysics of the presence, precisely as the "eternal now," preempts the possibility of the temporally engaged "moment" or "instant." Time is then "leveled off" or "covered up," as Heidegger says. It is seized in a concept, "'sighted' in the manner of the use of clocks" (*Being* 474–76). In this way it is "mastered" in the Platonic perspective. In being framed solely in terms of its regularity, time is brought to account, placed at our service, much the same as language. Indeed, Plato's conception of time marks his most cogent repudiation of poetic experience, because the timelessness of the Forms is the surest hedge against becoming and change, the very phenomena that is life, the stuff of pathos and tragedy of which the earthly poets do indeed sing. Borne of this addling mix of language and events transpiring in time, poetry literally speaks the mystery that there is time at all, that there is Being. And so the mystery is displaced by Plato's stark discourse of eternal truth, by this essential prejudice of philosophers.

Plato and the Rhetorical *Kairos*

In at least a couple of respects this examination of Plato is hardly controversial. One is the fact that "truth" is the cornerstone of his philosophy. Another is that the dialectic, whether we believe it precedes or is itself the essence of the rhetorical process, is the means of achieving truth and therefore a necessary condition of his rhetoric. Undeniably this is the case insofar as his rhetoric can be considered "philosophical" at all. And, in addition, another necessary feature of his rhetoric, for it to be considered "rhetoric" at all, is *psychagogue*. Here his attention is directed to the psychological component of our souls, that part attentive to "linguistic" rather than strictly dialectical considerations. Thus, to Plato, the "genuine" rhetorician must not only base his appeal on truth and reason but must, as well, understand how truth is best expressed and communicated in reckoning with others as creatures responsive to the effective use of language. But there seems to be a catch to this curious juxtaposition of truth and language, reared in Plato's own deliberations, that attests to something especially significant in his rhetoric. Given his belief that truth is non-verbal, the issue turns on the sort of creatures we by the nature of reason and the sort we are by the nature of language, of how we take our measure in regard to each. Indeed, it is not overstating the case to say our very souls are at stake in this matter, and Plato resolves it in the only way he can, on behalf of reason and thereby the soul's integrity. As the rhetor must be proficient in the art of leading other men's souls by means of language, he necessarily seeks, in the phrasing of a very critical line from the *Phaedrus,* the "qualitative correlation that obtains between speech and soul" (271d).

Of course, Plato's more immediate purpose here is to distinguish "types" of souls, matching them to types of discourse, but to achieve this correlation, as Michael Leff points out, Plato is inevitably preoccupied with an "analogical relationship" existing "between the genuine Forms of truth and forms available in language." As the Forms of truth are utterly non-verbal, the relationship is necessarily analogical because language, in contrast to the Forms, partakes of the world of appearances, functioning exclusively in terms of practical, human contexts (22). Richard Weaver's observation is equally cogent in this context. He explains the relationship by saying that in Plato "rhetoric appears" in the passage from "the logical to the analogical" (18), because "the soul is aware of axiological systems which have ontic status" (17). *Logos* remains supreme in any event, though it is here manifest as the soul's reasoning function, as rhetoric thereby makes its appearance on the analogical level as secondary, subject to this function. Within these restraints, the Platonic *kairos* is the moment when the practical situation is understood by the speaker, so that, in the descent of the logical to the level of the here and now, the situation might be exploited by the speaker through the appropriate choice of words. Rhetoric appears insofar as he wields language, insofar as he is its principal. In this way *kairos* may be the only sure warrant of Plato's rhetoric, its defining ingredient, but that hardly elevates his rhetoric to some august station in his overall scheme of things, as his ambiguous or otherwise spiritless depictions of it seem to indicate. What is at work here, quite simply, is the recognition on the part of the rhetor of a situation that he suitably responds to or masters through the appropriate use of language.

In Plato's dialogues there are any number of references to *kairos,* two of which are important to this discussion. The first of these is in the *Parmenides* (156) and will be examined later in this chapter, as it figures significantly in the discussion of the Gorgian *kairos* to follow in the next. A second reference occurs in the *Phaedrus*, is the better known of the two, and figures significantly in the discussion at hand. It develops the full rhetorical implications of *kairos* in Plato, providing in some detail the movement from the logical to analogical achieved through the agency of *kairos*. Moreover, since this reference is one of the most concise and complete statements of the idea to be found in classical rhetorical theory, and is the reference most often cited by rhetoricians when discussing the meaning of *kairos* in Plato, the passage is presented in full:

> Since it is in fact the function of speech to influence souls, a man who is going to be a speaker must know how many kinds of souls there are. Let us, then, state that they are of this or that sort, so that individuals also will be of this or that type. Again, the distinctions that apply here apply as well in the cases of speeches: they are of this or that number in type, and each type of one particular sort. So men of a special sort under the influence of speeches of a particular kind are readily persuaded to take action of a definite sort because of the qualitative correlation that obtains between speech and soul; while men of a different sort are hard to persuade because, in their case, this qualitative correlation does not obtain. Very well. When a student has attained an adequate grasp of these facts intellectually, he must next go on to see with his own eyes that they occur in the world of affairs and are operative in practice; he must acquire the capacity to confirm their existence through the sharp use of his senses. If he does not do this, no part of the theoretical knowledge he acquired as a student is as yet of help to him. But it is only when he has the capacity to declare to himself with complete perception, in the presence of another, that here is the man and here is the nature that was discussed at school—here, now present to him in actuality—to which he must apply *this* kind of speech in *this* sort of manner in order to obtain persuasion for *this* kind of activity—it is when he can do all this and when he has, in addition, grasped the concept of propriety of time—when to speak and when to hold his tongue, when to use brachylogy, piteous language, hyperbole for horrific effect, and, in a word,

each of the specific devices of discourse he may have studied—it is only then, and not until then, that the finishing and perfecting touches have been given to his science. (271d–72b, Helmbold and Rabinowitz trans.)

This is *rhetorical* theory of the purest sort. And of a particular kind. On both counts the idea of *kairos* makes it so. It can be understood specifically as "propriety of time," though, in a more general sense, the Platonic *kairos* designates the transition of theory into practice, the transformation of forms of truth into those of language stressed throughout the passage. In Plato's overall aim to influence men's souls, *kairos* entails a movement from "true knowledge" to "opinion," from *episteme* to *doxa*, where the rhetor, "through the sharp use of his senses" must obtain an "adequate grasp of the facts" as "they occur in the world of affairs and are operative in practice." This process of making "true knowledge" relevant to a given human situation is a point Kinneavy makes repeatedly concerning *kairos*, and his special contribution to our understanding of the rhetorical *kairos* is to make clear that in Plato's view the world of ideas, the timelessness of the Forms, "is brought down to earth by the notion of *kairos*"; that, indeed, "in Plato's system, rhetorical thought becomes effective only at the moment of *kairos*" ("Neglected" 88–89). Quite clearly, Plato does concede the particularity of a given situation that is realized and encompassed entirely by *kairos*. The passage provides, as well, the obvious warrant for those definitions of *kairos* given earlier. In the emphasis on the listener being present to the rhetor "in actuality," *kairos* is the adaptation of language to a particular situation, determining the sort of speech, the manner of its delivery, the particular appeal to be made to various "types" of souls—and, in fact, whether the rhetor should speak at all or remain silent.

The point is crucial. For truth as *aneu logou*, to be realized in this world, must pass through the portal of *kairos*, however damaging to truth that passage may prove to be. Just as crucial is that the passage requires the mediating presence of the soul, as the soul obtains its "qualitative correlation" with speech, and where Plato's emphasis is more on the soul as principal than agent, more on the speaker than listener. Therefore, in no sense does Plato say or imply that *kairos* stands in lieu of universal and binding first principles. On the contrary, as he depicts the function of *kairos* in persuading listeners, he is concerned primarily with truth. Dialectic, as a necessary condition of his philosophical rhetoric, remains in place. As Edwin Black says, "dialectic was Plato's general scientific method; rhetoric is a special psychological application of it" (182). Given this contention, which seems to very accurately reflect what Plato himself has to say regarding his rhetoric, truth is fixed but can be expressed in different ways. For instance, in a departure from his sweeping condemnations of rhetoric in the *Gorgias*, Plato offers in the *Phaedrus* a relatively detailed account of the correlation between speech and soul, saying the rhetor must discover the kind of speech that "matches" or corresponds to the nature of each type of soul. Once this is accomplished, the rhetor then "arranges" and "adorns" each speech in such a way "as to present complicated and unstable souls with complex speeches exactly attuned to every changing mood of the complicated soul," while the "simple" soul is presented with "simple" speeches (277c). Only by following such a process, Plato says, "will it be possible to produce speech in a scientific way" (277b—c). All else is subsequent and subordinate to this purpose.[21]

That truth is fixed and universal, but can be expressed in different ways to appeal to different souls, is indicative of a logocentric view of language. So here, most clearly, to bridge the gap between the nonsensible realm of ideas—from what Plato in the *Sophist* calls "the unuttered conversation of the soul with itself" (263e)—to the sensible, phenomenal world of appearances that is the realm of rhetoric, Plato necessarily relies on metaphor and analogy, on the "likenesses" that can be drawn between the two realms, so that the truth apparent in *aneu logou* can be mirrored, however darkly, through language in the practical "world of affairs." Language is thus analogical and the ability to use it effectively—"to produce speech in a scientific way" or "to give an account," as Plato so often says—is a consequence of our prior reckoning with the truth we garner in the nonsensible realm. In Plato, Glenn Morrow says, "the resources of rhetoric supplement and make effective the insights reached through dialectic" (342). Or, in Plato's analogy, the dialectician is like a farmer "who finds a congenial soul and then proceeds with true knowledge to plant and sow in it words" that grow themselves, to thereby nourish in that fertile ground the dialectic and cause the growth of yet other words. The analogy is extended, even convoluted, but in this context, as in all others, Plato claims truth comes prior to the rhetor's discourse, to his speaking or writing (*Phaedrus* 277a—c). Thus, this view of language ultimately puts the emphasis of the Platonic *kairos* on arrangement and style.

Though not itself a particular feature of either, the Platonic *kairos* is a means of determining arrangement and style, of how to best "arrange" and "adorn" the truth to make it most receptive to the souls of others, whether they are "complicated" or "simple." In short, the Platonic *kairos* is a device determining when to employ the specific devices the orator may have learned in school, such as "when to use brachylogy, piteous language, hyperbole for horrific effect." The only "sure warrant" of his rhetoric notwithstanding, in the end *kairos* is a device among others in Plato's philosophical rhetoric in service to the preeminent purpose of "truth."

This idea of *kairos* is also present in Aristotle's works. "The concept, if not the word, pervades the *Rhetoric*," Kinneavy says ("Relation" 15).[22] As previously noted, the presence of *kairos* is implicit in Aristotle's definition of rhetoric "as the faculty of observing in any given case the available means of persuasion" (1355b26–27). As we have seen, in the phrase, "in any given case," Kinneavy sees the "particularity and uniqueness" in given rhetorical situations that account for the "*kairos* element." For example, Kinneavy sees the presence of *kairos* in Aristotle's discussion of the functions of voters in the assembly and jurors in law courts given in Book I of the *Rhetoric*. As opposed to the "prospective and general" decisions required of legislators writing laws, the purpose of the voters and jurors is "to decide on definite cases brought before them" (1354b5). To Kinneavy, this "unique situational context" of the voters and jurors embodies the notion of *kairos*. Indeed, beyond these particular examples, Kinneavy says that Aristotle's incorporation of *kairos* in his *Rhetoric* is "completely in line" with the injunctions of Plato—and Gorgias—in these matters.[23]

That Gorgias's notion of *kairos* is identified with the Platonic *kairos* is, of course, one of the central concerns of this study, and will be addressed specifically in the next chapter. But there can be no doubt that *kairos*, at least as a feature of rhetoric, has been most closely associated with the Sophists, and with Gorgias in particular. The peculiarly Sophistic notion of *kairos* was quite possibly derived from a principle of Pythagorean "science" that structured reality on the basis of contrary arguments, on an "ontology" of dissociated concepts (*dissoi logoi*) perceived as inherent in the diversity of nature. *Kairos*, then, was the force that resolved the conflict in the context of a specific situation, resulting in a synthesis or in the imposition of one of the two alternatives. This aspect of *kairos* captures the sense of "rightness" implied in due measure, proportion, opportunity, and critical time. Thus, scholars like Untersteiner and Enos translate *kairos* as "balance," thereby to characterize the sense of harmony we achieve through *kairos* breaking up the cycle of antitheses that comprise all things (Untersteiner 110; Enos, "Epistemology" 44). This idea of *kairos* as balance or harmony reflects the mythological roots of the term, but in some crucial respects that idea has been left underdeveloped in rhetorical theory.

Nevertheless, a certain regularity prevails in the above. Kinneavy maintains that in its transformation from its source in Pythagorean science to a central feature of rhetorical theory, *kairos* retains a basic consistency in meaning that "runs through Hesiod, Pindar, Philolaus, the Pythagoreans, nearly all the Sophists, Socrates, Plato, Aristotle, and later Greek writers," extending even to the Roman rhetoricians, to include Cicero in particular ("Relation" 13; Kinneavy, "Neglected" 92). Kinneavy's is a viable interpretation, but one based on an understanding of *logos* as it achieves its hegemony more and more as the guidance of reason—as *ratio*, as *Vernunft*. Insofar as rhetoricians accept this interpretation of *logos*, *kairos* becomes essentially as Plato depicted it in the *Phaedrus*, as it is codified there by virtue of *res*, even as these rhetoricians say this understanding of *kairos* originated with Gorgias. Thus, in Augusto Rostagni's discussion of Gorgias, *kairos* is identified as the principle through which "the changeablity of speeches is justified and required by the need for adaptation to circumstance" (Untersteiner 119). Once again, both Kinneavy and Kennedy refer to Gorgias as the source of the rhetorical *kairos*, endorsing Funaioli's definition of the term, which is essentially the same as Rostagni's. So I return to the canonical definition given at the beginning of this chapter and reiterate: "*Kairos*, when resolved into the rhetorical skill" is what is "fitting in time, place, circumstance." It is "the adaptation of the speech to the manifold variety of life" and "to the psychology of the speaker and the hearer."

However, amidst this sheerest unanimity there are tidings from certain quarters that inspire second thoughts. In the course of his discussion of the rhetorical *kairos*, Mario Untersteiner, with care and good insight, provides grounds for a more enigmatic perspective on *kairos*. In his review of both Rostagni's and Funaiolo's explanations of *kairos*, Untersteiner claims that *kairos*, once it is endowed with "objectivity" and "cast into the rigidity of the rhetorical formula," is only a "feeble reproduction" of Gorgias's original "epistemological motif" that afterwards maintains its "imperfect survival" as part of the rhetorical *facies* in the "objective expression of a scheme" (196). After giving Funaioli's definition, Untersteiner adds that Gorgias did not "dwell pedantically" on "precepts" and that for Gorgias *kairos* meant, "even on its more rigid plane as rhetorical *kairos*, something living, a prompting which continually by means of irrationality overcomes recurring opposites" (197). And in his presentation of Rostagni's views, Untersteiner qualifies the former, apparently with remarks from Rostagni himself, to the effect that *kairos* not only justified and required the "changeability of speeches" to meet the need for adaptation to circumstances but that "in Gorgias it had a quite different meaning . . . that is, it touched the heart of the painful problem of knowledge" (119).

This alternative view of *kairos* will be explored in the next chapter, but needing emphasis at this point is the nature and disposition of *kairos* as it has survived, however imperfectly, and the manner of its survival.

The Rhetorical *Kairos* in Other, Various Guises

With Untersteiner's reservations duly noted, it can be said that the above definitions of *kairos* are as similar to one another as they are to Plato's, and surely they point to matters clearly vital to our understanding of rhetoric. But as the term "*kairos*" has gained currency in rhetorical studies, it is important to note that the concept has been always a fixture of rhetoric, if not invariably as *kairos* then in other, various guises. Untersteiner, for instance, associates *kairos* with *to prepon* (197). The latter is the Greek term for "propriety" or "decorum," and generally means "appropriateness" in matters of language and usage. Similarly, Kinneavy ("Relation" 16), Poulakas ("Toward" 41), and Kennedy also associate *kairos* with *to prepon*.

Kennedy's discussion of *kairos* is particularly pertinent. He places it under the canon of style, saying in that context it is "allied" with *to prepon*. "The two together constitute what may be called the artistic element in rhetorical theory as opposed to prescribed rules." *Kairos*, then, in its association with *to prepon*, "is largely restricted to the classical period," whereas *to prepon* "is more persistent and is the only provision for latitude and taste which found a permanent place in traditional rhetoric" (*Art* 67). Moreover, his definition of *to prepon* is essentially the same as his definition of *kairos*, as the latter is to Kennedy, once again, "the principle which governs the choice of organization, the means of proof, and particularly the style" (67); whereas, he says, "propriety ([*to prepon*], *decorum*) discusses the adaptation of the style to the circumstances of the speech, the character of the speaker, the sympathies of the audience, and the kind of speech" (276).

This survival of *kairos* as *to prepon*, propriety or decorum, is especially evident in Cicero and Quintilian. Both lacked the highly speculative and engaging intellect of Plato and, seemingly in alliance with the Sophists, sought a more pragmatic, pedagogical development of rhetoric geared to the needs of the "orator's province," to the dynamics "of human life and conduct," as Cicero says (*De Oratore* I.xv.68). Both possessed a well-grounded appreciation for the power of language and shared an attentiveness to the potentialities of style that was in marked contrast to the position adopted by Plato in the *Republic*. Moreover, to both, philosophy and rhetoric were interdependent disciplines, with Cicero in particular locating his model for rhetoric among the Sophists, especially Isocrates. Yet neither endorsed notions of language as equal in station to reason, each subscribing instead to Plato's logocentric view. In fact, both Cicero and Qunitilian explicitly denied the role of language as anything other than secondary to reason, and their views on "propriety" attest to the fact. The essential ingredients of the Platonic *kairos* are reflected in their ideas of propriety, specifically in the separation of subject and object, an understanding of language as logocentric in that sense, with the speaker perceived as the master of language; and even, indeed, logocentric in terms of a most extreme instance of the "metaphysics of presence," as time itself is perceived as separate from our being, something that by virtue of propriety we also master and put to use.

Cicero, for instance, defines elegant speaking in a manner to suggest the fundamental distinction between *verba* and *res*. In the persona of Crassus, he claims eloquence to be one of the "supreme virtues," having the most "beauty and distinction in outward appearance," which, "after compassing a knowledge of facts, gives verbal expression to the thoughts and purposes of the mind in such a manner as to have the power of driving the hearers forward in any direction in which it has applied its weight" (*On the Orator* III.xiv.55). This view is particularly apparent in a later work by Cicero devoted exclusively to style. In the *Orator* Cicero says, "Although a word has no force apart from a thing, yet the same thing is often either approved or rejected as it is expressed in one way or another" (xxii.73). From premises such as these, fully in accord with Plato's logocentric view of language, Cicero claims "the universal rule, in oratory as in life, is to consider propriety" (xxi.71). He defines it as follows:

> In an oration . . . nothing is harder than to determine what is appropriate. The Greeks call it [*prepon*]; let us call it *decorum* or "propriety." Much brilliant work has been done in laying down rules about this; the subject is in fact worth mastering. From ignorance these mistakes are made not only in life but very frequently in writing, both in poetry and prose. Moreover the orator must have an eye to propriety not only in thought but in language. For the same style and the same thoughts must not be used in portraying every condition in life, or every rank, position or age, and in fact a similar distinction must be made in respect of place, time, and audience. (xxi.70–71)

Nowhere in Cicero's work is there any mention of the word "*kairos*," but the above passage bears comparison to Plato's in the *Phaedrus*, and can also be considered one of the more likely examples marking this transformation of *kairos* to "propriety," to "decorum" in the Latin parlance.

Quintilian mentions *kairos* but once, and in a context having no clear rhetorical implications and seeming to reflect Plato's use of the term in the *Laws* rather than the *Phaedrus*.[24] Quintilian does, however, offer an explanation of decorum very similar to Cicero's. In fact, he explicitly endorses Cicero's views on "the appropriateness of speech" (XI.1.1), and later quotes Cicero directly: "*One single style of oratory is not suited to every case, nor to every audience, nor every speaker, nor every occasion*" (XI.1.4, his italics). Quintilian also employs the separation of thought and language that we saw in Cicero, with *res*, as thoughts or "purposes of the mind," given ascendancy over *verba*. To Quintilian every speech is composed of "matter and words." Matter is concerned with invention, and words with style (VIII.Pr.6). Words must be "well-adapted to produce the desired effect," but "nothing must be done for the sake of words only, since words were invented merely to give expression to things: and those words are the most satisfactory which give the best expression to the thoughts of our mind and produce the effect we desire upon the minds of the judges" (VIII.Pr.32). In this separation of language and thought, with "words merely invented to give expression to things," the logocentric attitude is epitomized. Indeed, one would be hard put to find a more complete expression of the idea, in antiquity or otherwise. Thus, as Plato and Cicero, words to Quintilian provide a speaker merely with alternative means for the expression of a single idea. The orator is presented with an audience, an occasion, and he must find the appropriate mode of expression to master the situation through his mastery of language.

That *kairos* survived in Cicero and Quintilian as propriety or decorum, and that it continues to survive under these guises, is given further credence in an early article on decorum and *kairos* by James Baumlin. There he defines *kairos* as the "epistemic counterpart of decorum" (179), claiming that decorum is based on "man's innate pleasure in and desire for harmony" (174–75) and that decorum's essential link to *kairos* lies in the emphasis that each gives to timeliness and sense of the opportune (177). As in all conceptions of the Platonic *kairos*, Baumlin's places a particularly strong emphasis on the role of the speaker, who shapes reality in wielding the power of language. Thus, his statement: "The observance of *kairos* becomes above all an interpretation of a mutable, contingent, temporal nature, giving the speaker or writer what amounts to creative control over the world he lives in and presents, by words, to others" (181).

Creative control is, of course, a critical feature of the Platonic *kairos*—is in fact its very nature—and is related in some fairly obvious ways not just to Plato's theory of language but thereby to his idea of time. As is the case with other conceptions of the Platonic *kairos*, there is little sense of time as sheerest "temporality" in Baumlin's, of the rhetor being involved *in* time, of the rhetor and his language being essentially determined by it. Rather, time is equated with occasion, and in that sense is the rhetor's to exploit as opportunity. Indeed, despite the emphasis on matters of a "contingent, temporal nature," in Baumlin's account it can only be one's mastery of time conceived as occasion, as a matter at the speaker's option to exploit by his wits, that would allow "creative control over the world."

The idea of time as occasion is given even greater emphasis in the doctrine of *stasis*, a stratagem providing a good likeness to *kairos* specifically in the manner that each pertains to time. As a concept in rhetoric, *stasis* involves a series of questions used to discern the central argument or crucial issue in a legal or judicial dispute.[25] It was present in an "embryonic stage" in Aristotle (Carter, *Art* 20) and came to full fruition in the Hellenistic period, particularly in the rhetoric of Hermagoras, Cicero, and Quintilian. But what is crucial to my purposes is the nature of the similarities *stasis* has to *kairos*. Michael Carter, for instance, makes the same point concerning *stasis* and *kairos* as Untersteiner, Baumlin, and others do concerning *to prepon* and *kairos*. He says that both *kairos* and *stasis* are associated with the beginning of discourse, serving as a means to prompt rhetorical action, and he extends the comparison by saying that "*stasis* may be the concept of *kairos* put into the form of a fully elaborated art" (29). Indeed, as both Carter and Otto Dieter suggest, the rhetorical *stasis*, like *kairos*, has its roots in Pythagorean science and, as a result, employs the same epistemological motifs. Or, to adopt Baumlin's terminology, *stasis*—like decorum itself—can be considered an "epistemic counterpart" of *kairos*.

Dieter explains the epistemological origins of *stasis* by contrasting it with *kinesis*. As generally accepted contraries in Greek thought, *kinesis* is motion or activity and *stasis* is "the rest, pause, or halt" occurring at the point of reversal between the opposite motions of *kinesis* (369). As the juncture between opposite motions—as the "transitory" state or "temporary standing in conflict" (350)—*stasis* bears resemblance to *kairos* in that each is generated by opposition: *kairos*, as we have seen, in the opposition of competing *logoi* and *stasis* in the opposition of the contrary movements of *kinesis*. Thus, both *kairos* and *stasis* are points or places of focus that transpire as disruptions of continuity. In this manner they incorporate "precisely the same imagery," as Carter says (27).

However, one of the more significant connections between the two ideas is given in Plato's explication of the "moment" or "instant" in the *Parmenides*. He writes:

> The word "instant" appears to mean something such that *from it* a thing passes to one or other of the two conditions. There is no transition *from* a state of rest so long as the thing is still at rest, nor *from* motion so long as it is still in motion, but this queer thing, the instant, is situated between the motion and the rest; it occupies no time at all, and the transition of the moving thing to the state of rest, or of the stationary thing to being in motion, takes place *to* and *from* the instant. (156d)

The importance of Plato's interpretation of the "instant" can be seen in light of the distinctions Carter draws between *kairos* and *stasis*. In *kairos*, Carter says, the opposition of *logoi* is resolved by an "irrational" judgment given in the impetus of the immediate situation; whereas in *stasis* the initial conflict is resolved on the "rational" basis of arguments "presented in response to the *quaestio*" ("*Stasis*"106). The distinction is one of the bases for Carter's contention that we might trace the art of *stasis* to the seminal idea of *kairos* and, thereby, to his contention that *stasis* is *kairos* carried to "a fully elaborated art." But in Plato's interpretation of the instant, this distinction can be seen as yet another similarity between *stasis* and *kairos*. Not only does Plato remove the sense of the instant from the irrational flux of temporality, he repositions it within a logical order, where the instant is no longer "in time" but is fixed in a concept, a nexus between past and future that is itself outside of time. It is a place of eternal presence and focal point of consciousness, ultimately a marker for the soul itself. Once again, the organic nature of time is precluded, so that the essence of human life is not manifest in time, nor even involved in it, but stands apart. As a result, time as *kairos* is accordingly taken as an abstraction, separate from the mystery of our being and put at our disposal in terms of the opportune moment, enabling our creative control of language by virtue of time as occasion. As *kairos* is thus stripped of its complicity in the "irrational," its place in the flux and exigencies of "event" is eliminated. It becomes instead one of the more curious illustrations of *nunc stans*, the "standing now," as it is transformed literally into *stasis*, the inanimate instant, frozen and formalized. It would seem, then, that on this basis *kairos* is made amenable to the elaborated art of *stasis*, and that, as a result, the latter is all the more clearly another instance of *kairos* persisting under yet another guise.

I conclude with one additional explanation of *kairos* that summarizes in practical ways some of the basic features of the Platonic *kairos* provided thus far. In *The Sophists* Guthrie says *kairos* was essential to Gorgias's art of persuasion, and quite typically he identifies it as "the sense of occasion or opportunity" which requires the speaker to "adapt his words to the audience and situation" (272). To illustrate the point, Guthrie offers an updated version of the idea, citing the case of Benjamin Disraeli who knew that the "opportune in a popular assembly has sometimes more success than the weightiest efforts of research and reason." In the same vein he also mentions the example of Bertrand

Russell who suppressed telling the truth to his countrymen of the cruelty, persecution, and poverty that existed in Russia soon after the First World War. In view of "the sort of people they were" and under the conviction that "the time was not opportune," Russell chose to say only favorable things about the Soviet government. Guthrie concludes that each example is a "good illustration of the Gorgian attitude to truth and *kairos*" (272).

The point I would make is that both illustrations only express what becomes of the Gorgian *kairos* once it is projected through Platonic epistemology and metaphysics, becoming then, as a result, the Platonic *kairos*. It is the same point I would make concerning the interpretations of *kairos* by Kinneavy, Kennedy, and other rhetoricians mentioned in this chapter. In their explanations *kairos* is conceived as a more or less critical feature of a larger conceptual scheme, a rhetorical skill brought to bear by a rhetor who appropriately responds to a situation by means of language. It is, as well, in this sense that the Platonic *kairos* has attained status precisely as the rhetorical *kairos*. But as I have suggested, it may be a far too narrow if not mistaken view of *kairos* insofar as it is assumed to have originated in the Sophistic rhetoric of Gorgias. As does Untersteiner, John Smith points to the narrowness of *kairos* in its rhetorical context, saying that it represents no more than a "human standpoint" and quite ignores its "ontological" dimension and other, corresponding implications that attach to the importance of things "not of human devising" (54). Here the argument can also be made, from Untersteiner among others, that the Gorgian *kairos* is possessed of any number of intriguing possibilities precluded by the Platonic interpretation that could deepen our appreciation of rhetoric, especially Sophistic rhetoric. This perspective on *kairos* would be based on the precedence of *verba* rather than *res*, and would thereby affirm *kairos* in the Gorgian sense, in a way to realize the measure of "the right thing at the right time" in a more dramatic context, one to give voice to the "original forces" inherent in language. It is to this notion of *kairos* that I now turn.

3 The Gorgian *Kairos*

... of Gorgias, a man to whom as to a father we think it right to refer [to] the art of the sophists.

—Philostratus

Plato and Gorgias are adversaries of a peculiar kind, being so closely linked in their differences that any well framed censure of Plato might serve in its own right to rehabilitate Gorgias and his rhetoric. If, as a result, this chapter seems to be as much about Plato as Gorgias, my effort is to compare and contrast, to achieve some initial insights into the nature of the Gorgian *kairos* even as a more complete analysis awaits development in subsequent chapters. I will attempt to come to these insights by emphasizing the role that language and style play in Gorgias's rhetoric, as each is subject in various ways to Plato's reproach. Secondly, I will apply and develop further some of the arguments introduced in the last chapter, especially those concerning the differing natures of the soul and *psyche*, so that the contrast between the Platonic *kairos* and Gorgian *kairos* can be made more distinct on that basis and the latter thereby brought into clearer relief.

Through Mario Untersteiner and others, I have related some of the difficulties associated with studies of Gorgias and the Sophists, among them the fact that we most often see the Sophists through the eyes of their adversaries. And Gorgias has a particularly bad reputation, even for a Sophist. The charge that he was an immoral opportunist with little or no regard for truth is categorical in Plato's condemnations of Sophistic rhetoric, and Aristotle criticized Gorgias in a number of places, generally for putting too much emphasis upon memorization and declamation ("Sophistical" 183b35–184a5). In matters even more pertinent, Gorgias has been condemned especially for the style of the oratory he espoused, for his use of the metrical patterns of poetry in his treatises and discourses, an effect achieved through the abundant use of tropes, schemes, antitheses, and various other rhetorical structures and devices. Aristotle says in his *Rhetoric* that Gorgias's use of metaphor was not just inappropriate but, most egregiously, was in "bad taste" (1406b5–10), and only a few lines earlier he claims Gorgias's "oratorical prose" was directed to "uneducated people" who "think that poetical language makes the finest discourses" (1404a25–27). And there are others. Diodorus Siculus says Gorgias's figures of speech, though once thought worthy of acceptance because of their "strangeness," are basically "far-fetched and distinguished by artificiality," and "now seem tiresome and appear ridiculous and excessively contrived" (Diels 82.A4). And Athanasius and Longinus found Gorgias's style not only affected and self-indulgent, but fit only to be scorned by laughter (Diels 82.B5a).[1]

Yet Gorgias was the most famous of the Greek Sophists, due in no small measure, ironically enough, to his numerous adversaries and the weighty reputations enjoyed by many of them, then and now. He was indeed the Sophist who most unnerved Plato, even more than did Callicles, Thrasymachus, and others of their sort whom Plato goes to great lengths to vilify as doltish, spiteful, and despotic. That Gorgias was foreign born may account for some of the hostility directed toward him. He was, in any event, an itinerant teacher of rhetoric who, according to many accounts, lived to be over a hundred years old. It is said that he gave exaggerated estimates of the worth of his pedagogy and extracted exorbitant sums of money from his students. It is also said that he had a statue erected of himself at Delphi that he personally dedicated in his honor. This in itself was not without precedent, but Gorgias's statue was not gilded as were the others but of solid gold, so great was the wealth he acquired from teaching oratory. Even Isocrates, once a student of Gorgias, said of the latter that he lived his long life making money, never spending any of it for public benefit, and never was a resident of any city and therefore never paid a tax. By some accounts, like that of Isocrates', Gorgias never married; by others, he had a wife but ignored her, enjoying instead an excessive fondness for his maid. A story in Plutarch's *Advice to Bride and Groom* has it that after Gorgias delivered a speech concerning concord among the Greeks, one of his listeners complained that "this fellow advises us about concord, though he has not persuaded himself and his wife and his maid, only three in number, to live in private concord."[2]

These statements are gathered from sources of questionable reliability, and I offer them as much as evidence of attitudes toward Gorgias than out of any intent to provide an historical account. I have no idea what Gorgias of Leontini was really like, but I suspect he was in attitude and acumen essentially as he is depicted in Plato's dialogue to which his name is given. That is to say he was self-satisfied and superficial, and relatively naïve in philosophical matters despite his abundant years and wealth. I also suspect that he was Plato's intellectual inferior in all things. Yet, whatever might be the case, the pitch of Plato's criticism of Gorgias is seldom directed toward the man himself. It is not the substance of the intellectual challenge that Gorgias personally brought to bear that is Plato's concern, but more clearly the world view that he so roundly personified—even though, in Plato, Gorgias is always at a loss to articulate it in any adequate way. What does emerge from Plato's dialogues, usually on very subtle levels, is the character of Gorgias as a man remarkably attuned to the workings of the prevailing *doxa* of ancient Greece and ingenious in the ways of exploiting it by means of rhetoric. Thus, many of the most crucial rhetorical implications of Plato's world of ideas lie in its encounter with the world of *doxa*, one manifest, as we have seen, by the whole of the early Greek *paideia* that Plato committed his life and philosophy to undo.

Not only was Gorgias in many respects grounded in the pre-Platonic way of life and thought, but in all instances he was clearly one of its chief beneficiaries and vendors. In this regard, of course, he was remarkably unlike Socrates, given the latter's austere life and unseemly demise. It is therefore especially important in these days of the Sophistic revival to remain unbeguiled by accounts that characterize Gorgias as brilliant and innovative, as suspicious of "conventional wisdom" and driven, indeed, by a revolutionary zeal that associates *kairos* with a "pure force of will," a "will to spontaneity," a "will to invent." I believe that such perspectives are based on misreadings of Plato's quarrel with the Sophists and are fashioned by doctrines of subjectivity, those mandating an autonomous self defined by concepts of will, ego, or soul that have little to do with Gorgias.[3] What we know from Plato on these matters, if not from Gorgias himself, is quite apparent. It was the intellectual and moral climate of the early Greek *doxa*, given so powerful a presence on the most practical and prosaic levels by Sophistic rhetoric, that alone made Gorgias such a difficult and dangerous adversary to Plato.

Another, related problem concerning a study of Gorgias is the dearth of primary sources, of evidence concerning Gorgias and his rhetoric given by Gorgias himself. Only two of his compositions have been preserved, the "Encomium on Helen" and the "Defense of Palamedes," and his authorship even of these is in dispute.[4] Though two additional works by Gorgias, "On Non-Being" and "The Funeral Oration," are available in relativity complete summaries written by others, short and scattered fragments of his works are otherwise all that remain. Nevertheless, it is possible to isolate the main features of his philosophy, and my effort to recover his notion of *kairos* begins with an examination of his only definitive philosophical work, the treatise "On Non-Being," available in the version given by Sextus Empiricus.[5] In sharp contrast to Plato, the treatise underscores Gorgias's recognition of the importance of *doxa*, as it also provides a good, workable means of arriving at a coherent picture of the rhetorical implications present in his other works.

"On Non-Being" and the Tragedy of Knowledge

Being is not manifest unless it succeeds to appearance, and appearance is powerless unless it succeeds to Being.

—Gorgias[6]

The main epistemological features of Gorgias's philosophy bear the influence of the Eleatics, Zeno and Parmenides in particular, and of Empedocles, once a teacher of Gorgias. Correlations with Gorgias seem to be explicit in Zeno's dialectic method of paradoxes and antithetical reasonings, in Parmenides' distinction between truth and opinion, and in Empedocles' theories of sense perception. "On Non-Being" opens with the statement containing Gorgias's three primary epistemological tenets, his famous trilemma: "First and foremost, nothing exists; second, even if it does exist, it cannot be known by man; and third, even if it can be known, it cannot be communicated to another man" (Diels 82.B3).

The statement is often associated with *homo mensura* doctrine (the Greek *"anthropos metron"*) affirmed by Protagoras, that man is the measure of all things. In Gorgias's specific epistemological formulation of the idea, the external world exists only insofar as it is created by our sense perceptions. In short, it lacks an essence of its own, a structure that is an analogue to concepts of the mind. Thus, the first tenet, that "nothing exists," is best understood as "phenomenalistic" rather than nihilistic (Untersteiner 159). It stands in opposition to Plato's idealism and Parmenides' "proof" of Being, and moves on the basis of antithesis, suggestive of the process of Zeno's dialectic that reduces opposing theses to absurdities. Gorgias argues that the "nonexistent" cannot exist because this is an absurdity. But, then, neither can the "existent" exist because existence and nonexistence, Being and Non-Being, are attributes of one another, though they are at the same time in opposition (par. 67). Thus, Gorgias concludes, they are the same thing: namely, nothing at all. Undeniably the issue "to be" or "not to be" was prevalent in the disputations of other pre-Socratics, and, as in this case, it was surely grounded in the realization that Being can obtain its definition and integrity only from an equally viable notion of Non-Being. But seldom was the option taken to favor the latter without qualification, as Gorgias seems to have done. Indeed, his position is in full counterpoint to that found in the fragments of Parmenides, where Non-Being can neither be conceived, experienced, nor expressed, and therefore cannot "exist."[7] Here, in ways as accomplished as Parmenides', Gorgias turns the former's doctrine of Being into its opposition, indeed demonstrating if not the proposition that "nothing exists," then at least the contradictory nature and ultimate fruitlessness of this kind of speculation (Versenyi 40).

Nevertheless, the proposition abides, seeming to be the work of some primitive, very perverse deconstructionist. It is therefore important to note that the conclusion, "nothing exists," is not a resolution but a claim advanced by "the weapons of the most uncompromising rationalism" (Untersteiner 158), that logic is here confounded by its own intrinsic paradoxes, that such paradoxes—antitheses, *dissoi logoi*, or *logoi* in opposition—are to Gorgias what structures our understanding of the world. Hence, the first tenet, as is the case with much of the treatise, is an example of the eristic logic for which the Sophists were famous, and Gorgias applied it to less esoteric matters in his "Palamedes," arguing that Palamades, accused by Odysseus of treason, can be either condemned or absolved through the application of the rational *logos*.[8]

The second principle of "On Non-Being" would seem to rely on either the Protagorean doctrine of man or on the very ancient theory of sense perception maintained by Empedocles, that "effluences" emanate from objects and enter various, appropriate "pores" of the sense organs of the body.[9] But here, too, Gorgias proceeds in the manner of

Zeno, reducing even the contents of sense perception to absurdities, thus excluding man as the measure. He claims that mental concepts or "things considered" can reflect the nonexistent as readily as they can existing things. He begins by arguing that the absence of immediate physical referents does not exclude mental concepts, "for Scylla and Chimaera and many other nonexistent things are considered in the mind" (par. 80). As such, mental concepts are not linked to the perception of existing things. But this thesis also falls back on its own contradiction, because external substances, perceived as things seen or heard, can only have their existence in concepts of the mind. "If someone were to consider in the mind that there are chariots racing in the sea, even if he does not see them, he should believe there are chariots racing in the sea" (par. 82). Given this analysis, objects of sight or sound exist for us only insofar as they are thought, yet, at the same time, thought that is apart from experience has no referential existence beyond the imaginative extrapolation of the thinker (Enos, "Epistemology" 47). Thus, according to Gorgias's rationale, neither by means of thought nor by means of sense perception can the existent—assuming it "exists"—be apprehended by us. The argument is based upon dubious premises and is otherwise far from elegant, but it would seem to be formally valid.[10] Untersteiner summarizes it simply as so: "Just as thought has not given existence to all possible experiences, so experiences cannot give existence to thought. In Gorgias, neither has absolute validity" (155–56).

To Gorgias this result is the inevitable effect of reason, of the rational interpretation of phenomena in which it is just as easy to prove "is not" as "is." Through appeals either to reason or sense experience, there can be no warrant for claims concerning the correctness of knowledge, nothing to guarantee in any conventional sense an accessibility of the knower to the known. *Logos* as reason is undermined. On a philosophical basis it thus follows that everything is resolved solely in terms of its irrationality. At the heart of Gorgias's epistemology lies what Untersteiner calls the "tragedy" of knowledge, tragic in the sense that the claims of reason cannot be satisfied or justified; that in all concrete situations in which we are involved, we are faced, often invariably, with *dissoi logoi*, each rationally posited and therefore each with an equal claim, rationally, to our allegiance. Thus, Untersteiner says that Gorgias is neither "a sceptic nor a relativist but a tragic philosopher and an irrationalist":

> Man cannot escape the antithesis. His thought discovers only opposite poles in all propositions which try to explain reality philosophically. The reality reached by dialectic expresses only *aporiai:* this is the conclusion of Gorgias himself, who makes us feel above all the drama of the continual clash and counterclash of the extremes, into which every attempt to arrest the mobility of the *physis* is resolved. In the increasing intensity of a close-locked struggle all human experiences, taking dramatic form, are brought to a standstill in the face of reason, which can no longer decide anything and therefore ends by denying on a rational basis every relationship between man and man, and finally all coherence within the individual himself. (159–60)

A peculiar merging of philosophical categories is taking place here. To the extent that the tragedy of knowledge is tragedy at all, it is ontological as well as epistemological, precisely because "human experience" remains in ways closed off, unmediated by views of reality established on rational bases. Experience is immediate and direct, takes dramatic form, and can be understood on an ontological basis in contrast to Plato's metaphysical conception of stable and enduring truth residing in the ultimate reality of "Being."

This Gorgian perspective provides a context for the discussion of the third tenet of his treatise and, for my purposes of uncovering some initial insights into the Gorgian *kairos*, the most important: "If anything exists and is apprehensible, still it is incapable of being expressed or explained to the next man" (par. 66). Gorgias's argument here is easily enough stated. We communicate with another through *logos*, but *logos*—our speech—is not the "substances and existing things" of external reality, but is *logos* and that alone. Given the first two tenets of the treatise, the emphasis in the third is necessarily on the medium of communication, on words alone and not their referents. As Gorgias says, "just as what is visible cannot become audible and vice versa, similarly, when external reality is involved, it would not become our *logos*, and not being *logos*, it would not have been revealed to another" (par. 84–85). Simply put, words are not things. We communicate with words about experiences, not the experiences themselves. In one sense, Gorgias, like Plato, establishes two realms, one of the reality and one of the discourse, though in Gorgias there can be no analogue or correspondence between the two. The connection we assume that exists between words and things is severed, and that is the nature of the tragedy of knowledge, the very problem that Plato's Forms were meant to overcome. But in another sense, the distinction is even more radical. There is an irreducible difference between Gorgias and Plato; indeed, in that "reality" is explicitly denied by Gorgias, all that can be known is revealed through *logos*—for *logos*, of and by itself, is all that we have, being the very ambiance in which we dwell.

Therefore, we must not draw the conclusion that communication is precluded altogether by Gorgias, for the central importance of "Non-Being" is hinged to the power of language, to *logos* as it is manifest in the immanence of the here and now. In the most general sense, the word was used in antiquity to designate the lawfulness that abides in the flux and in our way of knowing it, functioning in all respects as the means of discerning the cosmos from chaos by way of an "ideal commensurability of thing and speech," as Thomas Rosenmeyer says (229–30). In Parmenides, for example, *logos* was a means to establish a universal ground by way of an ontological identification of Being, thinking, and saying—an idea Plato carried to its ultimate conclusion with the Forms serving as objects of linguistic meaning and reference.[11] The Forms therefore became a means of mediation between human being and reality, a way to sanction our access to knowledge. But what we witness in Gorgias's epistemology is the irony of an utterly rational, systematic deconstruction of the concept. In Gorgias's treatise, says Untersteiner, *logos* as either thought or perception ends in the "annihilation" of *logos* as *logos*. "There remain only external experiences without the existence of a means of communicating the thing known" because *logos* "has not been given the power to escape from its own reality in order to establish the relation between object and its expression. It can impose itself only irrationally" (158). To put the matter another way, *logos* is free, its own master.

However one puts it, this "radical departure" of Gorgias, says Rosenmeyer, "completes" Heraclitus, even as it refutes Parmenides. Heraclitus laid the foundation of a "cosmic" *logos* as a medium capable of transforming itself into a human *logos* so that reality finds its appropriate formulation in speech, whereas Gorgias "utterly separates the cosmic *logos* from the human *logos*." And to translate the former, "or any part of it, into a human *logos* is not only difficult, it is impossible," because in Gorgias there is no measurable relationship between the two (230–31). One implication of this perspective has already been mentioned: reason is taken as an untenable means to knowledge, negated in a dialectic that ruptures any correspondence of *logos* with "reality." Another implication is that in Gorgias's specific epistemological formulation, the external world lacks an essence of its own or one that can be known by us, and, as a result, reality exists in Gorgias's scheme only to the extent that it is revealed in phenomena created by language. The cosmic *logos* as reason or means of mediation between human being and reality, the view upon which Plato builds his metaphysics, is undercut; the human *logos*, understood exclusively as speech—or, more aptly yet, as *listening*—is given preeminence. The critical lines from "On Non-Being" are, "For that by which we reveal is *logos*, but *logos* is not substances and existing things. Therefore we do not reveal existing things to our neighbors, but *logos*, which is something other than substances" (par. 84). Here, *logos* is sheerly *logos* as such, present in its primary meaning as language.[12] In such a case, the object is not only separated from the word, *res* from *verba*, but all that remains of the former are phenomena revealed, driven out of hiding or given appearance by the latter—by language, *logos*. As Gorgias begins with the premise that words are not things, in his formulation they ironically become, in the end, precisely that. If nothing exists, if reality is first annulled, then words alone become the "plastic" medium through which "reality" is brought to presence, through which it is, in effect, created.

I will return to these matters presently, in context of a discussion of Gorgias's style. But it is the irrational power of *logos* acting in lieu of the cosmic *logos* that I am interested in pursuing for the moment, for in this dynamic flux of contrasting *logoi*, in the "anxiety" provoked in the "close-locked struggle," the Gorgian *kairos* becomes most apparent.

Gorgias's "Funeral Oration"

If the parable of the cave provides a fitting expression of Plato's philosophy, the funeral oration of ancient Greece does much the same for Sophistic rhetoric. The orations were meant to praise the sacrifice and valor of those who gave their lives in defense of the *polis*, to offer solace for the loss and sorrow of all concerned. But they were far more, attesting to our earthly origin and fate, to *that* "mobility" of *physis* of which we are so much a part we can have no control or mastery over it. Thus, homage is paid in the funeral oration, affirming in the acknowledgement of this mysterious process and power the congruence of language, mortality, and native soil that are the crucial constituents of a rhetoric of the earth. In this sense the funeral oration is a repudiation of the abstract and universal principles of philosophy and philosophical rhetoric, providing in their place a most profound expression of the tragedy of knowledge that is apparent not only in the belief that there can be no correspondence between language and reality but that there is no "rational" reality whatsoever. Such is the nature of the Gorgian *kairos*, and it should then be no surprise that its finest exposition comes in Gorgias's "Funeral Oration."[13]

Indeed, of all of Gorgias's works, the "Funeral Oration" best illustrates the nature of the "dramatic form of human experience" that Untersteiner says is central to Gorgias's epistemology and ontology (159). In it, Gorgias eulogizes the Athenians killed in war, praising them but more saliently dwelling on the "tragedy" they faced. What he describes is a situation that confronts the "habits" of a given world, breaking the fixed and formulaic processes of thinking and speaking with the force of a moment—a *kairos*—that is new and different, of *doxa* redeemed through "para-*doxa*." In the course of the "Oration" Gorgias asks:

> For what did these men lack that men should have? What did they have that men should not have? Would that I could express what I wish, and may I wish what I ought, avoiding divine wrath, shunning human envy! For these men possessed an excellence which was divine, and a mortality which *alone* was human. Often, indeed, their choice was the kindness of the truly just rather than the arrogance of positive right; often, rather than the rigor of the law, the perfection of argument; for they believed that it is a universal divine law to say and refrain from saying, to do and refrain from doing, the right thing at the right moment. Above all, they cultivated two needed qualities of mind and body, the first for deliberation, the other for accomplishing; helpers of those in undeserved adversary, chastisers of those in undeserved prosperity; bold for the common good, quick to feel for the right cause, checking with the prudence of the mind the imprudence of the body; violent towards the violent, restrained towards the restrained, fearless towards the fearless, terrifying among the terrifying. As evidence of these things, they have set up trophies over the

enemy, an honor to Zeus, a dedication of themselves: men not unacquainted with the inborn spirit of the warrior, with love such as the law allows, with rivalry underarms, with peace, friend of the arts; men showing reverence towards the gods by their justice, piety towards their parents by their care, justice towards their fellow citizens by their fair dealing, respect towards their friends by keeping faith with them. Therefore, although they are dead, the longing for them has not died with them, but immortal though in mortal bodies, it lives on for those who live no more.[14]

The "Funeral Oration" is, of course, a panegyric, but a world of discord is depicted here, tragic in the sense that any confidence placed in the order and rationality of the world is dashed. The warriors were caught in a moral dilemma of some consequence, their lives and honor hinging upon decisions of flight or fight. Gorgias tells us that their courage came from God, but from "nature" they possessed a will to live. Thus, the tragedy that Gorgias depicts fits the classical Greek mold; that is, the warriors were overwhelmed by forces they neither controlled nor understood, yet they were exalted in an "excellence which was divine" by giving their lives, ultimately the only resource at any mortal's disposal that suffices in such a circumstance. And to Gorgias the dilemma was, at the same time, logical. How the idea is presented in the "Funeral Oration" is apparent in the opposition of competing *logoi*, the imperatives in conflict. He casts it in terms of the age-old plight of warriors. Their choice was between the sanctity of their lives, made so purely in their desire to keep on living them, and the equally imperative and no less sacred purpose of saving the *polis* from its enemies.

In this light Gorgias's idea of the tragic is grounded in his more basic belief concerning Being and appearance. As he suggests in his trilemma, ideas of Being based in reason are found wanting in their dramatic "testing" in specific contexts. In fact, he tells us elsewhere that "Being is not manifest unless it succeeds to appearance, and appearance is powerless unless it succeeds to Being" (Diels 82.B26). That seems to be specifically the precept that applies here. At any rate the concurrence of Being and appearance would seem to come about in the exigency faced by the warriors, as their dire circumstances blur distinctions we habitually make between the two. Indeed, insofar as these distinctions remain in place, tragedy is latent and encountered outright, as Untersteiner tells us, when the "elegiac speech of Being" does not attend to appearance and appearance to Being (121). Such is the nature of the "tragedy of knowledge," consummated in the sublimation of appearance or, what is the same, in the immanence of Being that is earth itself. The resolution, if such it can be called—the *kairos*, as Gorgias presents it—comes in this concurrence. The event marks the "perfection of argument" of which he speaks, of the "glory" achieved by the warriors that, ironically, we will come to call *doxa*. Here, in the immanence of Being there is no deliverance from earth, as the glory obtained by the warriors lies only in the reaffirmation of their tragic fate. Tragedy will have it no other way. Most simply the event marks the paradox of an immortality achieved only by virtue of mortality.

And seen from yet another angle, the tragedy of the Athenian warriors was, in effect, a disrupted time sequence. The predictable and familiar, the things bestowed in *doxa* construed merely as "opinion," were shattered; and the chaos beneath an otherwise well ordered world surfaced, immanent in the decision the soldiers faced. Time became organic, of the essence of their lives, as Being itself was focused in the "moment." As a result, the new awareness, this insight that is provoked, is hardly an act of the will but sheerly a matter of broken routines, foiling those habits of mind allowing them to think of themselves as separate from their circumstances at all. Thus, Gorgias tells us, the decision came not from the static, logical formulation of the cosmic *logos* or even from some pale reflection of it in the real world—not from "the arrogance of positive right" nor "the rigor of the law"—but was manifest in the "perfection of argument," in the irrational *logos* that has as its purpose the "right thing at the right moment." As Untersteiner describes it, the "unpredictable novelty or special combination of circumstances" presented in the "individual case" prompts the *kairos* (178). Represented by Gorgias as a force inherent in the situation, it is a divine and universal power, even though irrational, that imposes the decision. As such, those who died to preserve the polis were guided by the persuasive force of an irrational *logos* made explicit by the "inexorable logic" of *kairos* (122) in the context of a specific situation.

In this matter the more fundamental opposition can be seen not as competing *logoi*, but as an antithesis of the cosmic *logos* and *kairos* that in the thick of the confrontation releases the irrational *logos* at the instant of decision. In this sense it is a warrior's *ethos* these soldiers embody. Being neither instruments of the state nor even of their individual wills, they are altogether taken by the situation *in extremis*, by a force or power beyond them that is manifest in this context as an "irrational" *logos* expressed through them in their openness to it, the soldiers being then at the command of this "divine" mandate. Thus, given the "dramatic" situation, the emphasis is entirely on what is immanent, and *kairos*, as "the right thing at the right moment," seals the connection of *logos* with life, giving it a human contour, giving flesh to the word and putting it at one with the tangible, material world, with the very earth itself. The Platonic *kairos* to the contrary, we are not dealing here with what Untersteiner calls "a mere theoretical precept enjoining the observance of *kairos* on orators" (180–81), but a phenomenon comprised of the urgency and turmoil of purely time bound eventualities. Man is not the master of the situation, Untersteiner says, but in given critical situations is a focal point of anxiety, who "suffers in anxious regard, at a time full of stress" (122, 143). Indeed, this notion of *kairos* contains vital connections to time and language that have no place in the conventional understanding of the term given in Plato. And these connections seem to be based themselves on peculiarly Gorgian notions of *logos* and the nature of human being that this notion of *kairos* would itself seem to invoke. It is to these that I now turn, before coming back later to a fuller exploration of this perspective on the Gorgian *kairos*.

THE GORGIAN *LOGOS* AND PRE-SOCRATIC SENSE OF THE *PSYCHE*

I have discussed Plato's philosophical rhetoric in terms of truth and *psychagogue*, of attaining the former through the dialectic and then leading the soul to truth by means of language. In contrast, Gorgias's art of rhetoric is rooted in the essential "verbal element" of language alone, depending wholly on the "power and efficacy" of words themselves, as Plato himself has Gorgias admit (*Gorgias* 449d–e). Gorgias thereby seeks to exploit the psychological impact words have on us, in the "magic" they work, and does so in ways vigorously condemned by Plato. So it follows that Gorgias's art of *logos*, as Charles Segal says, "belongs more properly to the poet of the sixth and fifth centuries than to the philosophic logician of the fourth" (112). In contrast to Plato's, his rhetoric favors figurative language, the kind that affects immediately and directly, allowing little room for reflection (Poulakos, "Possible" 223).

Yet on another level the contrast is connected to something more significant yet, to the differing perspectives Plato and Gorgias have on the nature of human being, perspectives so pronounced as seemingly to designate different things entirely: the soul and the *psyche*. As I have indicated, the meaning that Plato attaches to the word "*psyche*" is not only different from Homer's meaning, but quite evidently different from Gorgias's as well. In fact, prior to fourth century Greece there was no doctrine of the soul or *psyche* in the manner it was later conceived by Plato. Rather, as John Burnet says, the *psyche* was something "extrinsic and dissociated from the normal personality . . . and altogether dependent on the body" (252). In this sense the *psyche* can be seen as the soul's pre-philosophic prototype, the result of the "archaic" and "rudimentary" views of the pre-Socratics. But in another sense the distinction between the soul and the *psyche* persists, lying at the center of conflicting but equally viable perspectives on human being, language, and rhetoric. With this qualification in mind, I follow the practice of using "*psyche*" to refer to what comes prior to Plato and "soul" to refer to what comes after, so that the pre-Socratic sense of the "*psyche*" is distinguished in this way from Plato's doctrine of a fixed and enduring essence of the self called the "soul."

There are various, often very diverse details in the accounts of the pre-Socratics concerning the *psyche*, but productive generalizations can be made nevertheless. In sum, the notion of the *psyche* was a necessary idea to mark the terrible transition from life to death or, indeed, to account for the consummate mystery that there is life at all. In this sense the *psyche* fulfills that same intellectual purpose as the soul, the critical questions being in what ways and to what ends. If the soul embraces the desperate hope of overcoming death, the *psyche* directs our attention elsewhere, compelling our awareness of what Erwin Rodhe calls the "arbitrary abbreviation" of life.[15] Specifically in this role the *psyche* was the élan *vitale* (Furley 17), a life force but one associated purely with physical functions and invariably of the same nature as the body, a "blood-soul" as Cornford perceptively calls it (191). Insofar as body and spirit could be separated at all, it was simply that in death the body remained, and the *psyche*, the life force, was gone. To the early Pythagoreans, for instance, the *psyche* was a "fluid-like substance, invisible and intangible, contained in the body during life but leaving it at death" (Furley 16). It is often portrayed as such in the *Iliad*. When Diomedes slays Pandaros, the latter's life and strength are "scattered" (5.296); and later, when the Trojan Prince Hyperenor is slashed through and through by Menelaus, his "fierce soul came rushing through the wound" along with the entails (14.518). Most momentous is the death of Hector. The essence of his being remains with the body, as his spirit streams free of trunk and limbs and down to the undergloom, leaving his youth and manhood behind (22.360–63).[16] Indeed, in many cases in the *Iliad*, whatever "spiritual" reality the *psyche* possessed would escape in the vapor rising from a wound, in the death rattle, or with the odor given off in decaying flesh. Nevertheless, however ethereal its portrayal might be in these contexts, the *psyche* is otherwise depicted exclusively in physical terms, such that the nature of the relationship of body and soul, certainly their distinction, could never arise as an issue. And if the *psyche* survived physical death, it did so only as a "feeble and witless thing" (Burnet 246), "a purposeless fluttering to and fro" (Rodhe 55), or as a wraith or shadowy double of the deceased that would, at best, abide momentarily to visit those who survived in their dreams and memories, and then would eventually dissipate entirely, to gather in some nether region as a shade, specter, or wisp of smoke or dust.

As a life force the *psyche* was often linked to specific bodily organs or physical reactions, such as the heart or respiration, or, more concretely, to pride or courage as each was expressed in specific situations by physiological turmoil, as labored breathing or rapid heart beat. The idea is particularly evident in the brawls and blow-ups of the *Iliad*, where gore and bloodletting are graphically portrayed. No truer battle scenes are to be found in literature, for here the violence is literally mindless, the warriors at the mercy of powers or passions seemingly not their own, doing only as these powers bid them. And, even as the *psyche* was at times identified with the mind, or *nous*, it was never considered to designate cognition, and definitely not extended cognition. At best it was associated only with momentary insights or instantaneous appraisals (Furley 8–9) given under the stress of external circumstances, where the pitch of the moment registers its impact in terms of immediate, visceral reactions that are very often manifest explicitly as language. In fact, the *dramatis personae* of the *Iliad* act (or react) at the behest of exhortations coming from without, as when Hector is forced to "speak forth what the heart within [his] breast urges" (7.68); or when Odysseus in dire straits carries on a conversation with his own valor (11.400–15). Significantly, instances such as these are not mere metaphors. Instead, according to D. J. Furley, a peculiar union of physical reaction and communication is at work, what he describes as a "seething turmoil of the blood" that in some cases in Homeric poetry is substantial enough "to give orders and to talk."[17]

Closely related to this peculiar aspect of the *psyche* was its connection to various ruling principles of the cosmos. To Heraclitus the *psyche*, like the cosmos, was made of fire; to Anaxagoras it was mixed with the cosmos as air, and for that reason he thought that all things are full of gods. In this manner the unity of the *psyche* and cosmos was conceived by the pre-Socratics to be implicit, as is apparent specifically in the early Greek conception of *physis*. We translate the term as "nature," though at this juncture, to use Heidegger's phrase, it is more gainfully understood as the "creative surge of Being"—of nature conceived primordially as "earth"—a process expressing the coalescence of human being and nature for it is ultimately the source of each. Here the "objective" stance is precluded, as *physis* is not simply the physical environment nor some amorphous mass in which we are enveloped, nor is it in any sense a backdrop for our exploits; and it is surely much more inclusive and embracing than the way it is conceived in terms of its customary contrast to *nomos*. In the decidedly Heideggerian emphasis that I am employing here, *physis* is earth, "the emergent and finite phenomena of the natural world" and is thereby rife with the life force manifest within the same, utterly corporeal context as the *psyche*. Given this perspective, apprehension, especially as it is manifest in language, is not a faculty belonging to the *psyche* but a process that engages it—"happens" to it, as Heidegger says—through its reciprocal bond with *physis* (*Introduction* 141). Though it is convenient and perhaps even necessary to separate what is within us from what is before us in order to isolate and thereby appreciate the idea of the *psyche*, doing so quite ironically distorts it, as it does the idea of *physis* upon which it is based. The mystery holds sway. To properly understand the nature of the *psyche*, as both Snell and Mansfeld claim, we must understand that the "borderline between inside and outside, between subject and object, is vague if not wholly absent (Mansfeld 4); or more simply, as David Claus says, the *psyche*, as the life force, occupies "an ambiguous position between the mind that knows and the objects or ideas known" (59).

It is hardly necessary to unearth and decipher arcane texts to encounter the idea; as I have indicated it is as near at hand as the *Iliad*. And nearer still. The Navaho have a word, "aho," that has come to be an inter-tribal expression of greeting. But it means originally both "breath" and "spirit," being as the Hawaiian "aloha" in that respect. The latter has also become an expression of greeting, though it is as "aho," meaning in that very capacity "to share the breath of life." Biblical references combining breath and spirit abound as well, and the single meaning of the two is apparent in the Latin *spiritus*, the Greek *pneuma*, and the German *Geist*. So it is with "aho." The word connects the Navaho people with the sacred on the most physical level, as earth itself is animate, as *physis* is here manifest as *Nilch'i*, as the air we breathe. And it is more. *Nilch'i* is the "Holy Wind" that "suffices all of nature," as James McNeley says, giving "life, thought, speech, and the power of motion to all living things," the "means of communication between all elements of the living world" (1). Thus, to breathe is a scared act; in this merging of the physical and spiritual, it is to pray. In this perspective we might say that the *psyche* is even less than a "feeble and witless thing," because in and of itself it is nothing except as it achieves ascendancy as "aho," precisely the relationship of the *psyche* and *physis*, having precedence over the things related, being *prior* to them. With every breath we breathe we are that relationship, being by nature the "outsider within," where there is no inside or outside, subject or object, only "an ambiguous position" where "the borderline between the two are vague or wholly absent." Again the mystery. Such beliefs will loom in subsequent discussions of Heidegger and language, especially as *physis* is there associated with *logos*, as *Nilch'i* is here. In each, language is as *physis* in being the very ambiance in which we dwell, beliefs in many ways not all that far removed from the pre-Socratics.

In any event, such beliefs obscuring boundaries between internal and external are in large part what fueled the rhetoric of the Sophists. They are especially apparent in Gorgias, where the dynamics of his thought, including most particularly his ideas of rhetoric, can be understood as falling under the pervasive, and prevailing, influence of the Homeric ideal of the *psyche*. In Gorgias there is no concept of the soul or other essence of the self to provide a privileged position or perspective from which to take stock of the world objectively, but only the *psyche* belonging intrinsically to the world. Having no independent existence apart from *physis*, the *psyche* is determined and defined only in terms of *logos*, a *logos* that is existential, corporeally manifest as language. That is to say, the Gorgian conception of language never advances beyond the basic unity of *physis* and the *psyche*, and only through this unity can the power of language function as a revealer of the being of things. In the familiar lines from Gorgias's "Encomium to Helen," speech is precisely the "creative surge of Being" that I am attempting to explicate, the "great *dynastes*" that achieves "the most divine work," making any "impression it wishes on the soul [*psyche*]"—removing fear and grief, creating joy, increasing pity, causing its hearers to be transported "by means of its wizardry" (par. 8–10). Indeed, consistent with the idea of *physis* developed previously, *logos* is conceived by Gorgias only in terms of the phenomenal world of sense perception, and its effect is rendered in physiological as well as psychological terms, such that "the power of speech over the constitution of the soul can be compared with the effect of drugs on the bodily state" (par. 14).

Thus, what Gorgias accomplishes is a restructuring (or more aptly a "structuring" prior to Plato) of our common sense notions of communication, one linked to the circumstances under which communication takes place, and, more importantly, *how* communication takes place. In the Gorgian configuration, as Charles Segal and many others have noted, *logos* is an independent external power, owing its source to nothing higher and possessing a physical *dynamis* that exerts an active force impinging on the *psyche* from without.[18] Once again, the power of speech functions on the same natural level as the *psyche*, working directly on the *psyche*, having an immediate, physical impact on it, as would a drug. The emphasis is on *logos*, that in Gorgias's peculiar interpretation can impose itself only irrationally because it is removed from the transcendental governance of the cosmic *logos*. Rather than being fixed in some essence beyond itself, in either the mind or external reality, *logos* has instead the "autonomy of a separate artistic medium" (Segal 112); it is freed from a dependence on a higher reality, is its own master, and is immanent, free to function in the dynamic realm of tangible, temporal experience—in the fray of the "clash and counterclash of extremes."

Here, not only are the boundaries obscured between intellect and senses, spirit and nature, but also those between Being and appearance, as these notions coalesce in the revealing and concealing powers of language. To reiterate Gorgias's conviction, "We do not reveal existing things to our neighbors," but "that by which we reveal [these things] is *logos*" ("On Non-Being" par. 84). Here, as there is no fundamental, fixed reality at the basis of *physis*, absolutely none that we can know, all things are made manifest in appearances revealed through language. Thus, the things of the world are not pre-existing and separate from language, nor are ideas themselves in any sense, but the world is composed of appearances, of phenomena "revealed" or given presence by language, so that language organizes our apprehension of the world to the extent that it does not create the world outright. Gorgias claims that "nothing exists," but essentially Being *is* language in his deliberations, especially as it is conceived as *physis,* as that "creative surge" taking its measure in appearances that language prompts. One could say, as many have, that on this matter Gorgias is again being droll or that he simply lacks the philosophic wits to cleave the cosmic *logos* from the human *logos*, for here each is congruent in his sense of *logos* as the great *dynastes*. Whatever the case, *logos* is not only prior to things in Gorgias's reckoning—*verba* made superior to *res*—but for these reasons all things are, in a sense, "made of words."[19]

Such notions dispose some critics to claim that Gorgian epistemology constructs a "prison house of language." It is, I think, an unfortunate metaphor, but for now let it pass. The task at hand is to further clarify and shore up the idea of the *psyche*, specifically the idea that in its oneness with *physis* the *psyche* lacks an essence of its own, its status as subject thereby resolved on a practical level wholly in terms of the power of language. We fancy human beings as free and autonomous, as "interiorized" selves and thus users of language. But in this different perspective *logos* itself is free, and human being is nothing until shaped by language. Thus, the latter is most adequately understood not only as the *logos* we speak but more essentially as the *logos* which speaks to us, and through us, ultimately the *logos* that we are. The *psyche* can then be seen as the place, or clearing, where language becomes manifest, where the power of *physis* is focused by virtue of language, in a strange equation of hearing and speaking, just as the idea is repeatedly illustrated in the *Iliad*. In this manner, then, we are also, in a sense, made of words.

The uncommon allure of this idea is but an extension of the otherwise simple notion of the primacy of language, of it ultimately determining the essence of who we are. Indeed, its effects are presented clearly enough in Gorgias's works and fragments. To read of the magic, incantations, and casting of spells in "Helen" and other works is to understand that the rhetor rides the back of the tiger that is language, and he is accordingly defined in terms of the radical closure of *psyche* and *logos*. Here, language—and more specifically rhetoric—can be understood initially as the defining ingredient in what Calvin Schrag calls the "communicative praxis," where, as I have suggested, any sense of self is a creation of rhetoric as much as it is a foundation for it. But in Gorgias's more extreme presentation of the idea, the rhetor wields a power he neither originates, fully controls, nor even comprehends. He at best harnesses or channels it, as would any magician or wizard. It would be hard to imagine a more forceful ascendancy of rhetoric over philosophy, but it provides a very likely means to appreciate Gorgias's contention, even in the context of Plato's condemnation of it, that rhetoric fulfills its function "entirely through words." Indeed, hardly one to take delight in fine nuances, Gorgias in the same context claims that rhetoric is "any art connected with words" (*Gorgias* 450b–451d). Given his view of language, and the seemingly restricted role of the rhetor essential to it, it is hard to imagine a more appropriate definition of an art of rhetoric than the one he offers.

In any event, Gorgias's view of the "art" of rhetoric, his *techne*, is inherent in the connection of *logos* with *doxa*. A necessary connection, it would seem. Inasmuch as *episteme* ("true knowledge") is unattainable in the Gorgian perspective, *doxa* is not only the ordinary state of communicable knowledge but in lieu of *episteme* is expressly mandated in his epistemology. Moreover, in that there exists no objective correlation of words to reality, due to the obscurity or complete absence of the latter, the speaker necessarily "deceives" his listeners. As a result, rhetoric inevitably involves the use of deception (*apate*). As *logos* reveals things that are otherwise hidden, it necessarily makes things seem what they apparently are not. It is, once again, the great *dynastes* that "achieves the most divine works by means of the smallest and least visible form" ("Helen" par. 8). No doubt the belief can be considered in accord with the ways of Sophistic rhetoric that Plato indicts in the *Phaedrus*, where he says of Gorgias and his ilk that they "can make trifles seem important and important things seem trifling through the power of language" (267b). However, insofar as we ever assume a reality apart from words, this Gorgian sense of *logos* simply mandates the essential truth that words are inherently deceptive. In fact, the deceptiveness of *logos* is, paradoxically, its virtue in rhetoric. With language freed from a necessary correspondence to some fixed reality, its use becomes a *techne*, an art which employs words not to find truth but as a power to be tapped and directed in the interests of persuasion. As Bruce Gronbeck states the Gorgian view, rhetoric then becomes in its essence a *"techne* of *apate"* (33).

Here the focus is on style, where the application of this *"techne"* is plainly evident in language itself, as the rhetor works through the conjuring power of language to achieve the desired effect on the audience. The style of the "Funeral Oration" provides a suitable example, where Gorgias seemingly attempts to spellbind or enchant by words alone. The inspiration for his particular strategies and techniques of style is open to dispute. Kennedy says that Gorgias's fondness for antitheses can be seen as either the result of his obsession with the antithetical structures native to Greek syntax or as a reflection of his belief that truth is relative and therefore requires expression of contrasts and alternatives (*Art* 64–65). George Thompson, however, speaking particularly of the *"Satzparallelismus"* of the "Funeral Oration," says that the Gorgian style is liturgical in origin, native to Sicily, and that Gorgias "divested it of its ritual setting, and secularized it as an art form; and that it was this sensational novelty that Gorgias brought to Athens" (81). In yet another view, Scott Consigny says of the "Funeral Oration" that "the somber sonorities, repetition of sounds, and rigorous formality of Gorgias's style is paradigmatic of the Athenian funeral oration rather than idiosyncratic; and what many critics identify as peculiarly 'Gorgian' might better be described as typical of these works" ("Styles" 47).[20] Whatever the case, parallel structures and figures of antitheses dominate the "Oration," apparent in expressions such as "the kindness of the truly just rather than the arrogance of positive right," "the perfection of argument rather than the rigor of the law," "to say and refrain from saying, to do and refrain from doing." There are other parallel structures as well, examples of the scheme of parison, seen in Gorgias's compiling adjective phrases with similar sounds in similar places, as in "violent toward the violent, restrained toward the restrained, fearless toward the fearless, terrifying among the terrifying." In fact, the full impact of the style of the passage is lost in translation, for the "Oration" abounds in the use of homoioteleuton, the parallelism of sound and syllable number, that accounts for Kennedy's contention that "on Gorgias's lips oratory became a tintinnabulation of rhyming words and echoing rhythms" (*Classical* 29).[21]

Moreover, the Gorgian figures of style do not in themselves mark the end of the matter, for the Greek language was itself a vehicle perfectly attuned to enchantments cast by words.[22] Gadamer is of the same mind as Nietzsche in claiming that the "domination of this 'most speakable of all languages'. . . over thought was so great that the chief concern of philosophy was to free itself from it" (*Truth* 378). The point reiterates one previously made, and one to be made again in this study, that the transformation of Greek culture from one based on the dynamics of poetic expression to one maintained primarily on the conceptualizing thought of metaphysics was as much a syntactical revolution as it was one of ideas (Havelock, *Preface* 259–63). As Gorgias's obsession with style reflects the power of poetic experience, Plato's injunctions concerning the "neuter" style set down in the *Republic* reflect the concern of philosophy, most definitely Plato's own. It is not unusual, then, that Socrates himself claims he was subject to the enchantments of language. In the opening lines of the *Apology* he says of his accusers, "I know that they almost made me forget who I was—so persuasively did they speak; and yet they have hardly uttered a word of truth" (17a). A more apparent concession is made in the *Menexenus*, where Socrates says specifically in reference to funeral orations that they "steal away our soul with their embellished words," until "I become so enchanted by their words that I imagine myself on the Islands of Blessed and not until the fourth or fifth day do I come to my senses and know where I am (235b–c). Socrates would seem to be ironic in both instances, as he surely seems to be in the remainder of the *Menexenus*, where he presents his own funeral oration. But Kennedy claims that in either case Socrates is not.[23]

If Kennedy is correct concerning the enchantment of Socrates, it can hardly be construed as an expressions of respect for Gorgian rhetoric. Elsewhere, Kennedy maintains that the stylistic peculiarities of the "Funeral Oration" were contrived and self-serving, and Gorgias's alone, the result of his penchant of borrowing for no clear purpose the techniques of poetry and applying them to prose (*Art* 64). Kennedy's argument is well taken. In translation or not, the sounds of Gorgias's elevated style appear to us now just as Kennedy has described them, or, for that matter, just as Diodorus Siculus and Anthanasius—and Plato and Aristotle—described them in antiquity. We might perceive his style as perverse and ugly or, on the contrary, as fully exploiting the potential of the Greek language. Yet either way, the crucial point depends upon its effect. It is "an intellectual display of skill," de Romilly says of Gorgias's style (20); and she continues, extending her argument to make it not only more generous to Gorgias than Kennedy's, but also more complete by virtue of its connection to a larger rationale. She takes Gorgias at his word, claiming that the very principle of his art of speech was to stir the passions, influence "the frailty and uncertainty of human opinion," and thereby deceive (25), and in that regard the effect of her interpretation is that Gorgias saw the irrational power of speech as invested with mystical significance, akin to magic in its capacity to evoke divine intervention.

Charles Segal makes a point similar to de Romilly's, giving particular emphasis to the force of fate or necessity, the "divine power" (*ananke*) that in Gorgias's view was expressed by *logos*. Specifically in this context the Gorgian *techne* possesses the critical aspects mentioned earlier, those requiring the rhetor to be both a "linguistic technician" and a "magiclike charmer" (117). And here the Gorgian *techne* is concerned expressly with style, for in style alone lies all we can know and control of the great *dynastes*. As a linguistic technician, the rhetor applies as a matter of conscious design the metrical patterns and other schemes and figures of style to suit his knowledge of their effect on the *psyche*, his own as much as those of others, for ultimately there can be no distinction. Thus, in his role as magician, the rhetor employs these same strategies of style, as they are his only means to ply the powers of language that are otherwise beyond his conscious control, if not beyond his means to tap by virtue of style. He is, then, a conduit, a means of passage for this power rather than its source. Gorgias's *techne* of *apate* is thereby an art dealing in style, a system to deploy words in ways designed to most effectively harness *ananke* and bring it to bear in service to persuasion.

The irony of this idea of *logos* is that from its conceptual development in Tisias and others, *logos* has been conceived as the alternative to physical force or violence, an idea that seems fully accomplished in Plato, where the influence of the physical was restrained and the appeal of *logos* was to the intellect, to the spiritualized essence contained in the soul. But in Gorgias, *logos* retains its nature as something violent, albeit, as Laszlo Versenyi says, a "superior form" of violence (45). Nevertheless, the effect of violence is a central feature of the great *dyn-*

astes. Versenyi goes on to say that *logos* in this Gorgian sense moves in an altogether separate realm from that described by Plato, and perhaps has nothing to do with knowledge, intellect, or reason (45)—at least, I might add, little or nothing to do with Plato's conceptions of these faculties. Indeed, *logos* in this sense necessarily impinges from without, for there is no "inside" to the *psyche* to house knowledge, intellect, or reason to which *logos* might appeal or from which it might originate. These "higher functions" are best conceived not as the effects of the mind—*nous* in that sense alone—but as attributes of language, and we take part in them only by virtue of our relationship to *physis*, only to the extent that both the *psyche* and *physis* are "actualized" by language, as each, in this sense, is made of words.

In this manner, then, the Gorgias's *techne* is essentially as he claims it to be. It is a "linguistic *pharmakon*," having the effect of a drug, producing palpable physiological effects on the hearer and, by extension of this logic, on the speaker as well. Here, once again, the *psyche* betrays its basically physical nature, as does *logos*, bearing as it does the effect of physical stimuli. Here, as well, Gorgias reflects the thinking of Empedocles, of his belief in continuous "effluents" flowing from objects to various pores of the body. But such a primitive notion of sense perception should not obscure the more important point upon which Gorgias's *techne* is based, that the only reality that can be affirmed "lies in the human *psyche* and its malleability and susceptibility to the effects of linguistic corruscation" (Segal 110).

If language is not "reality" in this perspective, their boundaries are blurred, coinciding in ways to constitute the character of magic precisely. The magic not only explains Gorgias's fascination with the effects of style but utterly confounds views of *techne*, especially as we customarily have them in rhetoric, as analytical and systematic treatments of subject matter that summarily exclude charms and enchantments associated with magic. Just as crucially, style is to Gorgias not merely a part of the external dress of language, as it tends to be in traditional views of rhetoric, but provides the surest access to the power of language, a feature of language that can be garnered and resolved into a *techne*. Nor is language conceived as a means to a larger end, as an instrument to express some extralinguistic reality or meaning beyond itself. In this "art" of rhetoric, words possess, embody, and realize. Words are words alone rather than means to whatever we assume them to reflect or represent. Language is experience in this view, a region where, Gadamer says, the "ideality" of the meaning of experience lies in the word, in language alone (*Truth* 377).

Plato's Dialectic and the Doctrine of the Soul

Nevertheless, applying the term *techne* to Gorgias's rhetoric had best be hedged with qualifications. To see the Gorgian *techne* in contrast to that elaborated in the well wrought wholeness of Plato's thought gives us reason indeed to appreciate Plato's complaint that Gorgias's art of rhetoric is no art at all. What is apparent in the contrast is that Plato's *techne* is based on critical judgment. It delineates both the "knowing what" and the "knowing how" of any given activity, to include a clearly defined goal for that activity, specifically stated principles governing the route to that goal, and, most importantly, the reasoning needed to apply these principles in a manner to best accommodate that goal.[24] Plato spells out the nature and some of the implications of an exclusively rhetorical *techne* in the *kairos* passage quoted earlier from the *Phaedrus* (271b–272b). Here, *techne* obliges not only a set of principles and a thorough understanding of these principles by the rhetor, but Plato describes as well the method constituting *techne*, the rational procedure essential in the production or accomplishment of a specific goal, all of which amounts to a "*logos* of the soul." Thus, what is also apparent in this conception of *techne* is that the means to rhetoric, the power to effect change through language, resides with the rhetor rather than coming uncertainly, obliquely, and in some collateral fashion by way of the rhetor's homage to language as a higher power, as a great *dynastes*. Two features of this conception are readily apparent, both vital to Plato's rhetoric. The first is that *techne* involves a rational method, a process of critical reasoning that constitutes, in essence, Plato's dialectic. The second is that by virtue of the dialectic the focus of Plato's rhetoric is overwhelmingly on the speaker. He is, as Plato says, the "father" of *logos*. The conception marks explicitly a most significant transition. Rather than the speaker being the means to express the power of language, he wields power over language, a turn of events that betokens in all respects the ascendancy of the soul over the *psyche*.

Surely the soul is one of the most important inventions of Western philosophy, in many respects the very basis of it. In Plato, the soul is also one of the most complex. He attributes qualities to it that seem to make the soul a sort of intermediary between physical objects and the idealized essences of the Forms. But he is clear in claiming that in its kinship with the Forms the soul is immortal. In fact, as the soul exists prior to the body, it knows the world of the Forms immediately in a "prenatal" or anterior state of existence. Thus, once the encrustations of bodily interests and cares are left behind, and "no things of sense trouble the soul," the vision of the Forms is grasped only by reason. Plato depicts it as a process of anamnesis, "a remembering of what the soul once saw as it made its journey with a god, looking down upon what we now assert to be real and gazing upward at what is Reality itself" (*Phaedrus* 249c).

Yet this splendid image of the soul's journey is contingent on the dialectic, for the latter exists ultimately as a means to recover, to "recollect" this knowledge embedded in the soul. Undeniably the dialectic takes its name from "dialogue," and therefore has been defined as the art of adroit questioning to elicit provocative and insightful answers, a method of knowledge that Plato says was brought to perfection by Socrates, a method that some would have us believe is recapitulated in his dialogues. However, the dialectic remains above all else the science of first principles, its purpose to reach the highest and ultimate source of knowledge residing in the Forms. That this is not achieved through dialogue with another—the dialogue form notwithstanding—seems to be the substance of Plato's position. This is definitely the case in the later dialogues, where the dialectic becomes more and more insular and internal, a process that transpires wholly within the soul. Perhaps the surest way to this understanding of the dialectic begins by revisiting Plato's attack on Sophistic rhetoric, especially as I have presented that rhetoric above, through Gorgias, with its reliance on deception and the odd, attending circumstances of magic and witchcraft.

Ironically, the idea of deception, clearly in terms of our most common understanding of the word, is in large measure contingent on Plato's view that there exists perfect truth and that language is essentially ill-equipped to capture it. And if the Greek word *apate* is rife with nuances allowing various shades of meaning to apply in certain contexts or circumstances, in most accounts it seems to mean "deception" pure and simple. But the word need not be associated with falsehood in a pejorative sense, certainly not with epistemological values where issues of right and wrong, true and false, are deduced objectively or methodologically. As Untersteiner tells us, prior to the fifth century *apate* designated a creative process associated with the very ambivalence of Being, as was so often apparent in the caprice of the gods themselves. In this way, he says, it was associated with a "particular tone," neither moral nor otherwise but drawn simply from the general ambiguity of things, that "can give a thing a certain meaning or its opposite (108–09). Any number of authorities, to include Havelock, Steiner, Wheeler, and Pratt, have come to very similar conclusions, claiming in essence that the early Greeks tended to assume a fatalistic or even sporting view of deception, often taking it as a means to power that could be used to very good effect. Not only can this be surmised from Gorgias's treatises and fragments themselves, but by one account he goes so far as to claim the deceiver and the deceived should be more highly esteemed than the nondeceiver and nondeceived; the deceiver because he succeeds in his effort to persuade and, what is more telling yet, the deceived because he is the wiser for being affected by the pleasure of words.[25] But Plato's criticism is relentless and always very imposing. He begins by denouncing the methods of deception. He condemns magic and witchcraft and the incantations on which they are based, and would sentence to death those who practice them (*Laws* 933d–e). His remedy for those who practice Sophistic rhetoric is graciously less severe, though his contempt is very bit as intense. In fact, in the *Euthydemus* Plato considers rhetoric inferior to witchcraft, for the practitioners of the latter can enchant beasts, insects, and other "vermin and pests," while the rhetorician can charm only juries, parliaments, and various other "collections of crowds" (289b–90). Otherwise, Plato places Sophistic rhetoric on the same level as witchcraft, each as sham wisdoms involving deceptions of a similar nature.[26] No less than the sorcerer, the Sophist summons phantoms. By means of deception he makes the weak case seem stronger and the strong case weaker, and, in effect, makes people believe in things that do not exist—or, more properly, in things that did not exist prior to the work of deception.

However, as pointed out in a prior reference (*Timaeus* 51d–e), true knowledge (*episteme*), according to Plato, shelters one from deception. As he separates the super-sensual realm of Reality from the earthly realm, he also separates *logos* from the *physis* of tangible experience, projecting it as something beyond language, essentially as reason itself. This is indeed the philosopher's state of grace, of a soul possessed of true knowledge and thereby delivered from the ambivalence of *physis* and the lure of persuasion, at least of the sort practice by Gorgias. In his dialogues Plato describes the soul as "self-moving" and "absolutely complete," the psychical source of physical movement, and he sharpens the contrast by repeatedly vilifying the body. He would thus sever, as best his philosophy could, the soul from communion with the body, eliminating any sense of the pre-Socratic *psyche* altogether, "for in attempting to consider anything in the company of the body [the soul] is obviously deceived" (*Phaedo* 65b). The specific rhetorical emphasis of the idea is given in the *Phaedrus*, where, in contrast to the Gorgian notion of the ascendancy of *logos* over the *psyche*, Socrates says that "every body that is moved from without is soulless, and every body that derives its motion from within has a soul," so that "the soul has charge of all that is soulless" (245e–246a).

The distinctions Plato makes between the body and the soul are first given extended treatment in the *Phaedo*. Appropriately, the conversation turns on whether Socrates fears his imminent death and the "common fear" that the soul, as was the fate of the *psyche*, "may be scattered to the winds at death" (77e). Simmias, a Pythagorean, believes in the pre-Socratic sense of the soul and therefore seems to give Socrates good reason to feel some apprehension. He contends that the soul is not stronger nor more lasting than the body but is a harmony composed of the elements of the body at the right tension. He then likens it to a well-tuned lyre, a harmony that vanishes once the lyre is broken or its strings cut or snapped (86a–c). As this harmony perishes, so does the soul at the death of the body, which itself remains only so long as it is burned or rotted away.[27] In response, Socrates offers the philosopher's argument, telling Simmias that he is not afraid of death because death is, in fact, a purification consisting in the separation of the soul from the body, that our immortal part retreats before death and goes away safe and indestructible (106d). In one of the more telling passages in the *Phaedo*, he argues that the body is the prison house of the soul and it is philosophy alone that delivers it from this bondage (82e). Verily, to Plato, philosophy itself is the achievement of immortality (Cornford, "Introduction" 27). In short, the human soul is an immaterial agent, considered by Plato as immortal and existing prior to the body, and in all instances impaired by the body in its role as our only emissary to the higher, spiritual Reality sought by the philosopher. Therefore, Socrates claims that the philosopher must hold the body in aversion, even despise it, because we are defiled by it and can be released from this enemy only in the wisdom of "the other world and nowhere else" (67e–68b).

Here the contrast of the soul and the *psyche* seems complete, and we are left to deal with the implications. Erwin Rodhe says that the *psyche* is a work of resignation, whereas the soul is that of hope (55), or, if we wish to be particularly ungracious, of a cowering in the face of non-existence. In any event, it is clear that in the hope of deliverance tendered by the soul, the tragic sense of life and definitely any sense of the tragedy of knowledge are put aside. The soul is quite literally an "antibody," as Derrida says, a *pharmakon* in its own right, dispelling fear of death, and fear of life as well (*Disseminations* 122–23).[28] To put the issue in a more usable context, we could say that the distinction between the *psyche* and the soul is resolved in terms of perspective, what Santayana in the following passage calls a "change of aspect": "The same thing looked at from the outside or biologically is called the psyche, looked at from within is called a soul. This change of aspect so transforms the object that it might be mistaken for two separate things, one a kind of physical organization and the other a pure spirit" (*Realms* 570).

Santayana's statement by all means stands on its own, but it is hardly intrusive to say simply the *psyche* and the soul are the same thing until we look at them differently. Then they are as far apart from one another as are our perspectives on them, meaning in this case worlds apart.[29]

The difference then seems far more pronounced than "mere" perspective would allow. For instance, to presume language is master, free and autonomous, it is all the same whether we perceive it as affecting us from within or without, for it precedes that distinction, thus preceding as well that drawn by Santayana between inside and outside. Here I appropriate Scott Consigny's analysis of Gorgias and for my purposes extend it further. Not only does language "denote all human action" (and reaction) to Gorgias but its fundamental element is the trope, the essential "maneuver" of the human drama. As the distinction between inside and outside is precluded by the power of language, so is that between literal and figurative language due to the absence of any reality existing prior to language upon which "literal" language could be based. The trope is then a "form of life" and the *psyche* a series of tropes expressed in social and psychological dispositions or moods (*Gorgias* 77, 121). It can scarcely be any other way when "nothing exists," when "language is language and nothing else besides" and *physis* the phenomena formed by language's fluid nature, a "plastic medium" indeed of process, flux, and change.

But by virtue of the soul this Gorgian world view is inverted. As the soul gains access to enduring reality behind the phenomenal world through the machinations of the dialectic, we come to understand appearance in the customary sense depicted by Plato, as a mere shadow show, figures cast upon the wall of the cave. And comprehension of all that is essential is effectively removed from the realm of appearances and is instead centered in the interplay of the soul and the dialectic as they procure true knowledge in association with the Forms. In general, the view of the world shaped by *physis* is not only foiled but seen as corrupt and corrupting, a source of despair and deceit. And *techne* is then no longer an extension of the power of *physis* but in Plato becomes part of an enterprise to usurp that power and use it against *physis*, a process bearing the mark of an analytic attitude epitomized by the dialectic itself. It is then a process used to separate the *psyche* from all that environs it, thereby to create a "soul" in its place.

Often it seems not to be the substance of Plato's ideas at issue but the manner of their presentation. Indeed, the dialogue form is at the center of many recent and worthy endeavors that take it as the essence of the dialectic, as a more pointed deployment of the Socratic method itself. Those under the influence of this belief, seeking to secure Plato's place in the rhetorical fold—to rehabilitate *him*—would deduce a Platonic rhetoric that is a perfect likeness of the Platonic dialectic itself, a philosophical rhetoric that is based on "mutual inquiry, trust, and assent," as John Gage says (33); a "seeking of fundamental truths through sustained critical conversation," as Virginia Steinhoff says (31). Of course, there are times when it is hard to imagine any such "mutual inquiry" taking place in any of Plato's dialogues, let alone in the *Phaedrus*. But Steinhoff's interpretation of that dialogue is an apt expression of the idea. She postulates what she calls the "Platonic Stance," a version of the dialectic that seeks what is "synthetic and artful," shaping unexpected material into new forms that are "suggestions of things not seen" (39). Teachers under the sway of the Platonic Stance would be "ethical performers who shape instruction with style, character, and language" leading to a dialectical enactment—indeed, reenactment—of the *Phaedrus* itself (41). As well presented as these arguments often are—so practical and down-to-earth—it would seem nevertheless, given Plato's doctrine of the soul, that his purpose is hardly so worldly, his dialectic seeming to be hitched instead to a far distant star.

In a sense the question of what comes first, the dialectic or the soul, is as fruitless as talk about chickens and eggs. But there can be no dialectic save for the knowledge of truth that the soul is born with, and there can be no soul without this knowledge, for that is essential to its nature. Moreover, in the *Phaedrus* and elsewhere in Plato, the soul is given precedence, serving as a first condition for the dialectic. Here, as I say, the dialectic is essentially a rational process that leads the soul to uncover truth rooted within the soul itself.[30] In the *Phaedrus*, Plato states that there are two principles of dialectical reasoning, classification and division. The former is a process of perceiving and bringing scattered particulars together in one idea, and the latter involves dividing things again by classes or species (265d–e). It is on this basis that Plato establishes his science of first principles, from whence it moves eventually to knowledge of the Forms. As elaborated in the *Republic*, classification and division express a double movement of ascending and descending. In the ascending phase the mind, or soul, seeks to grasp the first principle and then, having grasped it, proceeds "downward to the conclusion, making no use whatsoever of any object of sense but only of pure ideas moving to ideas and ending with ideas" (VII. 511b–c).[31]

In this process the Forms are, seemingly, not thoughts but objects of thought, the Platonic Ideas themselves, and for this reason they remain separate and distinct from the soul. Nevertheless, both the Forms and the soul are on the same level of reality, insuring the viability of their relationship and possibility of communication between them by way of the dialectic. They are, indeed, the only essential elements of this reality—Reality as such, to Plato—and they subsist as principles of a conceptual scheme rather than as entities or tangible fixtures of this reality. Though separate, the Forms and the soul are creations necessary to authenticate one another. They are coupled in the admirable consistency and organization always at work in Plato's thought, such that we can understand one only in terms of the other. In fact, to frame the issue in this manner is to understand the soul is no more a "reality" than the celestial sphere of the Forms. The intricate, inner workings of the one and the vast expanse of the other are analogies themselves, metaphors of one another to give an account of what is ineffable. As such, the domain of each is *aneu logou*, and either domain, within or beyond, is as "unsayable" as the other. The power of language is overcome here, and all that is timely and terrestrial is transcended temporally, spatially, and culturally. To say that reality is something other than language, being beyond it, is to be well on the way to saying it is absolute reality and not simply another narrative, another "mere" perspective, and generally that is how we have come to accept the Platonic perspective.

Nevertheless, in this perspective the Forms and the soul are essentially the same, as the Forms themselves are lodged in the soul. In this perspective "dialogue" is at best a useful analogue for what Plato envisions as taking place among the Forms or within our soul for the sake of knowledge of the Forms. Indeed, in this context "knowledge" itself is too frail a term, for in these issues concerning the soul and the dialectic we are speaking of philosophy, of the soul's longing, its transfiguration in "the splendor of the light of reason," as Plato says in the *Republic*. And in matters of rhetoric—as in all others in Plato—emphasis on the soul is consummate. Every implication of a rhetorical nature that is broached in the *Phaedrus*, every idea, statement, or suggestion having anything to do with rhetoric, has its basis in Plato's doctrine of the soul, to include most emphatically the Platonic *kairos*.

As the soul is presented as a unity in the *Phaedo*, at odds only with the body, in later dialogues it is more complex, becoming tripartite in nature. Thus, there is an evolution of the soul in Plato's works, beginning with the simple dualim of body and soul in the early dialogues to the way the soul is presented in dialogues such as the *Phaedrus* and *Timaeus*, where the soul coalesces with *nous*, the mind, in a tightly woven dialogue of the soul's parts that constitutes the dialectic itself in its movement toward pure reason.[32] For our purposes the *Phaedrus* is much more critical to this argument, as it is at once Plato's enduring showcase of the soul and, accordingly, most complete statement of his rhetorical theory. Here the doctrine of the soul unfolds in the myth of the chariot, the steeds, and their driver. The charioteer is the soul's pilot, reason itself, and must keep in check the two winged horses. Thus the soul, in contrast to how it is presented in the *Phaedo*, has three distinct parts—reason, courageous spirit, and willful appetite, a division Plato maintains through the rest of his works, to include the *Laws*. And as John Moline argues, these parts are best seen as agents, not faculties; that is, these parts are not mere proclivities or inclinations but are independent of one another in that they stand in direct opposition, vying for control and thereby establishing the soul's susceptibility to internal conflict (51). A fully "philosophized" *psyche* results, in that capacity being every inch the soul. As such, it remains alien to the body, and a healthy soul is a harmony of its own parts, just as Simmias claimed in the *Phaedo* that the *psyche* is a harmony, though of bodily parts. The soul is given autonomy, and just as importantly, the harmony is achieved not when the three parts or agents of the soul are in balance but when reason, the soul's pilot, is predominant, directing and controlling the soul's two less judicious elements by means of the dialectic, the "dialogue" occurring among them. Thus, in Moline's analysis the unity lost in the tripartite nature of the soul is regained by the dialectic, as the soul seeks to achieve true knowledge in its ascent to the Forms.

This idea of the soul leads to the critical if much diminished role that rhetoric, and language as such, have to play in Plato's philosophy. Given the soul's tripartite nature, all decisive utterances in Plato are interiorized, resulting in a dialectic that is a conversation of the soul with itself, a conversation, as Plato says in the *Theaetetus*, complete with questions, answers, affirmations, and denials (189e). This is precisely what one would expect of a theory that deems the soul composed of agent-like parts, where the "conversation" is internal and takes place among distinct parts of the soul, providing in the interface required by the dialectic a way for the soul to "watch itself" (Moline 108–09). Thus, the ascent to the Forms is more of a journey inward in search of these structures, these intelligible realities, ideas, or essences of things apart from the matter which they reflect, for they are, at long last, wholly of the divine scheme of anamnesis, of "pre-natal" memory of true knowledge that Plato posits. This sense of the dialectic, of reason as internal dialogue, is the basis of philosophical rhetoric, at least Plato's version of it, and is also in accord with what Hannah Arendt describes as the twofold approach of Greek philosophy as such. First comes the activity of *nous*, consisting of contemplation and intellectual apprehension of the everlasting. This work is the soul's highest function, albeit a function that is itself *aneu logou*, "speechless," being as far removed from language as it is from sensuous apprehension. Then, secondly, follows the attempt to translate the vision of the everlasting into words (137). In the *Sophist* Plato makes the argument himself, suggesting that the thought of a person and the account given by the person are the same, thinking and speaking the same, with this very important difference: thought is the silent, internal dialogue of the soul with itself, while the account, the language reflecting the thought that comes after, is "the stream that flows from the soul through the mouth" (263e).

Rhetoric is thereby reduced to an act, and whatever art or *techne* is attached to rhetoric comes prior to the act. An interior-exterior dichotomy is established, one coupled with an initial distinction of body and soul—the first an image, the other the original, the first appearance, the other real—and consequently speech is relegated to a channeling of interior contents. The idea establishes categorically the congruity of the soul with the individual personality (Claus 179), a comprehensive personal self that relative to the pre-Socratic notion of the *psyche* is fixed, stable, and enduring. It is this idea that lends special poignancy to Socrates' refrain to "care for your souls" and his resolve at the beginning of the *Phaedrus* to follow the Delphian injunction to "know thyself." As Guthrie maintains, both are invariably calls to exercise the mind, to maintain personal integrity, and both, I might add, are an absolute impossibility for the *psyche*, for in the transition of the source of *logos* in *physis* to its source in the soul, *logos* is no longer free and no longer language. It is reason, originating in the soul and remaining there, literally at the beck and call of the soul. In this scheme rhetoric is thereby based on the prior affirmation of the self as an interior mind standing over against an exterior world, an independent "privacy" struggling, by means of rhetoric, for externalization in the public realm. Insofar as rhetoric is genuine, hinged to dialectic, it is present only in the soul. Insofar as rhetoric is *psychagogue*, a practical psychology concerning how the soul can be influenced by means of language, it is corrupted as soon as it passes from the soul and into linguistic forms.[33]

Thus, what supremely matters in Plato is not, as Steinhoff and other apologists for Plato suggest, a dialectic as a process involving the participation of more than one soul but, given Plato's conception of the soul, a dialectic that is wholly internal, absolutely within the individual soul. It is a process of rational analysis through which the soul seeks and achieves true knowledge in a self-contained and introspective journey that requires, as Havelock says, an intellectual effort of oneself acting upon oneself ("Socratic" 7). Indeed, insofar as rhetoric has a place in Plato's scheme at all, it is secondary to the dialectic, or is otherwise consigned to the inner parts of the soul, a rhetoric of a very peculiar sort where the real audience is divine not human. We speak to please the gods, Plato says in the *Phaedrus*, not our "fellow slaves" (273e). Or, in the more exacting fashion of the *Sophist*, the mastery of the dialectic is the work of the philosopher, whose thoughts, by virtue of the dialectic, constantly dwell upon the nature of reality, even though this effort is arduous and often harrowing, for in these regions reality is so painfully bright. Here the philosopher in all respects is unlike the Sophist who "takes refuge in the darkness" of non-Being, "for the eye of the vulgar soul cannot endure to keep his gaze fixed on the divine" (253d–354b). Again, the contrast of earth and sky, the metaphor of the cave. Accordingly, we have the remarkable image of a soul with wings, and its quest for truth is achieved amid the incorporeal and ineffable Forms, such that the proper subject of the art rhetoric is paradoxically, as Michael Leff has stated, not language but the soul itself (22).

Thus, even as the nature of Plato's dialectic is always open to question, it nevertheless seems especially ill-advised to equate it too closely with dialogue. On a general level the issue is resolved simply in terms of the distinction between thinking and speaking, between two fundamental and perhaps irreconcilable notions of *logos* upon which the differing practices and speculations of philosophy and Sophistic rhetoric are ultimately fixed. The matter will figure more substantially in the discussion of Heidegger and language that is to follow, so here it can be addressed with some celerity. As Gorgias's sense of *logos* measures its impact on the psychical and even physical levels, Plato's noetic vision is bereft of the drama of tangible experience and moves vertically (or internally) to the Forms until sensibility is abandoned altogether. Apparent in Plato's attitude toward the poets is the belief that the sensual appeal of language corrupts the pursuit of truth. Indeed, in contrast to Gorgias's view, Plato's is invariably that language as a sensual medium can and must be renounced if the integrity of philosophy and philosophy's method, the dialectic, is to be sustained. Plato claims that the soul best attends to truth when the mind is "gathered into itself" and nothing of the senses trouble it, whether of hearing or seeing, pain or pleasure. As he explains, only when the mind takes leave of the body, is utterly rid of corporeal considerations, sense and desire, can the soul aspire toward true Being (*Phaedo* 65a–67b).

What remains, then, is one of the more conspicuous messages of the *Phaedrus*, that of an eager, naïve, and not very bright young man being lead by Socrates, an astute and well seasoned philosopher who knows the truth and implants it in the soul of his pupil, the epitome of tractable youth. In the end I doubt that Plato himself claims the dialogue's message to be otherwise. If rhetoric is at all an affair of language, then this imposition of the will of the stronger *is* rhetoric, such as it is, in the *Phaedrus*. In contrast, the essentials of the dialectic would seem to be captured in full by a tale related in the *Symposium*. It concerns Socrates, standing perfectly still and silent, transfixed in one of his celebrated fits of abstraction. The trance stretched from one early dawn until well into the next day, as the Ionians gathered in amazement on their mats and pallets in the open night air to watch this curiosity unfold, to unwittingly witness, perhaps, the dialectical ascent of a soul toward true Being (220c–e). We can only guess, for this image of Socrates so utterly absorbed in thought is all the more to which we are privy, though I suspect in all its imposing insinuation it is a much more exquisite example of the "Platonic Stance" than that presented by Steinhoff. Of course, the civilizing influence of that image is something quite wonderful, to me every bit as poignant as the Pietà, or the Buddha in any one of his various attitudes or postures. The world will forever be in desperate need of such concentration, such magnificent intellectual endurance, that there can be little wonder why Plato was so enamored of this man. But, then as now, we can only see Socrates' behavior at some point far removed from rhetoric, for the essential character of what transpired within his soul is beyond rhetoric, is left unsaid and will forever remain impenetrable by words spoken or written. On this note it seems fair, even fitting, to supplement Leff's contention, to add that a rhetoric having the Platonic doctrine of the soul as its essential feature is at best a very feeble rhetoric and quite possibly no rhetoric at all.

Of Time and *Kairos*

The episodes of Socrates spellbound for hours on end in contemplation of the Forms attest to the mystical emphasis that attends the dialectical ascent. Indeed, as de Romilly says, there is a "divine quality" about Socrates as he is seemingly possessed in such moments by the "demonic voice" of inspiration (36). The manifestation of Platonic love is replete, the soul able to affirm its divine nature during these moments, to gain its wings and complete itself in union with the beloved. There is thus a sense of self-surrender to the ecstatic vision of true Being, a "divine madness" that affirms the presence of a cosmic *logos* securing on a metaphysical plane the coincidence of our being with Reality itself. However, as the passage from the *Symposium* suggests, the demonic voice is not a force that encroaches from without, but is enclosed entirely within the confines of the soul as an interior monologue, present as divine reason, as *nous*, as "the mind gathered into itself." The point de Romilly makes concerning the power expressed through Socrates is that it is nothing less than magic itself, that of "implacable truth" instigated in reasoning and, if abetted by discussion with others, ultimately secured in the direct and pure apprehension of the Forms. It is, then, a magic comparable to that of the Sophists.[34] As Gorgias represents the deceiving power of speech at work in the "dramatic fray" of human experience, the divine madness depicted by Plato "consists of the apprehension of concepts in and of themselves, without mediation by physical or linguistic embodiments" (Swearingen 314–15). The Forms, comprehensible to the mind alone, are alien to "physical utterance" (Leff 22). They are "entirely separate from the structure of the lived world" (Poulakos, "Possible" 221).

If the direct apprehension of the Forms is the destination, and the dialectic is the means to the Forms, anamnesis—Plato's theory of remembrance—marks the route. Significantly, remembrance is not a recollection or reconstruction of things that have occurred in the past of one's life but the way to recovery of "pre-natal" knowledge embedded in the soul. Once cleared, the way leads to a coming together, closing with the eternal that transcends time in what Stanley Rosen terms an "instantaneous vision" occurring "in the interstices between moments of time" (61). The phenomenon is thereby apparent, as the temporal is transcended, in terms of *stasis* as much as *kairos*, but more on that presently. I again raise these issues here not just to heighten the contrast of Plato's celestial Reality with the nature of human experience depicted by Gorgias but to frame each in terms of differing perspectives on the nature of time. In Plato's theory of remembrance, the ecstatic vision of the Forms is rendered in majestic stillness beyond time and place, while the Gorgian *logos* is joined to dramatic earthbound situations, very often bearing a violent, well-nigh physical impact upon the *psyche*. However, each view, even Plato's, is related to a concept of time, and, in the respective interpretations of the nature of the moment that emerge from each view, there is apparent one of the more obvious distinctions that can be made between the Platonic and Gorgian *kairos*.

In the last chapter I cited an example from Plato's *Parmenides*, of another of his conceptions of the "moment" or "instant." He defined the "instant" there as a category of transition between rest and movement (156d–e), essentially as that "instantaneous vision" of which Rosen speaks. The dialogue is concerned with the difference between words and reality, and thus bears directly on the point Gorgias addressed in his treatise, "On Non-Being." More specifically, the question confronting Socrates in the *Parmenides* is as follows: Once the truth of things is known, what is the mechanism, beyond mere magic, that makes its possible to *speak* the truth? Gorgias's answer to the question has already been discussed. Truth does not exist, and if it does we can neither know it nor communicate it to another. It is hard to imagine a state so utterly removed from projections of pure and perfect truth, of a more thorough resignation to the verbal environment of this earthly realm where to speak at all is necessarily to deceive. But the matter is not so easily settled in Plato because, given the existence of a cosmic *logos*, a means must be devised to translate it into a human *logos* to bring about its expression in speech. In short, *res* must be transformed into *verba*.

In this impasse, if that is what it is, Brice Parain locates what we might call the metaphysical basis of the Platonic *kairos*, of which the rendition of *kairos* given in the *Phaedrus* is but an extension. In Plato's conception, Parain says, if truth is to be spoken at all, it must pass through a moment when it ceases to be uniquely "what is" and becomes "what is said." In the absence of such a transition, it would be impossible to speak the truth, or for "a philosopher or serious person of any sort to speak at all" (102). Thus, he says of Plato's idea of the moment that "we see 'what is' in a flash, and in that instant 'what is' is astonishingly transformed into what must be said about it" (105). He concludes his argument, incisively, by claiming that "in the strange mixture in the same instant of being and nonbeing, we would find the essence of the word" (106). Or, at least, we find the word in Plato's view of things.

If such a process accomplishes the transformation of *res* to *verba*, speech is then indeed one with truth. Language is created—or, at least, has its source—in the realization of the moment. But it is an event exclusively divine, transpiring in a celestial never-never land that, of course, Gorgias claims does not exist. Thus, in this transformation speech is as thought itself, instantaneous and trans-temporal. A self-identity of the two is achieved through the work of *kairos* on the metaphysical plane, where they are immediately self-present to one another, a perfect singularity of dialogue and dialectic so that, ideally, "speaking is considered a loud thinking, and thinking, a mute speech," as Felix Cleve says. Yet it is important to realize that in Plato's design thinking arises in our perception of the Forms, these "patterns fixed in the nature of things" (*Parmenides* 132d), and not from the impact of words, sentences, and sounds. This is the philosophic view of the matter, and it clearly accounts for the transition of *logos* from an issue of language to one of reason. If it were otherwise, Cleve says, *logos* "could never have arrived at the meaning of reason" (42). Moreover, insofar as this coincidence of thinking and speaking occurs on the metaphysical level, the possibility of a temporally situated moment is denied and, by the same token, so is temporal duration or, indeed, any manifest significance to the moment whatsoever. In Plato's conception, the moment is "exempt from the impingement of the bite of time," as Schrag says (*Existence* 138); it is not "in any time" as Plato himself puts the matter in the *Parmenides* (156d).

What is apparent, then, is that in essence the Platonic *kairos* has a wholly transcendental constitution, abstract and sterile, and is at best exemplary only of the solitary mental life unique to the soul.[35] In Plato's conception of *kairos*, meaning is self-present to the subject, but only to the subject, and for that reason incapable of being communicated in a literal fashion to another. Thus, speech is but the "moving image" of the stable reality of eternity, as Plato concedes it to be, while truth itself remains "rigorously nonverbal" (Cascardi 225). Except by means of analogue or metaphor, there is no way truth can be turned outward and made available rhetorically. To the extent that Gorgias eliminates the "reality" of the cosmic *logos*, Plato, in effect, eliminates the human *logos* by denying the stamp of authenticity to the contingencies of the "lived world." Not only is thinking an inner contemplation ruled by the dialectic, but in the timelessness of Plato's celestial world speaking is wholly subsumed by a process that inevitably has silence as its goal. Plato's "genuine rhetorician" is thereby a contradiction in terms. "His objective is to guide the auditor to a point where he can escape the prison house of language and turn his soul toward a direct vision of the Reality that language can suggest but not encompass," as Leff says. Hence, to the extent that the genuine rhetorician is truly genuine, "he destroys the very medium in which he works" (22), and we are left with the spectacle of Socrates, alone among the pristine thoughts of his innermost soul. A prison house in its own right, however radiant and divine.

In this light the distinction between the philosopher's word and the rhetorician's word is all the more compelling. The former is intelligible and internal, as Samuel Ijsseling says, while the latter is external and sensible (16). The distinction marks not only the difference between Plato's and Gorgias's notions of *logos* but also the implications of the metaphysical basis of Plato's conception of *kairos*. In Parain's lively speculation concerning the moment, the identification of truth with words is secured by *kairos* wholly within the internal sphere, confined to the relationship of one's ecstatic vision of the Forms. Therefore, as Parain himself stresses, in basic agreement with Leff, the dialectic always ends in silence (78). The point Derrida makes, specifically regarding Plato, is that *logos*, in its expressive form as language, is a secondary event superadded to the stratum of sense—"something supervenient upon the absolute silence of self-relationship."[36] Thus, once *kairos* is removed from its metaphysical basis and applied on a rhetorical level in dealings with the external and sensible word, it becomes, like expressive language, a "secondary event" itself. In all events it is derivative, an inferior, indurated rendering of an original insight. It "survives imperfectly," as Untersteiner says of the transformation, and is necessarily "hardened into a formal value" (196). In this way, then, the Platonic *kairos* can only be understood in its rhetorical application as the adaptation of language to a situation. As the rhetor confronts a given situation in the temporal world, he expresses a "re-presentation" of an already fulfilled mental content separate from the situation, for the rhetor himself, by virtue of his soul, is separate from the situation, invariably and always by the precepts of the Platonic *kairos*.

Indeed, in Plato's conception of the moment as nontemporal and isolated, the "authenticity" of language is established in terms of *veritas*. As Heidegger would have us understand it, *veritas* is truth understood as "calculation," a measure of how well or poorly language corresponds in the phenomenal world to the "what is" of some transcendental realm. Ironically, then, its authenticity in the sense of a direct and immediate relationship with this transcendental reality is established only on the transcendental level and, just as Gorgias has told us, remains beyond the capacity of language to express in the here and now. So here, too, we witness the silence, a "Platonic stance" given full justice in Theodore Kisiel's contention that in such circumstances "consciousness always arrives too late to seize what seizes it, the immediate present" (94).

As I have attempted to establish, this idea of consciousness is in stark contrast to Gorgias's conceptions of the *psyche*, *logos*, and time. And from the outset I have used Kisiel's phrase "indigenous field" to designate the place language occupies in such a world view, though it is *physis*—earth itself—that just as properly and perhaps more specifically names the place where language, rather than any *dynamis* of the soul, constitutes the ontological structure of our being. Here, time itself, like *logos*, can be understood as earth, as the "indigenous field" from which we can no more escape than we can from the "prison house" of language. Thus, rather than temporality being comprised of appearances, mere shadows of reality, it comprises "reality" itself insofar as it can possibly be known. Most profoundly the funeral oration heralds this sense of temporality, of being born of the earth and returned to it. If earth conceals, it nurtures and sustains as well, not in some metaphorical sense alone but in the tangible, immediate presence of native soil. Here the physical and cultural are one, in that manner fostering the historical mode of the being of a people—their "*doxa*"—and Gorgias's "Funeral Oration" in particular provides the pattern, putting in place the essential ingredients of a world view through which the nature of that union is both identified and realized in the experience of *kairos*. Here the moment is utterly temporal rather than abstract and vacant, "instantaneous" in the Gorgian sense. As Gorgias expresses it, the "right time" is not only given presence in the context of a situation but is itself decisive, notwithstanding the purposes and plans of the rhetor.

These observations lead to a few initial conclusions regarding the Gorgian *kairos* that can be made now, prior to its fuller development in light of various considerations gathered from Heidegger. First of all, *kairos* as understood in Gorgias is not, or is not only, the application of language rhetorically selected and suited to fit the occasion or proper time. In contrast to Plato, there is nothing transcendental to the meaning of *kairos*, nothing of the non-temporal moment marking the passage from "reality" to language, from "what is" to "what is said." Instead, the Gorgian *kairos* maintains its integrity on a temporal level, where, through the force of an irrational *logos* prompted by *kairos*, language is not only created but is creative; or, as Rosenmeyer says, "a creator of its own reality" (232). Therefore, the purpose of Gorgias's rhetoric is to direct and utilize *logos* to whatever extent possible and thereby tap—or, more properly, provoke or create—the forces that shape the *psyche*. This purpose is achieved ostensibly by the rhetor's manipulation of the style of language, so that the Gorgian *techne* is concerned, at least on the surface, with metrical patterns and techniques of verbal elaboration discussed earlier. Thus, style possesses for Gorgias an "inventive" character which allows the rhetor to depart from the conventions of established language in order to exert the most telling effect upon the *psyche*. These tricks of style do indeed violate the "habitual" *nomos*, marking divergences from the pre-existing *doxa*, and can therefore be resolved into instances of "para-*doxa*."

Yet this sense of paradox is related in far more significant ways to the Gorgian *kairos*, these stylistic effects being but pale reflections of the far more significant place *kairos* has *as* the Gorgian *techne*. Here the irrational *logos* prompted by *kairos* can not only be seen in contrast to the cosmic *logos* but to the human *logos* in terms of the paradox of certain situations that seize one's consciousness, overcoming received opinions, and reshaping one's sense of self and social obligation. Segal discusses the physical manifestations of such situations on the *psyche* in terms of the *"ekplexis"* that prompts "a sudden yielding to an emotional and nonrational response" (107). In his "Encomium to Helen," Gorgias describes the *ekplexis* provoked in soldiers by the sight of the enemy in battle array, causing the "powerful habit induced by custom" to be "displaced by fear" (par. 17). The same condition of *ekplexis*, with a much different result, overpowers the soldiers of the "Funeral Oration"; though here, as well, there is a similar violation of *doxa* that overturns "the rigor of law" and "arrogance of positive right," to secure "the perfection of argument." Thus, in both cases the *logos* prompted by the situation overcomes *doxa*—overcomes it, at least, in the manner we have come to conceive *doxa* as conventional understanding, as mere opinion. Especially in the example of the "Funeral Oration," there is present a view of paradox that, as Michael Hyde puts it, signals the creative function of discourse that "moves present understanding beyond (*para*) an habitual, commonsense view (*doxa*) of truth that has been preserved in the language of present understanding" (153).

I place the influence of the Gorgian *kairos* in such a context. In the *kairos* prompted by their situation, the Athenian soldiers move beyond *doxa* to experience the "authenticity" of language in the "perfection of argument." The autonomy of *logos* is seemingly breached in a radical closure of *logos* with the *psyche*, such that, in its "self-revelation," language itself opens the world of the speaker, as the speaker speaks at the behest of language, creating an audience composed of himself as much as any other, as each is one in the power of language. All things are linked to language, never to individual insight, so that, for the *psyche* under the influence of language, awareness is language and speaks as language—an idea that comes to us now, I am sure, as some curious haunting from the world of the *Iliad*. But Heidegger speaks quite sanguinely of the phenomenon as a "primordial hearing," as the "call" of language exerting a claim to become manifest. To follow explicitly the Heideggerian perspective, I would suggest that what lurks behind Gorgias's notion of the situation is "Being"—not the *ousia* of Plato's cosmic *logos* where truth is transcendent and forever a function of *veritas* in this world, but Being in terms of the radical immanence that imparts the irrational *logos*, where truth is disclosed in this world phenomenologically, in the special sense of "event" and "happening" as we have them in Heidegger's sense of truth as "*aletheia*." Yet this recent, and very ancient, view of our being, beliefs, and language seems in large part contingent on retrieving, by way of Heidegger's thought, a Sophistic sense of *doxa*. Which, among a number of related things, is now time to explore.

4 Das Sein, Dasein, and Doxa: Attending to the Way of Heidegger's Thought

. . . it is wise to agree that all things are one.

—Heraclitus

At a time when philosophy seems to create as much confusion as it dispels, one could scarcely find a philosopher as cryptic as Martin Heidegger. The point is especially pertinent here, as his thought seems no less elusive than the pre-Socratics he strives so intently to explicate. In fact, Gerald Bruns identifies Heidegger's work, specifically his texts on language which figure significantly in this chapter and the next, as shaped by "a headlong retreat from critical reason" (111); he calls Heidegger a figure of the "counter-Enlightenment" who practices at the most crucial junctures a "rhetoric of bewilderment" (95). Nevertheless, Heidegger's work remains to all intents an unusually creative recollection of Heraclitus and others of his kind who expounded so obscurely on the nature of Being.

In some sense, then, the bewilderment Heidegger provokes is a consequence of his subject matter, being nothing less than the meaning of Being itself, obscure whatever the context. It is certainly a consequence of his approach to his subject matter. In this regard Heidegger preaches what he practices, to the effect that what he says and how he says it are inseparable, and the curious irony is that his message is all the more compelling as a result.[1] In a radical departure from the perspectives of representational, discursive thinking that has been the standard of philosophic thought since Plato and Aristotle, he shuns conventional methods and attacks what he calls the accumulated prejudices of centuries which hold that "thinking is a matter of ratiocination" (*Way* 70). In this vein he calls for a vast reordering of the solemn conceits of traditional philosophy set to prove what the world is really like. In his later works he specifically attacks the conception that language is, or ever can be, lucid and transparent as a sheet of glass under which truth can be sealed and displayed. In defiance of his role as philosopher, he would seek to use language to bewitch rather than explain. He "re-mystifies" and "re-enchants," as Bruns says, and to read him productively we must abandon the "steady state" idea of clarity and engage in a "meditative groping," where clarity is measured by the images gathered in "the slow dawning, the glimpse, the momentary insight" that then "darkens at once into bewilderment" (xxvii). Or, as Gadamer says, it is a philosophy based in the recognition that the "law of the day" is limited by "the passion of the night" (*Heidegger's* 189)

Indeed, to read Heidegger productively one must muster, then cultivate, a tolerance for ambiguity, and in the end an adequate reading is all that can be expected, never an explicit, comprehensive elaboration of his philosophy.[2] And a reading is all I claim to offer, one to the point of my purposes here. Even so, my excursions into Heidegger's texts surely have been among the most frustrating experiences I will ever have, and yet nothing beyond excursions is possible, frustrating or otherwise. They reflect, indeed, the very purpose of Heidegger's writing, as he himself would seem to have it. He invites us to attend to the "way" of his thought, to a path of thinking he likens to *Tao*, so that we might be allowed "to reach what concerns us, in the domain where we are already staying" (*Way* 93). If the way is wholly inaccessible by means of "calculative" thinking, and the "deadening" effects of philosophic language are spurned at every turn, Heidegger encourages us for these very reasons to be open and "guided by the hidden riches language holds in store for us" (91). In any event, it is the way that bears me agreeably to an appreciation of the pre-Socratic mystery as Heidegger presents it and, from this, to particular insights that can be gathered along the way into the nature of Sophistic rhetoric. It is a way that ironically begins with an idea seemingly dismissed by Gorgias, that of the mystery of Being.

Heidegger opens his *Introduction to Metaphysics* with the question that engaged Prince Hamlet: "Why is there something rather than nothing? That is the question."[3] According to Heidegger that has always been the question of metaphysics, yet the malaise he sees in modern thought is rooted in the "forgetfulness of Being," in our reliance on previous answers and strategies that, ironically because of metaphysics, are fashioned in ways to avoid the question itself. So Heidegger asks the question anew, and here the question concerning the meaning of Being (*Die Frage nach dem Sinn von Sein*) underscores his response to the Western philosophical tradition as a whole. The question is never a peripheral concern in Heidegger, exclusive to any of his particular works or to any particular period in the development of his thought, but embraces the breadth, and incredible depth, of his entire philosophical enterprise. And he frames the question anew. Claiming it cannot be fully grasped by way of traditional metaphysics, he emphasizes the essential circumstances of our "being-in-the-world," thus giving the question an existential cast. Heidegger says, "Man alone of all beings, when addressed by the voice of Being, experiences the marvel of marvels: that what-is *is*" (*Existence* 355). Typically, the question of Being is never answered in Heidegger. It remains a question, meant to tap a "primordial source" of additional questions, for as he says at the close of *Introduction to Metaphysics,* what we know "authentically," we know only "questioningly" (206).

Indeed, to even broach the issue of Heidegger's account of Being would seem to mire this analysis in matters beyond its scope, but to speak of Heidegger at all the issue must be addressed and, beyond that, as Being is taken in its "pre-philosophic" sense as "earth," it is precisely to the point of this analysis. I therefore begin this chapter with a discussion of his view of Being, limited and shaped by the specific purpose of bringing a Heideggerian perspective to bear on Sophistic rhetoric and the meaning of the Gorgian *kairos.* This discussion leads to sections in the following chapter dealing with Heidegger's interpretations concerning language that, in turn, lead to concluding sections on time, rhetoric, and "authentic" discourse that establish an interpretation of *kairos* in terms of Heidegger's notion of the *Augenblick.* In the course of this analysis I will introduce and address other elements of Heidegger's philosophy as they arise in context and relate to my explication of the Gorgian *kairos.* Particularly critical among these are the nature of our being and the nature of our beliefs. Each is addressed in this chapter in terms of the Heideggerian perspectives on *dasein* and *doxa.* As each pertains to Being, they pertain as well to the nature of "appearance," whether conceived as Being's opposite or its singular means of manifestation. As always, my approach is recursive, as I circle back to cover old ground even as I break the new, to recapitulate and generally shore things up.

The overall philosophic connections between Heidegger and Gorgias are scarcely explicit and often tenuous, but there are similarities in their thinking on significant matters that I will develop and apply whenever they can be established, seem relevant, and serve my purposes. As mentioned before, I proceed from the premise that Gorgias was in many respects representative of a world view that Heidegger found particularly congenial to his philosophical approach. In this context the most crucial connection between Gorgias and Heidegger is based on the fundamental role that language plays in their respective notions of *kairos* and the *Augenblick*—specifically as language is mysteriously bound up with Being in Heidegger's thought and how, in the case of Gorgias, all the relevant attributes of Being are attributed to language in any event, even as they are otherwise unexamined, unappreciated, and unacknowledged apart from language. Heidegger says that "in its essence language is not the utterance of an organism; nor is it the expression of a living thing. Nor can it ever be thought in an essentially correct way in terms of its symbolic character, perhaps not even in terms of the character of signification." Then, in typically cryptic fashion, though rife with meaning for my purposes here, Heidegger concludes, "Language is the lighting-concealing advent of Being itself" ("Letter" 206). It is at all times a mysterious and decidedly pre-Socratic conception of Being that he seeks to exploit, what he came to call "earth," and from the many implications he gathers from it, Heidegger deploys and in various ways explicates anew the entire lexicon of the very practical art of Sophistic rhetoric, to include novel perspectives on the nature of appearance and *doxa,* the meaning of human being, language and *apate,* and, most critically, the meaning of *kairos,* which in the end proves to be the measure of all of the above.

Being and *Doxa*: Earth as *Physis*

The central features of Heidegger's thought emerge from what he calls the "ontological difference," namely the distinction between "Being" (*das Sein*) and "things" (*das Seienden*).[4] The latter can be understood as "that which is" or "things that are." These are the physical features, traits, or characteristics that compose the world we inhabit. The "things," for instance, that were to Plato mere appearances whose true essences were found in the Forms, in the supra-sensible realm of ideas. The meaning of the term "Being," on the other hand, is not so easily settled. In English the word is a verbal substantive, functioning invariably as a noun, while the German *das Sein* preserves the infinitive root, and to Heidegger *das Sein* is always understood verbally (*Introduction* 42). In the most crucial sense, Being is indeed the marvel of marvels, what George Steiner calls "an epiphany out of Nothingness" (*Heidegger* 68). It turns on the enigma of "what-is *is*," on the difference between things and Being, between that which is manifest and the process by which it becomes manifest (Richardson 12).

Being therefore implies an active presence, a process vitalizing physical existence, surely our own, as it prevails to dispose us to the strangeness and wonder of the very "isness" of our being. Traditional philosophic conceptions of Being—understanding it as Idea, *energia*, subjectivity, will or will to power, or as some other celebrated category in a conceptual scheme—are in Heidegger's view the results of metaphysical perspectives based on principles of transcendence that are not only derivative of Being's immanence but work to exclude its effect altogether, preempting, as Calvin Schrag says, the "originative presence of Being" in the world (*Radical* 25–26). On the other hand, as Heidegger also contends, to whatever extent metaphysics sanctions the immanence of Being, the latter is seldom conceived in its difference from things but exclusively in terms of things, utterly identified with them, so that in this instance Being is forgotten as well. In the first case, as Stanley Rosen says, "Being is not thought in its own terms but as an abstraction" (58); while in the second, Being is once again derivative, categorized as "things that are." Either way, metaphysics precludes the possibility of thinking about Being *as* Being, which is the reason for Heidegger insisting upon the ontological difference in the first place, as it comes about in the effort to preserve this "wholeness" of Being, of its expression as dynamic process.[5]

For Heidegger, then, the decisive moment in the history of philosophy was the Platonic separation of the world of ideas from the world of opinion, manifestly truth from appearances, *episteme* from *doxa*—ultimately the ideal realm from the earthly realm—thereby eliminating the presence of Being from the world and destroying the fundamental ontological inseparability, the thoroughly mutual relationship, of *das Sein* and *das Seienden*. Thus, rather than one of transcendence, the difference between Being and things in Heidegger is one based on what Steiner calls a "radical immanence" establishing Being as the "very occurrence" of existence in all things. It is, in effect, not a distinction at all in the traditional, Platonic sense, where Being is "forgotten," banished from the world and lodged in some transcendental realm. Nor, just as critically, is it an understanding of the things of this world in terms of "the degenerate fragments of a Platonic sphere," as Heidegger "concentrates on the total *thereness* of these particular existentials" that furnish the "ontic" substance of our world as things and presences (*Heidegger* 65). In this treatment there is inevitably a mysterious "oneness" to Being, though ironically for that very reason still separate from "things" by vitrue of the ontological difference. It is thereby more properly understood as an event rather than a fixed reality, as all things are manifestations of the immanence of Being and can become separate and distinct as phenomena only in terms of their appearances in particular contexts or perspectives in which they are revealed or concealed through the process of Being itself.

Here, most significantly, through the agency of the ontological difference, Being and appearance are joined rather than opposed. In one sense, Heidegger retains an appreciation for the Platonic notion of appearance, understanding it as the "front" or surface features of Being that offer a mere "appearance" to be looked at. However, in one of the more splendid moves to be found in poststructuralist thought, Heidegger conceives of appearance in another, more important sense in terms of *phainesthai,* as a "self-sufficient emergence" (*Introduction* 101), where appearance emerges not as a mere copy of something more genuine but as authentic in its own right. It "brings-itself-to-stand" as Being, as the medium or manner in which Being is manifest, an idea vibrant in reiteration of the Gorgian precept that "Being is not manifest unless it succeeds to appearance, and appearance powerless unless it succeeds to Being" (82.B26). Thus, in the sense of *phainesthai,* appearance is not something subsequent to Being nor an issue of individual perspective. It is, instead, of Being's essence, altogether a matter of how things reveal themselves (*Introduction* 101). In the lithesome world that Heidegger conceives, appearing is a "showing" that belongs "side by side" with Being, "the one changing unceasingly into the other" (109).

Heidegger might well avoid any talk of immanence in describing his philosophy, yet he repeatedly invokes this "togetherness" of Being and appearance, especially as he explicates each on the basis of the pre-Socratic conception of *physis.* The term is most often translated as "nature," and in what Heidegger has called "the tragic flaw of Western thought," we have come to understand nature in terms of features of the physical world that exist apart from ourselves, features that we then objectify, measure, and describe by means of the physical sciences in particular. Heidegger rejected this sense of the word, claiming further that to the early Greeks *physis* meant Being essentially in the manner he depicts it in his own writings. The move is intrinsic to the process of "destruction," that "teleology in reverse" of which Gadamer speaks, as Heidegger's interest was never to seek some goal or principle on the far side of philosophy or metaphysics but always "the retrieval of the long forgotten Greek argument about '[B]eing'" (*Truth* 227). By virtue of this process *physis* is manifestly "earth," Being conceived in its "primordial" sense, free from the "objectivization of science" and other conceptions of sheer utility, of "permanent availability and usefulness" expressed in technology, as Otto Pöggeler says ("Being" 109). Free from these and seemingly all "derivative" conceptions of Being present in metaphysics and traditional philosophy. As we shall see, *physis* is ultimately the "event of appropriation" brought about *as* language in lieu of "objectivization," being coupled in that respect with appearance, such that we "dwell" in language as upon the earth.

From the likes of Pindar and Heraclitus, Heidegger derives this "primordial" sense of Being from *phyein,* meaning "to grow" or "to bring forth." As a result, *physis* can be understood as both Being and things, a "becoming" and a "to be"—a "realm of emerging and abiding" that is "intrinsically at the same time a shining appearing" (*Introduction* 100–01). The language may be especially confounding, as if the more Heidegger talks of Being the more his idiom has the ring of the pre-Socratics, both their insights and obscurities. But his interpretation is crucial, and on this issue very accessible. Not only does *physis* designate the things of this world but is inseparably the "elemental power" that brings them to presence, a "self-blossoming emergence" that is at work in the unfolding petals of a flower, the rising of the sun, and more poignantly, in the journey of all things from birth to death (14). It is, then, as "aho," in crucial respects being the process that binds with every breath we breathe what we are and what we so often presume to be other than ourselves.

In this interpretation *physis* is the "creative opening of Being," expressing the intimate affinity of the event of Being and the things of this world while at the same time doing no violence to their difference. Indeed, the process achieves its integrity as Being only by virtue of a phenomenology rooted in the ontological difference. Being comes to the fore in *physis,* and if the latter is not necessarily equated with Being in all contexts, it is the process through which Being emerges from concealment, manifest in the play of appearances that constitute the things of the world.[6] Here the true nature of the world is not hidden behind a veil of appearances but is appearances themselves. And it was this process of Being that Heidegger claims was altogether dismissed in Plato, where Being as idea is exalted. In Plato, what appears is no longer of the nature of *physis* but is "mere" appearance, marked in all cases by the deficiency of illusions, of copies or pale reflections of forms more genuine and essential. Either Heidegger's or Plato's notion of appearance could shape our understanding of Sophistic rhetoric, but so pervasive is Plato's that even as those who come to the defense of the Sophists most often do so with Plato's sense of appearance in mind.

Indeed, even as we are immersed in the thick of Heidegger's argument, our sympathies, along with our biases, seem to rest with Plato. His commitment to enduring truth, unaffected by "mere" appearances, ingratiates and beguiles, as do his corollary condemnations of those who seek to manipulate these appearances through the power of language. Those like Tisias and Gorgias who "go chasing after opinions" (*Phaedrus* 262c), making "trifles seem important and important things seem trifling" (267b). Especially as *doxa* is linked with language in these passages, as it inevitably *must* be, Plato's purpose might well strike a responsive chord in people of good sense and good will, who see in the Sophists the classical Greek model of their least favorite lawyer, politician, or adman. Or some likely other, say a poststructuralist philosopher or woebegone rhetorician endeavoring to make their way in the world through the artifice of language. What is more, it is a purpose that, for all its reliance on the lofty premise, seems to offer something of detail and substance in contrast to the "bewilderments" and "momentary insights" given in Heidegger's texts, those fleeting images no more distinct and lasting than the shadows cast on the wall of Plato's cave. But if appearances are to Plato necessarily deceiving, accounting for an understanding of *doxa* as mere opinion, the implications Heidegger draws from his contrary view of the relationship between appearances and *doxa* are no less compelling.

Doxa is usually understood as the practical wisdom inherited from tradition, that collection of beliefs we hold by virtue of subjective conviction, beliefs that are themselves anchored in some durable and enduring cultural system. The particular theoretical expression of *doxa*'s place in rhetoric is given by Plato, patently in the *Phaedrus* and *Republic*, where in the latter *doxa* is consummately of the cave, the vision gained in shadows cast by flickering firelight; and in each is presented as "mere opinion" in contrast to *episteme*, to "true knowledge." When we are aimless in our pursuits, failing to scrupulously follow the way of dialectic to true knowledge, *doxa* prevails by default, for opinion is blind, Plato says, and those who hold "true opinions" are not appreciably different from "blind men who go the right way" (*Republic* 506c). In this manner he associates *doxa* with the deficiency of appearances. *Episteme*, on the other hand, is wedded with reason and reality in Plato's scheme, and is the very élan of truth and flawless clarity that he undertakes to disclose and display in his dialogues. As *episteme* is joined with the attitudes of the gods and philosophers who have fathomed the nature of true reality beyond *doxa*, *doxa* itself is the medium of the rhetoricians and poets, and Plato likens it to the caprice and incorrigibility of subjective conviction that is bereft of knowledge, a circumstance that he says suits so aptly the "vulgar souls" of the Sophists (*Sophist* 254b). Though some might choose to be less derisive in their expositions of *doxa*, Plato's has stood the test of time. And here, most evidently, those who seek a rehabilitation of Sophistic rhetoric, and who otherwise can be quite inspired in their condemnations of Plato, consistently structure their understanding of *doxa* in terms of its contrast with *episteme*, where it is altogether the scoundrel issue of Plato's idea of truth and reality.

And, as I say, these features of *doxa* are hardly precluded in Heidegger. In fact, in some instances they are further entrenched. But like his conception of *physis*, *doxa* is both product and process in Heidegger, thereby moving well beyond the strictures given by Plato. In Heidegger's interpretation, *doxa* is not mere opinion in its essential sense but is necessarily prior to subjective conviction, pertaining instead to the "aspect" given in appearance—to how things, including human beings, "show themselves," to how they appear or enter "into the light," as Heidegger says (*Introduction* 103). Insofar as appearance is authentic, it participates in *phainesthai* as a self-sufficient emergence that "shines forth" in Being as "that which is." Insofar as appearance is inauthentic, it has no support in the thing itself and the aspect of appearance is "one we take and make for it" (104). Admittedly, the argument is circular, projecting Being only in terms of Being, where presumably we could ascribe to Being whatever we will, seeming to leave us none the wiser of what is authentic in appearance and what is otherwise. But to express Being as Being alone is a necessary consequence of this conception, for to do otherwise is to reduce it to something much less than itself, to think of it as an abstraction rather than on its own terms.

Just as crucially the question of Being is thereby kept open, forever remaining a question, as Heidegger would have it. Here the issue of appearance is recast, or surely our role concerning it is drastically changed, as we are deprived of the privileged perch that would provide the measure of authenticity. We are subordinate to the process itself, as the emphasis turns from the seer to the thing seen. The shift in perspective is critical: To Heidegger, *doxa* "holds" us—indeed, "beholds" us—rather than us holding it, as we would mere opinions. *Doxa*, then, is the event of disclosure "in the light that endows with permanence, Being"; and if appearance is "authentic and particularly distinguished," *doxa* ultimately culminates in "fame and glory," in "brilliance, esteem, and mode of highest Being" (*Introduction* 102–03). Here, with his purpose fixed on the uncanny, outward manifestations of Being, Heidegger would discern a process of interaction or even reciprocity in keeping with the dynamics of the ontological difference. In this manner *doxa* partakes of the same mystique as that given *physis* in Heidegger's thought. It is intrinsic to it, for *doxa* is an extension of Being outward into appearance, obtaining its most salient realization in disclosing and attributing, "though glory and glorification," the "regard in which a man stands" (102–03).

Heidegger insists that this conception of *doxa* is not to be mistaken for celebrity, arrogance, vainglorious display, or any other attribute of self-aggrandizement that puts the emphasis on the human subject rather than *doxa* itself. In fact, Heidegger's interpretation of *doxa* has one of its clearest parallels in Christian theology, as is borne out repeatedly in biblical references to the term. *Doxa* is "the glory of the light" that blinded Saint Paul on the road to Damascus (Acts 22.11) and the "marvelous light" that called the Jews out of darkness (1 Peter 2.9). It is apparent in the aura attributed to holy men and women, expressed by the lustrous, ever-present halos seen in depictions of the saints. It is most obviously apparent in the celestial beatitude of Jesus witnessed by the Apostles Peter, James, and John in the Transfiguration, when in sudden radiance Jesus was seen as the Christ, as the very immanence of God, and his face shone "as the sun" and "his raiment was white as light" (Matthew 17.2; Mark 9.2).[7]

Though Heidegger acknowledges the place of *doxa* in Christian theology, he claims that in similar ways it was just as much a fixture in early Greek thought. In a curious sense his interpretation of *doxa* is a recapitulation of Plato's metaphor of the soul's journey in transcendence, as it moves in remembrance to the celestial regions of Reality in contemplation of the Forms (*Phaedrus* 250b). The way is marked by the radiance of higher glory in each, but with the crucial distinction that in Plato the destiny of the soul is "true knowledge," which is most profoundly *doxa*'s opposite in Plato's metaphysics. Here the difference could not be more revealing, as the soul achieves its glory by means of transcendence rather than immanence; and as the mystique remains nevertheless, Plato would seem to usurp the glory of *doxa* for the purpose of bestowing it upon philosophic truth.[8] Indeed, the greater dynamism that Heidegger attributes to the idea of *doxa* was present all along, he claims, in the pre-Socratic understanding of the term, which was later brought to quiescence by metaphysical interpretations, Plato's in particular.

If *doxa* is manifest in the persona of saviors and saints in the Christian tradition, in pre-Socratic thought it is most clearly evident in the character of the tragic hero. In each, a person's nature stands out in the radiance of a higher ontological order, the crucial difference being that the deliverance of the tragic hero comes utterly in a process of the revealing and concealing power of Being, in immanence rather than transcendence; and, as a result, *doxa* in this pre-Socratic sense, in lieu of haloes and transfigurations, is apparent in the practical but compelling occurrences of the tangible, physical world, where the reality of things is discovered by the bodily senses and given its full integrity in appearances alone. One source expressing the matter most appealingly is Robert Pirsig's *Zen and the Art of Motorcycle Maintenance*. Pirsig deals with a profusion of things in his book, but his keynote throughout is his quest for "Quality," an elusive, very slippery concept that he claims lies at the heart of Sophistic rhetoric. He says that Quality defies definition, but one of the better illustrations he offers comes quite appropriately from a passage of the *Iliad*. In this scene, Andromache, the wife of Hector, tells her husband in a voice of dispassionate but most sincere persuasion that his bravery will be the death of him, that he therefore has no pity for his wife nor his little son, who will be left widowed and orphaned.

Hector's reply, in the Kitto translation used by Pirsig, is as follows:

> Well do I know this, and I am sure of it: that day is coming when the holy city of Troy will perish, and Priam and the people of wealthy Priam. But my grief is not so much for the Trojans, nor for Hecuba herself, nor for Priam the King, nor for my many noble brothers, who will be slain by the foe and will lie in the dust, as for you, when one of the bronze-clad Acheans will carry you away in tears and end your days of freedom. Then you may live in Argos, and work at the loom of Messene or Hyperia, sore against your will: but hard compulsion will lie upon you. And then a man will say as he sees you weeping, "This was the wife of Hector, who was the noblest in battle of the horse-taming Trojans, when they were fighting around Ilion." This is what they will say: and it will be fresh grief for you, to fight against slavery bereft of a husband like that. But may I be dead, may the earth be heaped over my grave before I hear your cries, and of the violence done to you.

Hector continued, and even as he conceded that he surely would be killed and that the Trojans would lose, the issue of victory and defeat—and even that of life and death—is to him beside the point. If he seeks death to avoid hearing her cries, his is hardly the coward's way out but preeminently an issue of sacrifice. After all, if he were alive to witness her disgrace, he would not have died shielding her from that fate, but with the enemy at the gates he fulfills by his death his obligation to her and himself as a warrior. So he held out his arms to his son, but the child shrank back, taking fright at the sight of the gleaming bronze armor and the horsehair crest of the helmet. He put the helmet aside, took his son and dandled him in his arms, and prayed to the gods, disclosing what seems to be expressly the point:

"Grant that this my son may be, as I am, most glorious among the Trojans and a man of might, and greatly rule in Ilion. And may they say, as he returns from the war, 'He is far better than his father'" (*Iliad* 6.440–62; Kitto 57–58; Pirsig 369–70).

Pirsig identifies Hector's dauntless bearing in this passage as the embodiment of *arete*, the classical ideal of excellence, and so it is. But at this point the distinction between *arete* and *doxa* is imprecise, if it can be made at all. The latter is surely the case once Heidegger's conception of *doxa* is taken into account. The glory that Hector desires for his son and would achieve for himself is what Heidegger associates with the "aspect" in which one appears, the *doxa* that is "the regard in which a man stands." Attendant to the issue, Heidegger draws his support from a fragment from Heraclitus: "The noblest chose one thing before all else: glory, everlasting abiding over against things mortal; while the rest are satisfied, glutted like cattle" (B29).[9] Indeed, as Hector's speech makes apparent, the Homeric hero was motivated as much by a thirst for glory, and fear of ignominy should he fail or falter, as by any patriotic obligation to his nation or city state.

To say nothing of familial obligation to wife and child. Surely there can be little doubt that the curious ways of an ancient warrior society are hardly the sort to sit well with modern sensibilities, seeming to pertain, as well they might, solely to brute force and the macho bluster underscoring it. We could be sorely tempted, as a result, to condemn these ways outright or, at best, consign them to the farce of that rueful dispensation that "a man's gotta do what he's gotta do." But if such societies are surely the domain of ardently believed and practiced male prerogatives, the issue is not for that reason alone so easily dismissed. Though the glory of the Homeric hero is self-justification, what Pirsig enthusiastically endorses as "duty to one's self," the justification is never self-serving; nor, more crucially, can it mean in this context to serve some enduring essence of the self external to the situation. It turns instead on how one stands relative to the light of *doxa*, to how one stands, ultimately, as an exemplar of the virtues that constitute the social fabric of the *polis*, and thereby *being* the light in and of itself. Here, *doxa* is never simply a body of beliefs inherited from tradition, for it is principally the manner in which these beliefs are vitalized in action, beliefs that must be continually sustained and proved anew through struggle and suffering. Witness the example of Hector. All recourse or appeal to some higher, transcendent power is denied, especially as that power is presumed to exist within the self by virtue of an immortal soul. It is the trial itself, the hero's embodiment in a dangerous situation, that accounts for his apotheosis, and the *Iliad* is virtually a cavalcade of such trials.

One of the most noteworthy comes in Book Eleven. Here, Odysseus's compatriots have fled the battlefield, and he is left alone to confront the Trojan infantry, many lines deep, who close in like hounds for the kill on a "wild boar." Encircled and hard pressed Odysseus "speaks" to his valor, his own "great-hearted spirit": "Since I know that it is the cowards who walk out of the fighting, but if one is to win honor in battle, he must by all means stand his ground firmly" (11.400–15). Odysseus does precisely that, and escapes unscathed to fight again and experience further adventures.

In a later scene, from Book Twelve, Sarpedon is not so fortunate, though his response at a time of great personal danger is unequivocally the same. Here the Trojans are about to breach the ramparts of the Greek camp, and in a critical moment of its defense, Sarpedon's words to his friend Glaukos capture as succinctly as Hector's the *doxa* of the Homeric hero—to wit, as life is precious for being so short, only in the quest for glory do we extract the full measure of its worth:

> Man, supposing you and I, escaping this battle, would be able to live on forever, ageless, immortal, so neither would I myself go on fighting in the foremost nor would I urge you into the fighting where men win glory. But now, seeing that the spirits of death stand close about us in their thousands, no man can turn aside and escape them, let us go on and win glory for ourselves, or yield it to others. (12.322–28)

This light of *doxa* giving luster to the peculiar paths of glory trod by Homeric heroes seems now to be every bit as faint as the many years to the ancient past can possibly make it. Yet the idea of *doxa* I am attempting to explicate is not so impossibly remote, nor can we yet be unmindful of its appeal. Indeed, in this final example it is as near at hand as a classic television series on the American Civil War. Given the immense waste and destruction of that war, and the prevailing historical and literary records we have of it, one could justly argue that in essence the Civil War was an event staged by profiteers, inept politicians and generals, and by brutal men filled with fanatical religious convictions and mindless ideals of gallantry. And given the likes of Fredericksburg, Cold Harbor, and many other similar places, some would say it was above all an event committing lambs to the slaughter. But these are all far removed from the principal motifs of Ken Burns's PBS documentary, *The Civil War*. While the carnage, villains, and other damaged souls are present and obligingly censured, what is most conspicuous in the documentary is the presence of a pure and inviolable nobility that Burns attaches to the soldiers who fought the war.

The courage and high-minded behavior of the soldiers, seldom so ably and plausibly depicted, is given such prominence that, in the end, we might surmise if the better angels of our nature could not prevent the war, they were assuredly in some complicated, mysterious way an active presence in the fighting of it.

A centerpiece of the documentary illustrates this motif in particular, and in a way the message of the series as a whole. It is presented in the recitation of a soldier's letter, a Union officer by the name of Sullivan Ballou writing home to his wife, Sarah. If the letter does not duplicate precisely the sentiments expressed by Hector to Andromache, it captures the idea of *arete* equally as well. The fate that Hector would crave as "most glorious among the Trojans," Ballou would seem to cherish as "honorable manhood" for himself and his children:

My very dear Sarah:

The indications are very strong that we shall move in a few days—perhaps tomorrow. Lest I should not be able to write again, I feel impelled to write a few lines that may fall under your eye when I shall be no more.

I have no misgivings about, or lack of confidence in the cause in which I am engaged, and my courage does not halt or falter. I know how strongly American civilization now leans on the triumph of the government, and how great a debt we owe to those who went before us through the blood and suffering of the Revolution. And I am willing—perfectly willing—to lay down all my joys in this life, to help maintain this Government, and to pay that debt.

But my dear wife, when I know that with my own joys I lay down nearly all of yours, and replace them in this life with cares and sorrows . . . is it weak or dishonorable, while the banner of my purpose floats calmly and proudly in the breeze, that my unbounded love for you, my darling wife and children, should struggle in fierce, though useless, contest with my love of country?

I have sought most closely and diligently, and often in my breast, for a wrong motive in thus hazarding the happiness of those I loved and I could not find one. A pure love of my country and of the principles I have often advocated before the people and "the name of honor that I love more than I fear death" have called upon me, and I have obeyed.

Sarah my love for you is deathless, it seems to bind me with mighty cables that nothing but Omnipotence can break; and yet my love of country comes over me like a strong wind and bears me irresistibly on with all these chains to the battlefield.

The memories of the blissful moments I have spent with you come crowding over me, and I feel most grateful to God and to you that I have enjoyed them for so long. And how hard it is for me to give them up and burn to ashes the hopes of future years, when, God willing, we might still have lived and loved together, and seen our sons grown to honorable manhood, around us. I have, I know, but few and small claims upon Divine Providence, but something whispers to me . . . that I shall return to my loved ones unharmed. If I do not my dear Sarah, never forget how much I loved you, and when the last breath escapes me on the battle field, it will whisper your name. Forgive my many faults, and the many pains I have caused you. How thoughtless and foolish I have often times been! How gladly would I wash out with my tears every little spot upon your happiness.

But, O Sarah! If the dead can come back to this earth and flit unseen around those they love, I shall always be near you; in the gladdest days and the darkest nights . . . *always, always,* and if there be a soft breeze upon your cheek, it shall be my breath, as the cool air fans your throbbing temple, it shall be my spirit passing by. Sarah do not mourn me dead, think that I am gone and wait for thee, for we shall meet again.[10]

As the letter concludes to the strains of maudlin fiddle music, the documentary's narrator declares that Ballou was killed a week later at the First Battle of Bull Run. More widows and orphans. But his death, and their fate, hardly makes the matter any more poignant, for the expression of *arete* lies as much in words as in deeds. In fact, the two cannot be separated, just as we cannot separate Ballou's sacrifice on behalf of the American *doxa* from his heroic status, his apotheosis as a paragon of that *doxa*. The letter alone, and its presentation in this film series, deflects accusations of Ballou's fate as just another lamb to the slaughter. For all we can possibly know, Ballou *is* his letter. In *doxa*'s web of meaning, he is his words, here and otherwise.

In one sense, then, Burns's documentary is a dramatization and complete narrative of Lincoln's Gettysburg Address, an ode to the glory of the men who gave the "the last full measure of devotion" that the nation might live and flourish by their example and sacrifice. Burns himself calls the Civil War our Homeric epic. It is not. The war was the war and nothing else besides, an event long since past that, in reality, is no more. Nor does the documentary seem to harbor any pretense of uncovering certain realities of the war that have thus far eluded us. In contrast to any conventional account purporting to tell what "really" happened, Burns's documentary carries the implication that an honest-to-God reality of the war and the soldiers who fought it is necessarily dubious. Rather, what it was, and who they were, is given in accordance with cherished political and cultural beliefs we take to be true, and then hold firmly in truth, so that in practical contexts of storm and stress these truths of the American *doxa* are presented as determining who we are as Americans. Maybe the heart of the matter is simply telling a good story well, rather than presenting a sound historical argument—if, indeed, that can be ultimately conceived as anything other than a good story well told. In any event, more than the Civil War ever could be in itself, Burns's documentary, whether myth or history, is a story of the American *doxa*, the story emerging from the war, and in that sense is more obviously our Homeric epic than the war itself. Steeped in the milieu and nuance of poetic experience, it is our *Iliad* in this sense, which in form and purpose, and even in style, it is so obviously modeled; yet it is shaped and promulgated, as it necessarily must be, by the modern American medium. Like those of the *Iliad*, the voices of those who fought the Civil War are seldom rendered in the documentary as expressions of hate and violence alone, nor do they offer only self-serving and inane proclamations of conquest and patriotism. On the contrary, they capture the essence of the American *doxa*, of the comportment of noble minded men and women holding to right conduct in the midst of the horrid brutalities of war.

The sense of *doxa* that emerges works to parallel in its vitality and effect the tribute Gorgias pays to the Athenian heroes of his "Funeral Oration":

> They were doubly exercised, above all, as was right, mind and body, the one in counsel, the other in action; helpers of those in undeserved adversity, chastisers of those in undeserved prosperity; bold for the common good, quick to feel for the right cause, checking with the prudence of the mind the imprudence of the body; violent towards the violent, restrained toward the restrained, fearless towards the fearless, terrifying among the terrifying.[11]

The "Oration" provides, in Gorgias, one of the clearest expressions of the Sophistic sense of *doxa* I have been attempting to explicate. Gorgias valorizes the warriors who proved their mettle, their prowess in battle, and if their paths of glory lead but to the grave, so it is with all paths. Moreover, their *doxa* is hardly restricted to them or their martial *ethos*. As is often the case in Homer, or in Burns for that matter, it transcends each. First of all, the trial or test is the important thing, and that hardly belongs exclusively to soldiers in battle, as our own history would attest. Though I acknowledge claims that modern, pluralistic societies may lack "a stable and coherent cultural tradition," as S. M. Halloran says (625), it does not require any stretch of the imagination to include such examples as Martin Luther King's glorious commitment to civil rights, Lillian Hellman's heroic stance during the McCarthy era, and a long litany of many others who through some sacrifice made good on the social contract explicit in Lincoln's injunction that this is a nation "dedicated to a proposition." It is only natural, then, that Lincoln's plea for union in his First Inaugural focuses on the American *doxa*; and focused there, he delineates its obscure nature, invoking "the mystic chords of memory" that stretch "from every battle field and patriot grave, to every living heart and hearthstone all over this broad land." There is an endurance in all this, some fitting deference to the cultural legacy in which we all share. This image of the "mystic chords of memory," that will "yet swell the chorus of Union," is the union indeed of those who have struggled and gone before with those about to face the tempest of civil war, extending even to those yet to come who will surely endure trials of their own for the sake of the American *doxa*.[12]

However, as significant as *doxa*'s reach, its intensity determines its nature as the embodiment of values and beliefs brought to presence through the sacrifices made by men and women for the sake of the community, *polis*, or nation. In contrast to analytical modes of understanding that would sever the believer from the belief and therefore discredit the latter as mere opinion, *doxa* in the Sophistic sense rests on prior sanctions of the social and cultural *ethos* that cannot be established independently of who we are. The idea of subjective opinion, in some sense the idea of individual consciousness altogether, is precluded in this configuration of *doxa*. Here the vital expression of the human personality, its authenticity, is achieved only as it participates in communal consciousness, in this peculiar way of believing. I. A. Richards says that our beliefs are a "veil between ourselves and something that other than through a veil we cannot know" (42). It is precisely so, even so much as to say that "nothing exists" behind the veil at all. We might then take him at his word, to discover the world as essentially a fabric of social conventions, and from that conclude that in this world we are virtually our beliefs. Again, *doxa* beholds us. We are realized—we "come to light"—in the vast, entangling web of social mores and conventions, of reality conceived wholly as the semantic environment of the here and now. *Doxa* beholds us and thereby occupies a mysterious region where the distinction between "what is" and "what is said" is particularly problematic, where *physis* comes to pass as language and where the distinction, say, between Homer and Hector is no more easily applied than that between the dancer and the dance.

Perhaps, like Socrates prior to his first speech in the *Phaedrus*, we should shroud our face in shame to say as much, for Plato places the commitment to such an idea under the onus of grievous sin. He would have us recant and make amends, and even then to risk the wrath of the gods. Indeed, the poet Stesichorus was struck blind for invoking the image of Helen as an adulteress whose allure incited the Trojan War, and it is the palinode of this poet that Socrates cites by way of atonement, to expiate the evil of his own irreverence: "It is not true, that tale/You never embarked in the well-decked ships/Nor come to the towers of Troy" (243b).

By these words of atonement Stesichorus regains his sight, and as Socrates repeats them, he proceeds, morally purified, in his instruction of Phaedrus. But in this explication of *doxa* it is best to keep faith with Homer just the same, blind though he may be. Here, traditional distinctions in structuralist theory between the "what" of a narrative—the "existents" of character, setting, and content of events—is never clearly discernible from the "how" of its expression in discourse.[13] In the context of poetic experience, distinctions between product and process, host and harbinger, are never sharply drawn in any event, with the result, here especially, that the congruence of word and referent, as Cassirer says, instills the illusion that comprehends reality. Thus, given the right cultural ambiance and storyteller, the difference between glory and glorification is undone in a context that excludes, as well, any negative connotations we might otherwise attach to each. The shift in perspective is crucial and perfectly thorough, such that conventional distinctions between words, reality, and appearance can neither take root nor even be apparent. In this regard Heidegger says that Plato's sense of reality is, in any event, but appearance arbitrarily stabilized in the changeless, supersensory realm of the Forms, a still life taken as the "sole and decisive understanding of Being" (*Introduction* 182). Here also truth is stabilized in terms of a correspondence, a "correctness of human vision" relative to the Forms (185). It is the template of Nietzsche's lament in *The Birth of Tragedy* that Heidegger invokes, where the Apollonian impetus is restricted to the "cocoon of logical schematism," the Dionysian to "naturalistic affects," and as one of many dire consequences of this division, the tragic hero is replaced by the dialectician (91). In Heidegger's terms, the "original sense" of *physis* as the "emerging and enduring power" is violated, reduced in Plato's scheme to a fixed and prosaic expression of natural phenomena. More regrettably, it becomes a mere "paradigm for man," where "human vision" rather than *physis* itself is decisive (*Introduction* 185).

In contrast, Heidegger says that the early Greeks did not come to know *physis* through natural phenomena but inversely, reflexively. "It was through a fundamental poetic and intellectual experience of being that they discovered what they had to call *physis*" (14). Such an account offers a view of intellectual experience where the creative act of "making" or "producing"—what the Greeks called *poiesis*—is the essence of *physis*, where indeed *creatio* is commensurate with *perceptio* in that we participate, are already involved, in the power of *physis*. This "revisioning" of *physis* was to Heidegger implicit in the pre-Socratic mystery all along. Hence, *physis* is both Being and becoming, the what and the how, embracing not just the barest physical features of our existence but the "doxic cogito" as well. Here the social, intellectual, and spiritual features of our being converge, expressed in this wholeness of earth, in the unfolding of "human history as a work of men and gods" that subordinates even the gods to its destiny (14).

Appearing continues to cloak and dissemble (109), but in the passage to immanence that Heidegger affirms, appearing "belongs to Being as appearing" (108), so that the early Greek effort was perpetually to "wrest" Being from appearance. According to Heidegger the idea is given its "supreme purity" in Greek tragic poetry (106), and is most apparent in the myth of Oedipus. In the Sophoclean version the tragedy is played out in a sequence of appearances, from Oedipus in the appearance of the august and arrogant king of Thebes at the play's beginning, to Oedipus in the appearance of the patricide living in profane union with his mother at its end. It may be that Oedipus's disposition, his singleness of purpose and the passion it occasions, is what is most memorable in the play. It is definitely what gives the play its meaning and marks it as drama. The struggle is always foremost, until it would seem that truth and reality, such as they are in this tale, are composed wholly of the struggle itself. Surely it is the measure of Oedipus's being, constantly in flux but marking the way to who he is, to his awareness of himself in the context of a situation. In this struggle, in his passion for the disclosure of Being, Oedipus "most radically and wildly" embodies the "fundamental passion" of the Greek purpose (107).

And quite significantly Oedipus achieves his apotheosis, his glory, in the irony of his blindness. Thus, rather than *doxa* being a matter of mere opinions belonging exclusively to those who individually hold them, here it embraces the entire community by way of the dynamism of poetic experience. *Doxa* is here, as Heidegger says, "authentic appearance," for it stands in the truth of Being, which is at once Oedipus's glory and tragic fate. And here, as well, *doxa* would indicate a radical departure from "human vision," for it possesses of its own accord that creative agency of Being, that poetic imprint of *physis*. Indeed, in a sense it is a radical departure from vision as such, from the "sight metaphysics" Heidegger so vigorously condemns. In matters to be discussed in the following chapter, "authentic appearing" has very little to do with seeing and, as I have already suggested, is essentially a function of language in Heidegger's phenomenology. It is a matter of listening, of paying heed. For the present, however, it is good to take notice of the recurrent theme in the above examples, that irony of blindness linked to insight given in poetic experience: of Homer being blind and remaining so, of Stesichorus regaining his sight only as he renounces Homer, and of Oedipus achieving awareness of the truth only by virtue of his blindness.

In the end, the issue turns on the divergent ways of transcendence and immanence, with *episteme* and all warrants of comprehension as a measure of the "correctness of human vision" securely joined to the former. But insofar as Being is conceived as immanent, as an active presence in the world—rather than seeing the world as a mere copy of Being that subsists in some great beyond—*doxa* is its obvious corollary. It is thereby fashioned by our passionate dealings with the world and is, at the same time, continually at their service, so that through the existential reality of lived experience thinking and Being are necessarily the same, realized in a manner both determined by and determining *doxa* (*Introduction* 103). No different that poetic experience, *doxa* is as well both product and process, the glory bequeathed and the bequeathing of the glory. In this context it participates in what Heidegger calls "authentic knowing," where comprehension is not a consciousness of something outside ourselves but a state of Being, or aspiration towards it, that was glory to the pre-Socratics and, to Heidegger, marked "the supreme possibility of human being" (103). This nexus of thinking and Being, thus replete in the excellence of authentic knowing, is the paradigm that fits so properly the personas of Hector, Sarpedon, and Odysseus, and can just as aptly be applied to Sullivan Ballou.

There is no great leap involved in moving from this conception of *doxa* to a perspective on Gorgian rhetoric and the Gorgian *kairos*, though it is but a glimpse for now. To begin, it is helpful to bear in mind, once more, that the art of *logos* espoused by Gorgias owes its pedigree more to the poets of the sixth and fifth centuries than to the philosophers of the fourth (Segal 112). It is with this suggestion, at any rate, that the connection Heidegger draws between Being and *doxa* is evident in the situation of the Athenian soldiers that Gorgias depicts in his "Funeral Oration." If Gorgias is no Plato in terms of intellectual prowess, he is even less a Homer or Sophocles in poetic inspiration, but the "Funeral Oration" nonetheless presents a tragic conception of human being that joins it to the work of the poets, and necessarily does so through a consolidation of thinking and Being that duplicates or at least foreshadows the idea of authentic knowing. As I have said, in his "Oration" Gorgias's praise is fixed on the warriors' horrific choice between life that is all the more worthy of reverence for being their own and the no less compelling purpose of defending the city from its enemies. Thus, as Hector and the others discussed, the warriors attained that singular glory of the tragic hero that, again, Gorgias defines as "an excellence which is divine and a mortality which is human" (Untersteiner177).

In just such an occurrence, the "supreme possibility of human being" is realized by virtue of this peculiar manifestation of *kairos*. Thus, to give the matter an Heideggerian cast, the warriors in this situation were hardly self-possessed and reasoning creatures, detached from the world, calmly contemplating their relation to it. They *were* their situation, utterly contained by a "thrown-ness" (*Geworfenheit*), a necessary involvement in the world, a "Being-in-the-world" from which they knew neither the whence nor whither. In such a situation the pure "that it is" shows itself, and the most fundamental question of human being arises, that of Being in all its overpowering immanence. In other words, the experience of the soldiers was not the stuff of mundane experience, of "average everydayness," as Heidegger would say.[14] Rather it is centered with profound emphasis on the moment, on a *kairos* compelling an awareness of human reality in that "close-locked struggle" of which Untersteiner speaks, that to Heidegger comes to pass as an "event of appropriation" consuming the traditional binaries of knower and known, Being and appearance, thinking and Being, and other categories of metaphysics which situate human consciousness above the tumultuousness of life. Heidegger has a term for this state of awareness as well: *Befindlichkeit*. Often translated as "state of mind" or "being in a mood," it is more literally and accurately "the state in which one finds oneself" (*Being* 172, 376), as "Being-in-the-world." In the anxiety and passion of *Befindlichkeit,* as Frederick Olafson says, "the world is most authentically uncovered as the world." The term directs our attention reflexively, for the place in which we "find" ourselves when in a certain state or mood is not at all the mind but the world itself, "since it is in the world that things happen or are happening that affect us" (106–07).

This is specifically the context of "authentic knowing." Here the very idea of "reality" is hedged with innumerable qualifications, being ambiguous at best, for it can neither be derived from ideals existing apart from the world nor from what Plato condemned as mere appearances or opinions that are, in any event, mere appearances and opinions only because of these ideals. Rather, prior to these ideals and whatever shadows they might cast in this world, and in place of all subject-object distinctions upon which either possibility is based, the warriors' role as men of *doxa* gives precedence to their prior involvement in the world, to their very embodiment in the world, making them intrinsically part of it. To Heidegger the phenomenon comprises Greek thought in its "original beginning," invariably leading to where it begins, to the "originary experience" he calls "earth." This is *physis* in its most essential sense, an immanent power manifest linguistically, as we are manifest ourselves, in *Ereignis*, the "event of appropriation" fusing "what is" with "what is said." Things are realized here in myth and legend, in language having its source in native soil, in that sense of *doxa* where heroes are literally born of the earth.

The *Homo Mensura* Doctrine and *Dasein*

This particular understanding of the human condition leads to another feature of pre-Socratic thought central to Sophistic rhetorical theory, that of the *homo mensura* doctrine of Protagoras, on whose authority we affirm "man is the measure of all things." Plato takes the doctrine to task, and, curiously, we have come to understand it primarily on his terms, as a sanction of relative or subjective truth, as knowledge gained by the equation of "appearing" and "perceiving." He condemns the doctrine specifically in his *Theaetetus* (155ff).[15] In feigned trepidation Socrates sets out in the dialogue to assail the "vast array" of those giving allegiance to the doctrine. He declares Homer to be their "Captain" and then proceeds to reduce the *homo mensura* doctrine to the contention that perception is knowledge. Going this far, he reduces the doctrine to an absurdity. If all things are exclusive to our perception of them, and if access to any eternal realm is thereby denied by the absence of what he would see as the higher functions of our soul, then we are no different than the brutes, for they perceive as well. Thus, Socrates asks, why not claim with equal justification that the measure of all things is the pig, or the baboon, or "some other sentient creature still more uncouth"? (161c).

Plato says a great deal more about the *homo mensura* doctrine, but no amount of trepidation, feigned or otherwise, can conceal his contempt for the idea that man is the measure of all things. Nevertheless, the doctrine would seem to be very congenial to the Sophists' purpose of a practical minded understanding of the world, and for that very reason it figures prominently in the thinking of current rhetoricians seeking a rehabilitation of Sophistic rhetoric. Indeed, as the doctrine would seem to foil all versions or patterns of truth not hewn to our measure, an obvious consequence is that truth is instrumental, relative to our place and purpose in the world. Thus, through the effect of the *homo mensura* doctrine all things would be reduced to a human dimension, as the testimony of Plato himself would imperatively have it, for we are neither pigs nor baboons, however uncouth we otherwise might be as human beings. This particular reading of the doctrine thereby affords an especially favorable ambiance in which rhetoric might thrive, for it places principal emphasis on what is practical in human life and conduct. On the strength of this doctrine rhetoric is indeed, as John Poulakos says, "an activity grounded in human experience rather than philosophic reflection." As such, it "does not strive for cognitive certitude, the affirmations of logic, or the articulations of universals" but is "satisfied with probability," lending itself in all cases to the "flexibility of the contingent" ("Toward" 35–37).

One of the more adroit extensions of this position is offered by Sharon Crowley in her "Plea for the Revival of Sophistry." She argues that the Sophistic method of inquiry focused upon what was "humanly desirable," upon pragmatic rather than theoretical matters, and that because of this focus Sophistry was above all a practical wisdom dealing with the very human if indeterminate and ungainly work of compromise, accommodation, political and social change. Hence, she claims, the teaching practices of the Sophists were dynamic rather than static, liberal rather than conservative. On this basis Crowley offers a curiously revisionist slant to the dialectic, arguing that the Sophists, because of their heuristic theory of language, "engaged their students in conversation, in a mutual give and take which was intended, unlike Platonic dialogue, to allow both learner and teacher to achieve new insights" (329).

If Crowley's rendition of the dialectic sounds suspiciously like those offered by Plato's apologists as essentially the nature of *his* dialectic, what most clearly brings her analysis in concert with the neo-Sophists is her embrace of what they perceive as most integral to the *homo mensura* doctrine: namely, the belief in the integrity of individual, subjective perception which, in turn, translates into the relativistic character of all knowledge and thereby its aptness to be applied in the social and political spheres. It is in this context, at any rate, that Sophistry would seem most clearly in harmony with rhetoric's aim to facilitate successful participation in the *doxa* of the *polis* by bringing disparate individual perceptions in line with communal belief.[16] It is specifically in this context that the *homo mensura* doctrine is in splendid accord with rhetoric's function of "adjusting ideas to people and people to ideas," as Donald C. Bryant defined the art so many years ago (211).

These arguments commending the relativity of truth are cogent and very satisfying, and our understanding of Sophistic rhetoric is all the more robust for the work of those presenting them. Heidegger, however, interprets the *homo mensura* doctrine differently, and his interpretation proceeds from convictions more in keeping with the mysterious side of the pre-Socratics. Yet they are hardly convictions that impugn aspects of "practicality" for the sake of "mystery." On the contrary, they regularly mark their equivalence, reflecting something more fundamental than what the modern day advocates of the Sophists would regard as the worth of "subjective relativism" and what Plato would regard as its evils. The "complete" fragment from Protagoras is "man is the measure of all things, of those that are, that they are, and of things that are not, that they are not."[17] Even in this form, perhaps especially in this form, the doctrine seems to reflect the priority of human being as subject, separate and distinct from the world—just as Plato would have us understand the doctrine, even as he condemns the epistemology that lies at its source and in particular the metaphysical implications he claims follow in its wake. In fact, Heidegger says of the doctrine that it seems as though Descartes were speaking through Protagoras, and due to the persistence of the metaphysical tradition we thus come to understand the meaning of the "human measure" as if it were an obvious foreshadowing of the Cartesian ego (*Question* 143; *Nietzsche* 4: 90).

But, once again, Heidegger endeavors to breach the tradition, and in this respect views the doctrine from what he calls a "uniquely Greek way" that summarily excludes Plato and Aristotle, because neither philosopher, Heidegger declares with characteristic aplomb, is "uniquely Greek." Accordingly, he informs us, we must take great pains not to "unwittingly insert representation of man as subject" into the doctrine (*Nietzsche* 4: 93). With prescriptions such as these, Heidegger says that "man" is not some "detached I-ness" but is determined out of a relation to Being as such; therefore "man" does not "set forth the measure to which everything that is, in its Being," to which Being, in return, "must accommodate itself" (*Question* 145–46). Here the Sophistic stance explicit in the doctrine is neither subjective nor relative. Heidegger interprets it as pre-subjective, and sees it as preserving the fundamental position of Heraclitus and Parmenides (145–46). To Heidegger, "all subjectivism is impossible in Greek sophism, for here man can never be subjectum; he cannot become subjectum because here Being is presencing and truth is unconcealment" (147).[18]

By virtue of this perspective, perception and knowledge are not functions of "human vision," of one's subjective consciousness; on the contrary, to perceive, and ultimately to know, is to face what is unconcealed by an object's outward aspect in "authentic" appearance. Here again, we understand Being in its immanence, of *das Sein* functioning as an active presence in the world, indeed as the "presencing" of "things that are" (*das Seiende*) in the world. Truth, then, is experienced as *aletheia,* the term meaning to Heidegger the "unconcealment" (*a-letheia*) or revelation of Being in the things of the world.[19] Within this compass the fundamental attitude of human being, insofar as it is authentic, is one of acceptance, of "unveiling" or openness to the "self-revelation" of things in their Being, "whereby man himself in his own way," as William Richardson says, "comes-to-presence with these [things] and thus achieves himself" (420). Perplexing as the process may seem, it is in clear contrast to conceptions of truth achieved on the basis of subjective relativism, or by means of the transcendental reach of a sentient center of the soul or ego. In either case human being is not only autonomous but preeminent, thereby provided with a vantage point to not just know "truth" but, indeed, to impose this "truth" upon the world. But in Heidegger's interpretation the emphasis is on "unconcealment," on the "self-showing" of Being. Thus, contrary to the conventional understanding of the *homo mensura* doctrine, "man" is not a representing subject who "fantasizes" the nature of the things of this world, but, as Heidegger says in one of the more compelling lines dealing with this matter, by virtue of unconcealment "fantasia comes to pass." It comes into appearance "as a particular something, of that which presences—for man, who himself presences towards what appears" (*Question* 147).

When all things are reduced to what we know, to what we can know, or to what we think we can know, then by virtue of the "knowing" subject alone the knowing subject is necessarily the measure of all things. But as the question concerning the meaning of Being prevails, and we retain an appreciation for the mystery of our being, then we are not, nor can we be, the measure of all things, for that capacity is vested in the mystery itself. Thus, in Heidegger's interpretation of the *homo mensura* doctrine the emphasis is not on what we perceive but how Being presents or reveals itself. Here the focus is as much on the truth "coming-to-pass" at the instigation of Being in its revelatory dynamism, as it is on human being and its openness to Being. More accurately yet, the focus is on what Heidegger calls the "co-constitution" of Being and human being. A relationship is established, in which the unconcealing of Being and our openness to it are necessarily reciprocal for there to be any relationship at all, which in contrast to a dichotomy of subject and object persists instead as a relationship preceding the distinction between the two. Thus, it is within the compass of this relationship that in any given instance man is the measure, the measure that the restricted radius, horizon, or region of his experience of the world makes available to him. It is the measure, then, not of man exclusively, if at all, but of the unconcealment of Being that comes to pass within the confines of this relationship. In a less cryptic if more abrupt account, Pirsig reaches a similar conclusion concerning the *homo mensura* doctrine, claiming that that which "creates the world emerges as a relationship between man and his experience." Human being is in this sense "a participant in the creation of all things. The measure of all things" (368).

Here, foundational thought is not only pre-subjective but is ultimately *non*human in that it cannot be intrinsically ours. In this account we would be the measure of all things without being opposed to them as subjects, for we would enjoy, as Heidegger says, an "ontological priority" over things as perceivers of Being. This is not to say that a notion of the self as subject—or of subjective relativism *in toto*—cannot be later abstracted and isolated from the world, but in Heidegger's perspective this process is a spurious if not destructive undertaking, thwarting the pre-subjective awareness of the world already in place when the subject-object distinctions are belatedly organized by representational thought. Indeed, traditional notions of consciousness establishing the self as an isolated ego, most of which can ultimately be traced to the Platonic doctrine of the soul, are abandoned in general by Heidegger. Surely these notions reach their apotheosis in the existentialist emphasis on consciousness as the distinguishing characteristic of human being and are themselves, I suspect, the basis of the neo-Sophistic position on the *homo mensura* doctrine. Here, in either case, the world acts as a material restraint on our spirit and, as a result, we then live pitted against the world inasmuch as our consciousness gains realization only in opposition to it. In any event, Heidegger goes to some lengths to disassociate himself and his philosophy from this attitude. He repudiates in particular the emphasis in existentialist treatments of consciousness as the primary criterion of authenticity. Integral to Albert Camus' philosophy, for instance, is the contention that "everyone begins in consciousness and nothing is worth anything except through it" (10). To Heidegger the position exemplifies a "subjectivity" that violates his idea of human being as fundamentally an "ecstatic" relationship with Being. In fact, his reproach of the existentialist view is most revealing of his own view, claiming "the higher the consciousness the more conscious being is excluded from the world" (*Poetry* 108).[20]

Here, once again, neither consciousness nor the world is given precedence in Heidegger's view, but prior to each stands the relationship itself, which is "more original" than that which is related (*Identity* 12). In the strictest sense, the nature of human being can be understood only in terms of this relationship, a matter of belonging or responsiveness—of "listening"—Heidegger says. We are "essentially this relationship responding to Being" and only this. But it is a belongingness that is mutual, as each is "appropriated" to each, "for it is man, open toward Being, who alone lets Being arrive as a presence" (*Identity* 31). Heidegger calls this relationship "*dasein*," a "being there," where, to reiterate, we are "thrown" into the world and "discover" ourselves already involved in it, such that the simple "that it is" forever veils the "whence" and "whither" in darkness (*Being* 173). Here again, our nature is not that of subject determined in contrast to Being as object. Indeed, thinking and Being are equated, blurring traditional distinctions between epistemology and ontology, between what is within us and what is before us. Rather than consciousness being subjective, conceived as a spiritual interiority housed in the body, Heidegger says "when dasein directs itself towards something and grasps it, it does not first get out of an inner sphere in which it has been proximally encapsulated." Instead, "its primary kind of being is such that it is always 'outside,' alongside entities which it encounters and which belong to a world already discovered" (*Being* 89). Thus, *dasein* is not a term used in place of consciousness ("Ground" 213), but a Being-in-the-world (*In-der-Welt-Sein*) that characterizes a relationship coming "in advance" of the ego or "I" that in this perspective is merely derivative and therefore "must not be taken as an isolated 'subject' or 'self' independent of the 'world'" (Brock 30).

And Being itself can be understood only in terms of this relationship. By virtue of *dasein* we dwell with things, rather than things being opposed to us as objects. *Gegenstand,* the basic German word for "object," means literally to be "in opposition to," "to stand against" or "to affront," and Heidegger rejects that explicit sense of the word to good effect, claiming that the idea of object is better understood as *gegenuber,* meaning "to be abreast" or "in relation to" (*Early* 34, 82). Not only does the distinction attest to the symbiotic relationship between thinking and Being, but the familiar scheme of the separation of subject and object is kept at bay, which Heidegger calls, in any event, a "modern falsification" of early Greek thought (*Introduction* 135). He says we tend to approach the pre-Socratics, who do not follow this schema, as if "they lacked training in epistemology" (136). We tend to understand objects as products that are at once inert and opposed to our being, so that we then assume to make forays into the objective world, to synthesize and accommodate, returning with our "booty" to the "cabinet of consciousness" (*Being* 89). On the contrary, he claims these things stand forth in appearance by virtue of the power of Being, as we do ourselves. Thus, in contrast to portrayals of our essential being in conventional interpretations of the *homo mensura* doctrine—in terms of interiors such as consciousness, ego, or subjective self—we can understand *dasein* scarcely as the measure of all things but more usefully as an openness to Being, as human being in awareness of Being, or potentially so, to the degree that it is then "co-constituted" with Being. In this way, Heidegger says, *dasein* can be understood as a place or location—a process, I would say—where Being occurs, where its truth is disclosed, temporally, both through us and to us.

Physis, Quality, and Doxa Revisited

Because "Being" is so well entrenched in the metaphysical tradition, Heidegger says the term has been altogether divested of meaning, that it has become an abstraction and "the emptiest of all words" (Seidel 36). So, as we have seen, Heidegger begins to use "earth" in place of Being in his later works, most often doing so under the rubric of "*physis*" to depict not just the fundamental connection of life to nature but of Being to material creation. Again, *physis* is what Heidegger claims to be the pre-Socratic understanding of Being, where Being is manifest in its most dynamic aspects as a force at work, verily accounting for the creation of the material world. *Physis* is therefore variously described by Heidegger as "the creative surge of Being," the "overpowering power," an "originative unity" emerging "out of itself" and "accomplishing itself as world." The relentlessness of the Heideggerian idiom is numbing, but another example from Pirsig might suffice to clarify things, for here, as in many other remarkable ways, Pirsig, unwittingly or otherwise, reprises a number of themes from Heidegger. As *physis* is "primordial Being" to Heidegger, it is both permanence and appearance, product and process, in all instances being prior to these differences. In Pirsig's more abrupt presentation of the idea, *physis* is what he calls "Quality." Or it is in many crucial respects Quality's paragon.

Though "Quality" is quite elusive in its own right, Pirsig claims that some of its aspects can be singled out, such as brilliance, vividness, and precision. He says that idea can be fully appreciated only as we are able to imagine its opposite, a world divested of Quality's wholeness, so that just a stark rationality of "purely intellectual pursuits remain" (210–12). Then we would live in Huxley's Brave New World or Orwell's 1984, he claims. It would be life as it was in ancient Sparta or, I surmise, as it would be in Plato's Republic. There would be no poetry and literature, no theater nor any football games and movies, and, most insufferably, "we would all wear GI shoes" (212). Though Quality itself defies definition, Pirsig says that there is no other reality, and that it abides wholly in the moment of vision before "intellectualization" takes place and separates the world into subject and object. He says that this "preintellectual reality is properly identified as Quality, the parent, the source of all subjects and objects" (214); that "the Quality which creates the world emerges as a relationship between man and his experience"; and that, in terms of this relationship alone, "man" is made "a participant in the creation of all things" (368).

If it seems as bothersome to deal with Pirsig's idea of "Quality" as Heidegger's of *physis,* implicit in both are a couple of things sufficiently lucid to be of use to us here. The first is that "reality" in each is pure process and never a static blatancy of the way things are, that we then abstract, isolate, and identify as reality. And secondly, Quality, like *physis,* seems to have no other reality beyond itself, no ultimate foundation that we can know by virtue of our separation and distance from it, or know by any other means except that of a profound appreciation that comes through our participation in Quality. In other words, Quality resides in "pre-subjective" awareness, as Heidegger might say; it is "pre-intellectual," as Pirsig says, and consequently owes its presence neither to ourselves nor to external reality but to a "primordial" awareness—pre-subjective or pre-intellectual—connecting and combining each. It is therefore an awareness coming prior to all divisions we would make between internal and external, subject and object, and despite Pirsig's condemnations of "purely intellectual pursuits," seemingly prior as well to divisions between rational and irrational thought or reflection. Notwithstanding his clipped analysis and scruffy tone that otherwise makes his book such a joy to read, one can only admire Pirsig's visceral, intuitive grasp of Sophistic texts and rhetoric. In the end, Quality is necessarily indefinite and well named by Pirsig, for it sums up in an aptly resourceful way the wholeness of all things, and moves irresistibly towards Heidegger's conception of the "originative unity" of *physis* that is Quality to Pirsig.

But the idea of *physis* remains unfinished until we understand that in both form and substance, as it is meted out in appearances and tallied to human awareness as language, *physis* is *doxa*. Heidegger says as much, claiming that *physis*, as the power of Being manifesting itself in unconcealment, is "dispersed" among the manifold of things that "display themselves as the momentary and close at hand," giving them aspects in appearance that we call *doxa* (*Introduction* 102). In this context *doxa* evokes the ambiance that constitutes "world" in Heidegger's specialized sense of the word, as that uniquely human web of relationship comprising all meaning.[21] In this context, as well, Plato's distinction between *episteme* and *doxa* appears far more profound than the difference between "true knowledge" and "mere opinion," for it completely vitiates the sense of *doxa* as *physis*, of *physis* as the power or facility of the semantic environment in the here and now. Here the distinction Plato draws is but an instance of the "ancient quarrel" between philosophy and poetry addressed in the *Republic* (X.603ff) and elsewhere, where, invariably, "opinion" is but the derogation of *doxa*. It thus preempts the Sophistic sense of the word, of *doxa* embodying the entire community through the dynamism of poetic experience. Again, in its coalescence with language, *doxa* is process in this Sophistic sense, being both the glory bequeathed and the bequeathing of the glory; in its coalescence with *physis*, it is as much the creative emergence of beliefs as the beliefs themselves. For thousands of years the Greeks turned to Homer and other poets to educate their young, in poetry that not only taught but embodied the virtues of Greek civilization. Once again, by virtue of *physis* a notion of *doxa* is projected that blurs the distinction between "to be" and "to be said."[22] Here, we are *dasein*, the relationship to Being, and as the latter is conceived as *physis*, utterly as poetic experience in this context, it marks our being solely in terms of those manifest social and intellectual features comprising *doxa*. Here, indeed, we ourselves are created and sustained in creation by words.

In a number of significant ways Pierre Bourdieu's work in anthropology helps to clarify these connections. He, too, describes a "pre-objective" world subsisting purely as a "self-evident and natural order." In this world nature can never be taken as the utterly separate and inert entity that in modern, industrial societies we presume to freely appropriate and use. In this pre-objective world Bourdieu claims that *doxa* expresses the "quasi-perfect correspondence" of the natural and the social, where the world of tradition is experienced as the natural world, their congruence in ancient societies going unquestioned, taken as self-evident (164). Thus, rather than nature being the end-product of the "long labor of disenchantment" occasioned by analytic, objective attitudes, in ancient societies nature expresses its coalescence with human beings in their reciprocal relationship with *physis*, where the "doxic mode" itself lies at the heart of the relationship. In terms of this relationship specifically, the "quasi-perfect correspondence" of *physis* and *doxa* is accomplished. Just as there can be no clear distinction between subject and object or thinking and being in such a world, there is none between *physis* and *doxa*. As Bourdieu says, in the "doxic mode," "the self-evidence" of the world's reality is recapitulated in the totality of ritualistic practices, discourses, sayings and proverbs about the world "in which the whole group's adherence to that self-evidence is affirmed" (167). In this mentality *doxa* is not only equated with *physis* but in their "correspondence" the phenomenon of earth is most apparent.

It therefore hardly carries us far afield to note that this idea of earth is not at all confined to the eerie mists of some primitive past. In Vietnam the correspondence of *physis* and *doxa* has a presence sufficiently real and exact to be given a name. It is called *xa*. The word translates to English as "village," but according to Frances FitzGerald it carries implications that constitute much more, designating for the Vietnamese a comprehensive and complete "politics of the earth." *Xa* signifies the ultimate wholeness of what we would otherwise see as separate features of earth and ancestors, village and villagers, the spiritual and material riches present in the land. *Xa* therefore designates, at once, the basis of the villagers' contract with society and with nature, their very identity being a function of the congruence of each. FitzGerald writes that "the sacred bond of society" lay "with the spirits of the particular earth of their village." *Xa* forms a "complete picture" of the village as a result and, in turn, the political design of a culture where all social, political, and economic relationships "appear there in visual terms, as if inscribed on a map" (144).[23]

And this phenomenon of earth is closer and more vital yet. N. Scott Momaday provides his own view of the wholeness of *physis* and *doxa,* an American Indian equivalent of *xa* that he calls "racial memory." In essence, racial memory is the commitment of a community of believers that comes to pass in a consciousness that is sacred, communal, and linked to an affirmation of the land. It is a splendid idea of earth, designating the perfect integration of the American Indian's mind and spirit with that of his people, a consciousness wholly realized in the rituals, legends, and beliefs of the oral tradition. And when Momaday speaks of nature, the land, or the earth, what he invariably has in mind is the concrete actuality of the physical world, the land and the landscape as a visible, tangible thing, manifest as such in myths and legends. It is through this perspective alone that community is established and preserved through the doxic mode of story and song, creating by means of language the cultural landscape through which the American Indian comes to achieve and acknowledge his being, to know who and *that* he is, as Momaday says.[24]

These ideas of *xa* and racial memory will be revisited soon, but for now we might surmise that difficulties we have in appreciating their significance are due to our eroded connections to places particular and unique—to nature as such—in the present-day, homogenized society of Wal-Marts and Burger Kings. These difficulties persist, along with the more essential and far larger impediment to understanding earth: namely, our heeding for thousands of years Plato's instruction in matters of language. It is important, then, to realize that *xa* and racial memory are not necessarily the staid, inherently conservative imperatives that we very often attach to "primitive" cultures. In the alchemy of *physis* and *doxa*, both persevere under the aegis of the oral tradition, where the essentials of Walter Ong's "psychodynamics of orality" and Eric Havelock's "self-identification with poetic experience" hold sway. In fact, we can understand both *xa* and racial memory as *doxa* itself in the pre-Socratic sense, as essentially the peculiar notion of earth I am attempting to explicate here. In each, life is lived flush with the earth, where "pre-subjective" awareness is the essence of the "doxic mode." Here an inviolable wholeness of nature and culture, manifest as language, comprises always the fundamental feature of our existence. The rituals, ceremonies, and beliefs of a culture are not only the stuff of *doxa* but constitute, as Heidegger says, the "Saying" of a people, apparent in a process that is nothing less than the event of Being itself. "Saying" (*Sagen*) is for Heidegger undeniably the definitive indication of *logos* as the word rather than reason. And it is more. It "rises" from the earth, he says, designating "native soil," consecrating it in the way of *xa* and racial memory. In "Saying" the connection with *physis* is consummate, and people "inhabit" language as a result. They "dwell" in language as upon the earth, in that sense being determined by it.

Therefore, language does not originate with the speaker, and with the poet in only the most qualified sense, for language "speaks from within its own speaking" and "not within our own" (*Poetry* 190). In the speech of a people, he says, "every living word fights the battle and puts up for decision what is holy and what unholy, what great and small, what brave and what cowardly, what lofty and what flighty, what master and what slave" (43). Here, in any event, *aneu logou* and its effects are precluded, as are all provisions of higher, nonverbal orders of reality to which our words would refer. And insofar as that reality is precluded, "reality" as we have it here is enveloped utterly in language. And language, like *physis,* has the character of event. In Momaday's world of the American Indian, for instance, there is no Western sense of the soul, of the isolated and unique self, only human being persisting as part of the fabric of the prevailing *doxa*, as "*dasein*" in that sense. Indeed, as will be examined presently, in Momaday's world the archetypal American Indian is the "man made of words," and if *dasein* is hardly identical with the concept, in ways concerning us here it is something very much like it.[25]

Dasein and the Psyche

This alliance of *doxa* and *physis* calls into question perspectives that structure reality in terms of the separation of human being and nature, society and nature, language and nature. The latter pair, that witchy mix of language and nature implied in Momaday's view, looms as the most basic of these, encompassing the others and ultimately being the one most pertinent to our understanding of the Gorgian *kairos*. It will be explored at some length in the next chapter, but for the moment it is necessary to consider further the relationship between human being and nature, only to see it now in light of the specific similarities that Heidegger's conception of *dasein* bears to the pre-Socratic notion of the *psyche*, for that, too, is pertinent to the Gorgian *kairos*.

The comparison is inevitably less than perfect. The concept of *dasein* in Heidegger's works was meant to repudiate among many other things the ill effects of "calculative reasoning," but *dasein* itself can be taken as the result of a calculative reasoning of the most rigorous kind, abstract and philosophical to the core. On the other hand, the idea of the *psyche* would seem to emerge in its completeness from the *praxis* of the early Greeks, implicit in the "natural attitude" of their way of life and therefore linked to specific relationships with *physis* apparent in their literature and poetry. In other words, the *psyche* is "pre-critical." And there are other obstacles as well. Chief among them, no doubt, is that to many the presence of a "turn" (*Kehre*) in Heidegger's thought pivots on this very issue of *dasein*. This reversal is marked, they claim, by a preoccupation with a "subjectivism" and the nature of selfhood in *Being and Time*, to concerns in his later works that deal not primarily with *dasein* but with Being as such, especially as he then relates Being with language in very direct and curious ways. I am not certain that a compelling case can be made to the contrary, though Heidegger himself disparaged "the groundless, endless prattle" about the "turn" in his thinking, and affirmed that he never abandoned the fundamental issue of *Being and Time*, namely to give shape to his basic and enduring purpose to query the meaning of Being.[26] In addition, it is clear that Heidegger never presented Being and *dasein* in terms other than their mutual implication or "co-constitution." Thus, there is a consistency to the meaning of *dasein* that never departs in any significant degree from the general sense of Werner Marx's contention that *dasein* is neither "substance nor subject" which would allow the "essence of man to be approached in a Cartesian way" (90). *Dasein* is not an "actuality" but is sheerly the relationship to Being that in its doing, making, and acting in the world comports itself all along in terms of its "ability to be," its "possibility to be" (88). "The question of Being and the self," as developed in *Being and Time*, "keeps ruling the philosophy of the later Heidegger," Paul Ricoeur says (*Conflict* 224). In fact, it seems to do so especially as Heidegger's turn to the coupling of language and Being not only further articulates his earlier formulation of *dasein* but very apparently and repeatedly recapitulates in that manner the disposition of the *psyche*.

Thus, even if Heidegger was concerned with "subjectivism" in *Being and Time*, there evolves in his later works an idea of *dasein* that is of the same lineage as the *psyche*. Insofar as his purpose to move beyond metaphysical perspectives is characterized by the effort to retrieve critical features of pre-Socratic thought, the resulting similarities of *dasein* and the *psyche* are found chiefly in their pointed contrasts to "humanistic" interpretations that postulate our being as supreme. Of these, Heidegger specifically condemns notions of *"animal rationale"* and "spiritual-ensouled-bodily being" ("Letter" 210), or, we might say, as both are present in equal proportion in Plato's doctrine of the soul. As the *psyche* precedes the philosophic advent of the soul, *dasein* comes later in defiance of the doctrine and all variations of it, surely of all ideas of an internal, unchanging essence of human being separate and distinct from the world. And if *dasein* is the product of theoretical reflection, it is at the same time the result of an effort to recapture the primordial sense of human being as "Being-in-the-world" that necessarily comes prior to theoretical reflection. Indeed, despite the rigid analytic procedures that some would claim underscore the hermeneutics of retrieval, the sense of *dasein* that emerges in Heidegger is based upon a relatively imaginative restoration of the *psyche* and the world from which it arose. It is itself "pre-critical" in that sense, for given the pre-Socratic sources that in good measure inspired it, *dasein* is a retrieval of the *psyche* in its own right, their differences matters of degree, never of kind. This is especially the case as each is positioned in place of Plato's sense of the soul to account for the immediacy of the concrete, explicitly lived existence that to Heidegger was the lifeblood of Greek thought prior to Plato.[27] Each, then, is shaped by *physis*, where the essence of human being is the result of its involvement in the world, its "Being-in-the-world" the sole warrant of its nature.

For instance, if the soul achieves its full integrity only in its separation from the world, the *psyche* and *dasein* are dependent on the world, seeming to owe their essence to things "outside" themselves. Their reality is situational and dynamic, each made up of moods rather than lasting, changeless properties. As the *psyche* occupies an ambiguous position between the subject that knows and the object known, its locus of being is always equivocal, seeming to shift between what is internal or innate and what is external or situational, otherwise seeming to manifest features of both at the same time. Again, it is often identified outright with specific bodily organs, so that virtues such as pride and courage were not conceived as spiritual attributes innately comprising one's identity; but for these and other attributes of character to endure as virtues at all they had to be externalized in action, having had their source in somatic functions, such as rapid, convulsed breathing or the flurried, agitated rhythm of the heart that comes about in the passion of critical moments. The concept of the spiritual is neither dismissed nor diminished as a result, but is joined with the corporeal sphere, so that it can arise only in the context of specific situations, where it evinces obvious, physical manifestations. The crucial point is that the *psyche*, in its bond with *physis*, is constituted in direct measure to the passion attendant upon its action, attesting to a relationship with the world that is reprised in Heidegger's conception of *dasein*. In particular, when Heidegger's presentation of the nature of feeling, passion, and anxiety is taken into account, the perspective that *dasein* brings to bear on the nature of experience parallels that of the *psyche*. Though the latter has a physical component seemingly absent in *dasein*, Heidegger maintained from the start that *dasein*'s essential state of "Being-in-the-world" is physiologically realized and conditioned (*Being* 234). In this connection he claims the very notion of *dasein* captures the essence of "feeling" as inevitably linked to specific situations rather than isolated, internal states that, apart from these situations, we might otherwise consider as intrinsic. It is here that the essential meaning of *Befindlichkeit* obtains its significance as the discovery of the "self" gained solely in terms of the situation. According to Frederick Olafson, we interpret the work of feelings as an uncovering, a "disclosure" that most often comes through feelings of anxiety, "where we are in the space of possibilities and actualities of the world" (106). In this manner the full importance of *dasein* as "Being-there" is apparent, for human being discloses itself more through attitude or disposition, what Heidegger calls "mood," than through intellectual reflection. In fact, the most significant implication of "Being-in-the-world," at least initially, is that we are not isolated egos or personalities but a succession of moods evoked by encounters in the world, by situations.

Time and again we witness the telling effect the idea is given in the dramatic action of the *Iliad*. Here the quality of the heroic is not borne intrinsically in the soul of the warrior, so that it can then be expressed, by and by, on the field of battle. Rather, in the quickening moment of its realization, the heroic cleaves to the needs explicit in the critical situations embroiling the warrior, thus obtaining its distinctive human shape in his persona, evincing his identity as hero in the aura of *doxa*. In this contest the heroic can only be as heroic does. The interrelationship of character and action is so complete in the *Iliad* that not only do traits of character spontaneously arise from the brunt of experience but materialize quite apart from the will of the warrior. *Doxa* is unambiguously *physis* in this context. At all times it is the prowess of nature on display, not that of the hero, whose *psyche*—his "being there"—is purely the means to express the power of *physis* in its "self-sufficient emergence." As Paolo Vivante says of the experience, energy and courage and passion as a matter of course are "drawn, like breath, from nature's innermost core" (51), in a world where "the burst of passion and the peal of thunder" are quite naturally considered as "two aspects of the same phenomenon" (167). The process preempts the fine distinctions we are prone to make between mind and matter, body and soul, especially as we take either mind or soul to embrace the constant and durable ingredients that alone would define human being.

It is therefore important to realize that these ways of apprehension associated with the *psyche* are not unique to Homer's account of the Trojan War but coincidental to an appreciation for the mystery and power of Being. The highly dramatized sense of the *psyche* that emerges in the *Iliad* is presented routinely, in sheerest delineation by Heidegger's concept of *dasein*. As the *psyche* subsists only through its coalescence with *physis*, so *dasein* cannot be separate from Being and still maintain its integrity as *dasein*. Nor can it be separate from *physis*, insofar as Heidegger chooses that designation for Being. Once again, the significance of *dasein*'s meaning as "Being-there" is apparent as the process through which Being is disclosed. Karsten Harries puts the matter of this relationship quite bluntly, saying that "dasein is not a fact but a nothingness; a relation, a gap, an in-between" (68). If *dasein* thereby seems, like the *psyche*, a "feeble and witless thing," it is such only as we presume it to be a thing, an independent "actuality." In fact, *dasein* is all the more significant in the terms given by Harries, for the relationship of which he speaks is more essential by far than the things related, as the latter exist only by virtue of the "in-between," the relationship of "Being-there." Indeed, we necessarily fail to appreciate the significance of this relationship in seeing it exclusively in Platonic terms, as somehow a combination or synthesis of what he separated, of a coming together of the physical and spiritual aspects of our being. But to see the relationship of *dasein* as prior to what is related, to see it then in the sense of "aho," the very act of drawing breath to the Navaho, is thereby to understand it as neither physical nor spiritual alone, but as a wholeness or integrity that is more than both, comprising by virtue of every breath we breathe the event of Being. If the act of breathing no longer transpires, *dasein* is no more.

In this vital relationship, Being is immanent by virtue of *dasein*, and Being *is* only as immanence. The ontological difference persists, yet "only so long as Dasein is, is there Being," Heidegger says (*Being* 255), and here we are not the "lord" of things but at best Being's "shepherd" ("Letter" 221). The event of *dasein* is thereby apparent not only in its relationship to Being, but by virtue of the "there" of its nature it grounds Being in the physical, temporal realm. In this context Being is *physis*, most conspicuously "earth." It is here as well that *dasein*, "Being-there," is itself manifest in the realm of action, time, and most especially as language, for it cannot possibly exist prior to its experience in the world, in this manner more than any other recapitulating the mysterious ways of the *psyche*. In this respect Olafson's definition of *dasein* is as telling as Harries', as he claims *dasein* is not simply the kind of entity that exists and also acts but that "its action, as well as the 'undergoing' and 'passion' that is the inevitable counterpart of action, is precisely what makes it the kind of entity it is." He adds that there is "no dimension of *Dasein*'s being that is prior to action and the kind of understanding its involves," as there can be "no locus outside the world or outside time that *Dasein* can be said to occupy" (103). At this point we might recall Olafson's definition of *Befindlichkeit* and appreciate more completely its implication that "the place we find ourselves when in a certain state is not the mind but the world" (106). As Plato interminably reminds us, we would dwell in the realm of the eternal forever and know its truth absolutely but for the burden of bodily cares and preoccupations that keep us fettered to the world. Here, by all accounts, to include Plato's in particular, action is the obvious measure of our involvement in the world, being especially apparent as we see it now against the backdrop of the soul's immutable stillness. Indeed, apart from its physical presence in the world, manifest always as action, neither *dasein* nor the *psyche* exists in any traditional sense at all. Otherwise, they are wraiths at best, mere shadows of the soul, such as those gossamer-like apparitions in the case of the *psyche* which dissipate soon after death and are gone.

And again, action is manifest not just in the bold antics of the Homeric hero but is evident in the barest physical circumstances of our existence. In this context we discover our being is real only to the extent that it is formed by temporality. Time, in itself, "is nothing," Heidegger says. "It exists only for the sake of measuring things," yet "time is that within which events take place" (*Concept* 3E). It is as *dasein* in this respect, gaining its significance only by virtue of a "gap," an "in-between," a relation. Here we must be aware of the closest possible connection in Heidegger between time and language, because both the *psyche* and *dasein* owe their nature, utterly, to this connection; and that, in turn, leads directly to an understanding of this perspective on the Gorgian *kairos*. For now, we can say language is the key, the touchstone of any adequate understanding of this perspective, for *dasein* and the *psyche* are events that transpire in time and are thereby necessarily resolved into narrative structures, into stories. "Being-human is *logos*," Heidegger says. "With man's departure into [B]eing, he finds himself in the word, in language" (*Introduction* 172). This uncommon formulation provides an express instance of the most extreme sort of Walter Ong's contention that language is, before all else, "a mode of action" rather than a "countersign of thought." And Heidegger offers a perspective on action that necessarily solicits a story, so that *dasein*, insofar as it is authentic, is manifest in "a rootedness in meaningful contexts of the past and its directedness toward some future end" (Guignon 8). This perspective not only serves to counter the "metaphysics of presence" but confronts us with a corollary implication, namely that the soul delivers us from the "corrupting" influences of the body only as it more crucially delivers us from the "corrupting" influences of language *that in itself is imperatively the relationship*, as "in-between" as the drawing of breath or beating of the heart. Thus, what is most essential in the make-up of *dasein* and the *psyche* is *logos*, as each achieves its most complete expression not only in language but, as I shall attempt to illustrate in the next chapter, *as* language. In this alone do they endure.

This perspective, so fundamental and expansive to embrace all things in terms of Being and the relationship to Being, cannot be reconciled to humanistic treatments that would exalt human being as sovereign. In this perspective, most emphatically, all things cannot be reduced to the human measure. On the contrary, the perspective invokes Being's mystery, and what makes the mystery as enchanting as it is redoubtable is that it bears elements not only of the obscure but the ominous, giving us a sense of something altogether alien to what we might otherwise perceive as our most favored place and purpose in the world. Therefore, we are mistaken to identify human being too precisely with *dasein,* for the latter is the relationship to Being, and Being comprises much more than we can know, or hope to know, or to manage and master. By its very nature, then, *dasein* implies the presence of the "nonhuman." As its essence lies in existence, this existence is marked by mystery, providing perplexing, very ironic, and wonderfully engaging twists to the contention that we are determined by a "whence" and "whither" that we cannot know, for by that very token we are not, humanly, what we are. To Heidegger the phenomenon means we dwell in the nearness of a god, that we are the "neighbor of Being" ("Letter" 222–23). In Gadamer's explanation, it means "humans, as humans, stand out in the open, that they are in the end more proximate to the furthermost, to the divine, than they are to their own nature" (*Heidegger's* 197). The idea will become more critical in the ensuing discussion of language, especially in the manner presented by Werner Marx, who claims, even as *dasein* cannot be identified as "anything human," this condition is essentially the "non-human enabling function of man" (213). At any rate, consistent in all these views is that Being and human being are never a relationship of autonomous entities, as one cannot exist without the other. In either case they are "co-constituted," their relationship preceding the things related, with *dasein*—"Being there"—the sole measure of each. The measure of our being.

Being, Gorgias, and Heidegger

That *dasein* evinces features of the "nonhuman" allows a Heideggerian cast to be placed not only on the idea of the heroic but on life in its more ordinary aspects, as it is lived far removed from the drama of epic tales. That is to say, in this perspective the particular expression of our involvement in the world is not primarily an issue of human will and aspiration but, heroic or otherwise, is supremely the "non-human" work of Being, of earth or *physis* in its most essential sense. However, to develop some of the rhetorical implications of this perspective, especially those relating to the Gorgian *kairos*, it is necessary to pursue certain connections that Heidegger's notion of Being might bear to the views espoused by the Sophists, by Gorgias in particular. Of course, one had best be wary to claim that Gorgias, or any of the Sophists, had ideas of Being that would match Heidegger's in terms of philosophical finesse or detail. In fact, the suggestion that a notion of Being lurks at all in the shadows of Gorgias's "philosophy" would seem to be dispelled by the very title of his treatise, "On Non-Being," seemingly the only work by Gorgias available to us where the question of Being is explicitly addressed.[28] But given his entanglements in pre-Socratic belief and thought, clearly he rejected Plato's notion of Being. Moreover, as I have attempted to illustrate through the analyses of Untersteiner and others, Gorgias was not a nihilist but a phenomenalist, and the particular notion of Being that he rejects is Plato's *ousia*, where Being is transcendent.[29]

The peculiar treatment of Being that does persist in Gorgias's thought is apparent in the very absence in his "philosophy" of any cogent distinction between Being and appearance. At least, this is where his treatment of Being begins. In the first tenet of his treatise, "On Non-Being," Gorgias claims that "nothing exists," and despite his notoriously primitive ideas of sense perception, he speaks of phenomena in a world of flux, only of "perceptibles" that we are able to know and communicate to others. To the extent that Being is addressed at all, it is in terms of this absence of permanence and stability, this process having its effect as an active if sometimes mischievous presence in the world, and very far removed from the array of abstractions lodged in the remote, celestial realm that Plato postulates. Thus, to take Gorgias at his word is to accept that the essence of the world is not hidden behind a veil of appearances but in essence is the appearances themselves. Here, the obvious effect of his analysis is to sever appearances from any metaphysical source in transcendental Being and thereby to see in appearances essentially what appears rather than reflections of things more essential yet. The result is a peculiarly vintage sense of *phainesthai,* the very archetype that Heidegger endeavors to invoke, of appearances, he says, in the "great sense" of epiphany rather than mere optics (*Introduction* 63). So, to draw out at long last the only practical consequences of Gorgias's notorious trilemma, the world can only be as it appears. To take Gorgias otherwise, as the complete nihilist as some have done, is to malign him for failing to answer the "why it is" of Being and to remain oblivious to his peculiar rendering, in all its compelling simplicity, of the more momentous "that it is."

Indeed, that there is something rather than nothing seems to be the sum of what can be said concerning a Gorgian and, in general, Sophistic sense of Being. Or, better yet, there is something only by virtue of being nothing. And if Gorgias does not provide a fully elaborated sense of *phainesthai,* the essence of the idea is nevertheless present in his belief that Being—or its absence, if you will—can only be manifest in phenomena, in its utter "thereness." In all instances it is a visceral awareness, but one where the "thereness" of Being is equivalent to its immanence. We accept so perfunctorily the analytic perspective, the separation of the knower from the known, that the notion of a "creative surge" of Being is apt to seem so unusual and remote as to be absurd. And "absurd" is precisely the word, as long as we insist on answers to the "why" of Being, or as we forget it entirely, rather than see it as the source of persisting questions. Or, as Heidegger says, Being is "uncanny." It is "fantasia" come to pass. The implication that necessarily follows is deeply rooted in the pre-Socratic mystery, that the world *is* simply because it is. Without a discernible reason or rationale in terms of cause, purpose, or goal, we are left bewildered and things are then as Heraclitus claims, "like a child at play" (22.B52). Heidegger invokes the fragment repeatedly, often by way of Nietzsche, to the effect that Heidegger's thought, for this reason and others, "strikes at the very foundation of Western rationalism," as Michael Zimmerman says (238). Surely we could say as much of Gorgias. And, in effect, Plato did.

With the emphasis wholly on "thereness," and nothing beyond, the question concerning the meaning of Being can only point to what is groundless. It begs itself. It is the idiomatic "*Es gibt,*" the phrase meaning "there is" or, quite literally, "it gives." We are left with an experience of Being that "is not God and not a cosmic ground. Being is farther than all beings [things] and is yet nearer to man than every being, be it a rock, a beast, a work of art, a machine, be it an angel or

God. Being is nearest. Yet the near remains farthest from man" ("Letter" 210).

I have mentioned the cabalistic turn that Heidegger's language frequently takes, and we definitely see it at work here. Theodore Kisiel calls it the language of "hide and seek" (102), a phrase particularly telling, especially for this word Being. Yet to attempt to fix the meaning of Being to anything more tangible and specific would be to subvert its significance. Again, the German "*das Sein*" functions as much as an infinitive as a noun, or, at least, it surely does in Heidegger's use of it. It is "a to-be" as William Richardson says (67). It is a force at work, a process—an activity as opposed to a thing or substance—and, as such, opposed most dramatically to Plato's "what is." To Heidegger, Being is "the quiet power of the possible" linked to this pre-conceptual or pre-subjective understanding, an enabling of "letting things be" (*gelassen*) apart from impositions of logic or metaphysics upon them ("Letter" 196).

We would then do well to understand also that Plato's idea of an immortal soul is more of a reason "why" for the way things are than a hope for a life to come. It is not just a rational agent meant to fashion rational answers to the question of Being but is more importantly an answer itself that, in effect, suppresses the question entirely. In contrast, the element of the "nonhuman" moves us forthwith and very far from this sense of the soul, and from there directly into the realm of mystery. Indeed, the things most vital in life, so unlike those ideas fashioned to transcend it, are never subject to proof and seldom even to rational analysis, and here, in Heidegger, that there should be life after death is of much less significance than that there should be life at all. In this perspective Being cannot be reduced to either a specific essence or a thing. It can only be encountered. Particularly in times of anxiety, in encounters with nothingness, the "meaning" of Being is disclosed. There are certain moments, says Heidegger, when this "nearness" of Being occurs—when Being is "illumined" for us in "ecstatic projection" ("Letter" 217 *passim*)—and, whereby, we achieve authentic existence and our resolve that there is, indeed, something rather than nothing. "Authenticity," of course, is the bromide of existentialists, though it appears in similar ways in Heidegger. In general it characterizes awareness following in the wake of anxiety, in Heidegger's thought specifically preempting habituation to things that preoccupy our attention, removing us from the grasp of our everyday experience. The central importance of *angst* in Heidegger's philosophy, the manner in which he describes its effects, places this feature of his thought in alliance with the "tragedy of knowledge" that Untersteiner says is essential to understanding Gorgias. In both cases, nothing is stable, nothing to offer us a firm stance or reliable sign by which to orient ourselves either by reason or habit. Here, in the specific context that Heidegger uses the term, *angst* is one of the primary instruments through which the character and context of everyday existence is "made transparent, rendered naked to, the pressures of the ontological." Steiner refers to this event as the "heuristic *angst*"; that is, to the extent that anxiety forces us to withdraw from our everyday concerns and confront our own being, it also makes it possible for us to open ourselves to Being (*Heidegger* 94–96).

In various context Heidegger refers to this event as "ecstatic projection," and most significantly cautions that the "projection" is not that of human being but of Being itself. And if, in one sense, our role in this projection is limited only to our openness to Being, in another its impact is so keenly felt as to be embodied, to become the flesh of our flesh. This "projection of Being" is *kairos* as we can now provisionally understand the term in Heidegger, as that time when Being is "nearest" to human being. In this projection there is little room for a Cartesian ego-subject, nor for any other genteel conceptions of a soul or consciousness that would have us believe we are born with the ingredients of our own validation in place. At issue is not just a disruption of "everydayness" but a disruption of a particular kind, abruptly bringing to bear a profound sense of the unusual or uncanny as integral to our lives. In "ecstatic projection" the "nonhuman" of Being is imposed to comprise the awareness of *dasein* in a flash point of revelation that, Heidegger says, "sends man into ek-sistence of Da-sein that is his essence ("Letter" 217). In this sense *dasein* is, decidedly, human *being*, the kind of being open to Being, an entity or event of temporal constitution "primordially" grasping authentic existence, just as the Athenian warriors, in their anxiety, became aware of "Being" through the tragedy of knowledge.[30] Invariably, within the ambit of the nonhuman, what we know, or can ever know, is far less than who we are. In his interpretation of Gorgias's "Funeral Oration," Untersteiner describes the warriors' lot in terms of the peculiar "ethics" of *kairos*, where the separateness of being and knowing, of "irreducible ontological-epistemological antinomies" are brought to completion only in their coming together in the press of extraordinary circumstances, where indeed the warriors' fate was most truly manifest in their possessing from God a courage which is divine but from human nature a destiny of death (176–77). To follow explicitly the Heideggerian perspective, they were "claimed" by Being—by *physis* or by earth—as we see a pre-Socratic variant of "ecstatic projection" in the distinguishing qualities of the hero's character that are "drawn like breath" from nature. Nevertheless, in that we are concerned with Being, the process is not one of being taken or possessed by that which is outside the self, but a "self-possession" that is neither inside nor out, but a return of the self "to its element"—indeed, to the "nonhuman"—and, thereby, to the wholeness of Being that comprises Heidegger's peculiar sense of "authenticity," of *Eigentlichkeit*.

Thus, authenticity, as Calvin Schrag puts it, "designates an existential comportment in which dasein appropriates that which is distinctively its own" (*Communicative* 203). In the conventional terms of metaphysics, the subject, human being, and the object, Being, are brought near to each other. Through the facility of Heidegger's idea of authenticity, they are, if never identical, resolved wholly in terms of one another. Self-possession is ultimately self-sacrifice to Being, as conversely self-sacrifice to Being is self-possession. Here, for instance, the believer exists by virtue of his belief and can no longer remain the believer, or exist in any authentic sense at all, once he no longer embodies his belief. Quite simply, what he believes—or more significantly, what he says—is precisely what or who he is at that critical moment. Or, as Heidegger says, moving us yet closer to a more exact Gorgian sense of *kairos,* "man and Being are delivered over to each other in an *event of appropriation*" (*Identity* 36). Thus, *kairos* is when Being comes to stand in *dasein,* appropriating us to its own purpose that, through the aegis of the propitious moment, is a purpose truly our own.

This application remains problematic, and if there exists a definitive word on the Gorgian idea of Being, I do not presume to offer it here. It would nevertheless seem to be the case that the connection between Being and Gorgias is as obvious, or as obscure, as that between Heidegger and Gorgias. Ultimately, Gorgias bears witness to Being in the manner of all good Sophists, as does Heidegger himself, through his steadfast belief in the power of language. If his treatment of Being begins with a phenomenology apparent in the play of appearances that is indeed the measure of Being, then language is where it leads, where our participation in the process, our "co-belonging," is manifest. In his phenomenology all things come to be, or cease to be, in the powers or proclivities of language, in the revelations and concealments it brings about. Here the crucial text is not "On Non-Being" but the treatise, "Encomium of Helen," where *logos*—language itself—is altogether an ontological power, extolled as the great *dynastes*. The momentous "that it is" is reared in language alone, and we can come to appreciate it only through language.

As I have stated, Heidegger came to a conclusion in his later writings that is essentially no different, due in part because it was derived from the sources that shaped Gorgias's beliefs. If, in Heidegger, language is not precisely Being in full measure, it is the "House of Being." Being is "aboriginal utterance"; it is the "Saying" of language, the "truth" to which we listen rather than speak, through which we discover and come to know ourselves not as creators but inevitably as creatures of language. Of course, a curious turnabout is at work here. Heidegger explains that "to bring to language" would seem to mean expressing oneself either orally or in writing, seemingly in either case the root of all theories of rhetoric; but to Heidegger the phrase means "to secure Being in the essence of language" (*Early* 77). It is an archaic, maybe chaotic, idea of language, but it provides a useful means of appreciating the subtle congruities of Being and language in Gorgias.

Our inclination is to understand things like language in terms of the parts composing them, dividing and sorting them out for individual study, the very process defined by the word "analysis." But here, in contrast, Heidegger says the "essence of language once flashed in the light of Being" when Heraclitus revealed the "Being of beings" in language, declaring that in *logos* "it is wise to say that all things are one" (78). In that fragment we are also told to heed the language that speaks through the speaker rather than the speaker himself (22.B50), an admonition upon which this interpretation of *kairos* turns. It concerns whether language is essentially a matter of how we use it or how it uses us. For the purpose of coming to a better understanding of Sophistic rhetoric and the Gorgian *kairos*, I have thereby proceeded in a virtually sleight of hand fashion in this chapter to maintain that Being is *logos*, most apparently so in its expression as earth, and therefore so is *doxa*, and ultimately so are we ourselves. Of course, the belief that all things are one remains conditional, especially under this rubric of earth where the nonhuman holds sway and the play of identity and difference is unrestrained. In this context it is qualified indeed by the ontological difference. Nevertheless, in the phenomenon of *kairos* two major themes comprising Heidegger's sense of Being come together by virtue of the ontological difference—namely, time and language. In Heidegger's conception of the latter, it may be wise to say that all things are one because language comes to pass *as* the ontological difference. So, to serve this purpose of understanding *kairos* it is necessary in the next chapter to turn to a closer examination of this idea of *logos*—of language—especially as we may come to understand it more completely by attending to the way of Heidegger's thought.

5 Heidegger and the Gorgian Kairos

When we go to the fountain, when we go through the forest, we always go through the word "fountain," through the word "forest," even if we do not pronounce these words and do no think of anything that would be of the order of language.

—Martin Heidegger, *Poetry, Language, and Thought*

The relationship we presume to bear to language in Western culture is both plain and apparent. Language is an agency of our will; we are, or would be, masters of language, so that rhetoric in this view quite properly concerns the apt use of language to achieve the particular ends of the speaker or writer. As a result, the most durable features of rhetorical theory can be seen as extensions of the idea that the speaker is the source and ultimate arbiter of language, that he is inevitably, as Plato says, the "father of *logos.*" However, as I have suggested, there is a less favored perspective, marginalized but apparent in a tradition stretching from Heraclitus to Heidegger, that puts the emphasis on listening rather than speaking. And here listening involves something far more significant than speaking's by-product, for in the end this emphasis attests to the priority of language over the speaker or writer. Indeed, this emphasis provides a way to begin to understand the idea of language's mastery over us, a view epitomizing the belief that in a comprehensive rhetoric we not only use language but in any number of ways language uses us. In this context it is not we who speak, at least not in any authentic way, but in every essential sense it is language itself that speaks and we are, artfully or not, its means of expression.

From the outset my observations have relied on Martin Heidegger's ideas concerning the nature of language, albeit in digressive and often random ways. In this chapter, then, I intend to deal expressly with his theory of language, and to do so in a manner to establish and extend the connections between Heidegger and the Gorgian *kairos*. All of these connections will be addressed in the course of the larger discussion of Heidegger's theory of language, developed specifically in separate sections of this chapter dealing with the role of listening in Heidegger's notion of "authentic discourse," his idea of "Saying" relative to time, situation, and the *"Augenblick,"* and throughout the function of language in his design of the "quiet power of the possible."[1] More importantly, each section relies in various ways on Heidegger's idea of "earth," and on that basis delineates terrain replete with rhetorical implications and associations, in particular those pertinent to this exploration of the Gorgian *kairos*.

Heidegger, of course, never presumed to construct a rhetoric, but much the same could be said of Gorgias insofar as we ordinarily think of rhetoric as involving a *techne,* a systematic body of knowledge embracing a set of principles and procedures that we impose upon language in order to use it effectively to serve some clearly established purpose. When Plato asks Gorgias what constitutes the field of rhetoric, Gorgias responds abruptly, and incautiously, "Words" (*Gorgias* 449b). Socrates seeks some elaboration or qualification but gets neither, and Gorgias reaffirms that rhetoric is any art concerned with words (450b—c). It is indeed a response to incite Socrates' most withering analysis, and true to the power and agility of his method, he extends Gorgias's argument and then summarily reduces it to an absurdity. Medicine, gymnastics, and arithmetic are equally concerned with words, Socrates argues, and, according to Gorgias's account, they are then instances of rhetoric as well. But Gorgias, out of well-disposed conviction or simple dull wittedness, does not relent. To him rhetoric is the power to produce persuasion, and its field is words and seemingly words alone. Thus, to Plato, Gorgias's art of rhetoric is no art at all. It is to the body what philosophy is to the soul. In his very familiar incrimination, it is flattery. That, and far worse besides. It is as cosmetics impersonating gymnastics, "a mischievous, deceitful, mean, and ignoble activity, which cheats by shapes and colors, by smoothing and draping, thereby causing people to take on an alien charm to the neglect of the natural beauty produced by exercise" (465a—e).

This is Plato's view of rhetoric, Sophistic rhetoric at any rate, and in the *Phaedrus* it is every bit as severe and categorical.[2] But given this assessment, from one who would take such great care to show Socrates always at his most masterful, and from one who himself expressed such concern about the dire consequences of the powers of language, there is in Plato's refusal to appreciate or even acknowledge these powers an apparent negligence that makes his reproach of Gorgias and language not only transparent but effete and insular as well, especially as we see it now, in light of philosophy's long transformations from Plato's time to the present. The heart of the matter is apparent in his effort to impose a *techne* upon language to provide not only some measure of our mastery over it but to serve some higher goal, some notion of truth existing apart from language.

As Gorgias blithely ignores or fails to comprehend Plato's effort, he would seem to accord language some mysterious or even mystical expression. If the move is anathema to Plato, who refuses the name of art to anything irrational (*Gorgias* 465a), it is to Gorgias a matter of necessity, and sheer common sense, to place language properly among the powers that befit its status as the "great *dynastes*." Indeed, not only is Gorgias's response far from unsophisticated but there is a peculiar wisdom to it, an awareness that comes in compliance to the powers of language to which Plato seems blind in this instance. Here we are concerned not so much with what we do with language but with what language does with us, and to us. Here, also, there is an appreciation for its powers, that they exceed our means to mold and use at our whim, a fact to which Gorgias constantly bears witness in his fragments, surviving discourses, and even as we see him made to play the stooge of Socrates in Plato's dialogue. And here, finally, to apply Plato's *techne* to the extraordinary sense of language embraced by Gorgias is not just illegitimate but well-nigh sacrilegious. Though Plato no doubt would seek to persuade us otherwise, what seems at issue is nothing less than the purpose achieved by the entire pantheon of heroes, saints, and gods described in the last chapter, that entire aggregation realized in the "doxic" sense of our being, in the vitality and unsparing geneses that language yields.

In the most immediate sense what is at issue concerns the nature of rhetoric, of it status as a *techne*. Indeed, what Gorgias presents, and so poorly defends in Plato's account that we often fail to see it as such, is not so much a theory of rhetoric but a theory of *logos*. Essentially the same point has been concerning Protagoras. Edward Schiappa says that the latter's "rhetoric" is "pre-categorical" and "undifferentiated," lacking altogether an organized and disciplined attitude toward language that would otherwise shape a well contrived and complete art of rhetoric (*Protagoras* 199–200). So it is with Gorgias. What remains is a rhetoric without a *techne* or, more aptly, without a "conventional" *techne*. Of course, to say that Gorgias therefore has no rhetoric at all is to state the idea in the most derisive way, just as Plato states it. But what is apparent, especially in this context, is that Plato's attack on Sophistic rhetoric is perforce wedded to his attack on language as such. In equal measure, what Gorgias's rhetoric might lack in terms of a well developed and cogent *techne*, it lacks of necessity, due to his view concerning the power of language. In the breach dividing the two are the perspectives of who—or what—is master, whether language or the speakers of language.

Once again, the paradigm from Heraclitus provides a suitable way to broach the Gorgian view. "Heed not me," he says, "but the *logos*, the language speaking through me" (22.B50). Or, as Heidegger says, "man acts as through he were the shaper and master of language, while in fact language remains the master of man" (*Poetry* 108). He affirms Heraclitus, extending the paradigm, so that its full expression lies in the belief that it is not we who speak authentically but language itself speaking through us. If these statements seem as arcane and cryptic as those of wizards, magicians, or other miracle-mongers, perhaps it is only because they violate certain expectations we have about language. In the dominant perspective, we are primarily concerned with the signifying function of language, where it represents, or "*re*-presents," a reality apart from language that, unlike language itself, we believe to be "true reality": stable, enduring, and ever-present. A reality that is the passion of Plato's own heart, and very often our own. One standing in contrast to words fashioned of the barest wisps of sound, so frail that they endure only fleetingly, and then are quickly gone. Nor should words be considered otherwise in this perspective, for here they are mere signs and ciphers standing in place of reality and performing even this function very feebly. Thus, insofar as we perceived rhetoric in the traditional sense of embracing an art or *techne*, we are quite properly—and necessarily—concerned with language as a means to extend our will, for we are the source of *logos*, its "father," and language is a tool of control or manipulation that we explore, exploit, and then apply by virtue of a rhetorical *techne*.

But a theory of *logos* is hardly bereft of significance, nor least of all a *techne*, merely for being a theory of *logos*. This is the case with Gorgias. Here the focus shifts more and more from a system of signs to communicate information, to a consideration of language as language and "nothing else besides," as Heidegger says, where language is not only the source of all essential relationships among human beings but determines even our nature as human beings. In Paul de Man's phrase, "rhetorical mystifications" transpire in cases such as these. And we might broaden this analysis, to claim that in this perspective we are in a sense realized in discourse, where our being, as either speaker or listener, is derivative of our relationship to language.[3] Thus, it follows that rhetoric is coincidental to that relationship, to the issue of where the greater emphasis is placed, to whether we harness the power of language through rhetoric or whether we are harnessed by language in ways that serve us most advantageously, through rhetoric in this sense as well. A *techne* can be applied in either case, even though the distinction between using and being used by language may ultimately be reduced to some very essential differences between speaking and listening. Indeed, the sense of a *techne* might be all the more significant for these differences. In opposition to Plato's process of "purification" from the corrupting influences of language, where truth is utterly nonverbal, all things in Gorgias—or, most assuredly, all things that we can know and communicate to others—conform to the measure of language.

The perspective severs word from referent, enclosing us within walls of words, that "prison house of language" Gorgias is said to construct. The perspective marks the ascendancy of language over ideas, social realities, and even physical realities, insofar as these distinctions are valid and can be made at all without the prior mediation of language. Moreover, in one of the more apparent instances of "rhetorical mystification," the specter of the irrational *logos* takes shape in the vital contradiction expressing *logos* as "irrational" in the first place, *logos* that would otherwise designate the lawfulness of the cosmos and our ability to know it strictly because of its lawfulness. But in this perspective language itself is an independent medium and, given its supremacy in lieu of all other realities, achieves its effect by *apate*, where deception is not only the better part of any rhetorical enterprise but is both necessary and essentially benign. And Sophistic rhetoric would simply make a virtue of that necessity.

Here, bedrock practicality is at once rarefied mystery. Here, as well, Gorgias's message is extraordinarily compelling, as simple as it is seemingly bizarre. We communicate by *logos*, he says, but *logos* is not reality, "not substances and existing things" ("On Non-Being" par. 84). Insofar as reality exists at all, it exists for us only as we create it with our words. Or, more accurately yet, as language itself, ever the master, creates reality for us. In this context the difference between practicality and mystery is ostensibly the difference between what we believe we control and what we believe controls us, and in Gorgias's art of rhetoric that is no difference at all. Indeed, in this scheme of things, where language is supreme, even to make distinctions between language and ourselves is problematic. It is a power in which we share and participate but a power that is ultimately not our own. What is at issue is not essentially a transformation of language but a transformation of our relationship to it (Dastur, "Language" 356). In such a context the Gorgian *techne* can be likened to a reverence or veneration of a very peculiar sort, an effort to comprehend and control what is either beyond us or so much a part of us that it exceeds our complete comprehension and control in any event. It expresses the mystery just as Gabriel Marcel describes it, in the end establishing a rhetoric of an uncommon but ennobling kind, drawn from the things that make us most human, coming about in that profound alchemy of what is in us and what is before us, in that relationship that stands prior to each. It stands, as well, as a fitting confirmation of our being and the all-consuming mystery that there is Being at all. It is this mystery, then, that Heidegger addresses from the outset and what seems to lead him inevitably to his preoccupation with language.

Indeed, the real "turn" in Heidegger's thinking occurs through his persistence with this question of Being, with the result that he is led to reverse, Dastur says, his former determination of language as a phenomenon belonging to human being to it being an independent power dominating human being (363). Hence, the later Heidegger endeavors to recollect or recapitulate the early Greeks, who, he claims, "dwelt in language" (*Early* 77). If, in the end, Heidegger does not have a theory of rhetoric, he constructs a theory of *logos* that supersedes it insofar as rhetoric is conceived exclusively as an epistemological affair, a means to arrange, classify, and control language. But here, as was Gorgias's intent, *techne* is not essentially a power we wield and impose from the outside but a power implicit in language itself. It emerges from a view of language apparent in Heidegger's purpose to think what the Greeks thought "in an even more Greek manner." And, as we have seen, it is a view of language clearly manifest in his conception of *Ereignis*, whereby "man and Being are delivered over to each other in the *event of appropriation*" (*Identity* 36). And finally, it is a view that displays a reverence for language and its powers that has now all but gone out of our experience, but one that I trust leads most directly to a Heideggerian interpretation of the Gorgian *kairos*.

The Alliance of *Physis* and Language

The idea of *ereignis* provides a likely access to the critical function that language plays in Heidegger's philosophy. As he says, "in the event of appropriation [*ereignis*] there vibrates the active nature of what speaks as language" (*Identity* 37). In a process of revealing and concealing, phenomenon continues to be "that which shows itself" or "that which is manifest," but in the later Heidegger what is manifest in phenomenon follows in the wake of language. Indeed, Heidegger's phenomenology is contingent on the prominence given language, for any reality beyond language is beyond knowing in his analysis, as language itself is invariably reality's final arbiter. And Heidegger is unequivocal concerning the mandate. Language is inextricably joined with Being. "Words are not mere wrappings in which things are packed for commerce of those who write and speak," he says. "It is in words and language that things first come into being and are" (*Introduction* 11). He invokes the poet Stefan George's refrain and puts it more bluntly still: "Where the word breaks off, no thing may be" (*Way* 61).

In this perspective language is no more a possession of human being than Being itself, for language, insofar as it is "authentic," does not originate with human being. We are not its "father" but are at best it means of expression. Nor even in its aspect as rhetoric is language essentially a matter of grammar or rules of elocution, tactics or strategies we devise to persuade or communicate. On the contrary, Heidegger applies a phenomenology where, as Werner Marx says, language "behooves" the essence of our being, projecting a realm of an "'essential-for-each-other' of Being and man" (207). Here we are "always already" involved in our life world, that verbal milieu in which we live, move, and have our being. We can therefore only encounter language as we do Being, inevitably *in medias res*, already in the midst of things.

In this ascendancy of language the role of human being is diminished only to the degree that we reckon its stature as independent and autonomous, in those terms, say, fostered by Plato's doctrine of the soul and its eventual heirs, such as the Cartesian image of the modern man of reason, and beliefs more recent still that reality is essentially gathered and contained in the locus of human consciousness. And here, of course, the transition from Plato to Heidegger is significant, and truly severe once we consider Plato's idea of the soul is in part a result of his fear of skillful and compelling displays of language, such as those practiced in poetry and rhetoric. And always, it seems, Plato's concern was with the power of language as such, rather than with those who so skillfully applied it. In contrast to his doctrine of the soul and, consequently, in contrast to our role as masters of language, our being in Heidegger's interpretation is grounded ontologically in language and, as *dasein*, is an expression of Being manifest as language itself. Therefore, language to Heidegger is not only what first brings things into being but is the source of our being as well, for "it is language that first brings man about, brings him into existence" (*Poetry* 192). The statement not only provides a further insight into his view of language but, as well, an insight into the dignity of our being that is closely connected, as I will argue presently, to the irrational *logos* of which Gorgias speaks: a *logos* quite fittingly as irrational and mysterious as our own being. There are two conceptions of language that are of importance here. The first is Heidegger's and the second is the logocentric attitude discussed in previous chapters. To Heidegger and many of his critics and disciples, Aristotle offers the definitive statement of the latter in his treatise, "On Interpretation":

> Spoken words are signs of the soul's experiences, and written words are signs of spoken words. Just as all men have not the same writing, so all men have not the same speech sounds; but the soul's experiences, which they immediately signify, are the same for all, as also are those things of which our experiences are the images.[4]

What is evident in Aristotle's definition is that language is an activity of expression, a propagation of words having their source in the "soul's experiences" and moving outward from there, to be expressed in the external world. In this view language is an event that originates and transpires exclusively within the compass of human being, where it is shaped and controlled through a culling of signs to signify things in the world and ideas in the mind. Heidegger says that this conception of language is manifest not only as a "phenomenon that occurs in man" but is experienced directly "in the sense of something that is present," so that we thereby encounter language solely as an act of "speaking" through the "activation of organs of speech, mouth, lips, tongue" (*Way* 96). Moreover, Aristotle's is clearly the commonly accepted view of language that Heidegger contends in its various forms has "remained basic and predominant through all the centuries of Western-European thinking" (115–16).

More specifically, as language is here construed as a process of externalizing something internal, it is also a means of representation, a functioning of brains and vocal chords to transmit a message from sender to receiver, just as the process is so often diagrammed in models in writing and speech communication textbooks. It is regarded as an expression of experience that follows experience—"a copy, as it were, of an experienced essent [thing]" (*Introduction* 87). As I have suggested, this conception of language, as a thing or skill we possess, fosters the understanding of *kairos* as a rhetorical tactic the orator deploys in order to shape, control, or otherwise respond to a situation through the selection and application of words appropriate to the occasion. In contrast to the vitality that Gorgias attaches to language itself, here the emphasis on our mastery of language accounts in large part for this conventional perspective on *kairos*.

In this same vein, Heidegger claims that Aristotle's conception is apparent as language is used in "average everydayness," the normal state of *dasein* as it is absorbed in the "public realm." In *Being and Time* Heidegger refers to language used in this way as *Gerede*, the "idle talk" of "they" (*das Man*). The latter designates *dasein* in its "fallen" state of estrangement, estranged from Being but more essentially from language itself insofar as language is "authentic" in its connection with Being. "Idle talk," in turn, is discourse that has either lost its relationship with the entity talked about or talk that has never achieved such a relationship in the first place (*Being* 212). But, more crucially, in such discourse the word's "primordial belongingness" to Being is violated, Heidegger says, so that "words are thrown around on the cheap" and "worn out" (*Thinking* 127), as language falls into the service of "expediting communication" along routes cleared by "objectification." A "leveling" of language occurs, characterized by "the uniform accessibility of everything to everyone." In this way, then, language is placed "under the dictatorship of the public realm which decides in advance what is intelligible and what must be rejected as unintelligible" ("Letter" 197). Thus, as *dasein* maintains itself in idle talk, language itself is wasted, often muddled away, even as it remains utterly in our possession, precisely *because* it remains in our possession, and as a result *dasein* "floats unattached" in its willful estrangement. In its refusal to listen to language, to be claimed by language in that way, *dasein* is alienated from language, and therefore from Being itself. Of course, none of this is to say that "idle talk" is in any sense false or illicit. For the most part it is necessary, being what we expect language to be, as it is often used in politics, other advertising campaigns, the classroom, the discourse of our daily lives, in such cases accounting for the label of "mere rhetoric," even "mere opinion." Yet it is far removed from the idea of *doxa* developed in the last chapter and most clearly from the way Heidegger himself developed it in his *Introduction to Metaphysics*.[5] As it is far removed from the rhetorical implications of the Gorgian *kairos*.

In effect, Heidegger's view of "authentic" language inverts the ordinary, linguistic tendency to make the operation of speaking primary. The relationship between word and referent, whether the latter is object or concept, is to Heidegger not an external relation between two independent entities as Aristotle's conception of language would indicate. The word itself is the sustaining relationship, for it calls things to the fore in Being, thereby occupying the place that holds everything in Being. "Relation is thought of here always in terms of appropriation," Heidegger says, and is "no longer conceived of in the form of pure reference" (*Way* 135). Thus, rather than language being a reflection of our nature, conceived in isolation from Being, language is pervaded by Being insofar as it is "authentic," and therefore our relationship to language is such that we "dwell poetically" in the world, as Heidegger says. As was the case in the Gorgian world view, we cannot escape or suspend the relationship but are enclosed linguistically in a world that is literally Being itself; and we are then unable to transcend it, this "prison house of language," to some non-verbal dimension. Here, the primacy of language is mandated, made explicit in the act of appropriation; and Being is manifest as "the simple nearness of an unobstructive governance that occurs essentially as language itself," so that "it is proper to think the essence of language from its correspondence to Being" ("Letter" 212–13).

We move, then, from an epistemological conception of language, Aristotle's, to Heidegger's decidedly ontological view, where language has its correspondence to Being, not entities or things—not to Plato's or Aristotle's renderings of "what is"—but to the simple "thereness" of our existence in the world. In this disposition Heidegger proposes a "fundamental ontology" from which, he says, "all other ontologies take their rise." This idea, first set down in *Being and Time*—but which Heidegger never abandoned and would continue to pursue relentlessly—led him ultimately to his contentions involving the ascendancy of language, to its alliance with Being. Though fundamental ontology is rooted in the "ontic priority" of *dasein* (34), the idea of immanence vitalizes this approach altogether. In Calvin Schrag's pointed and oft-quoted statement of the conception, "the ontological is not a separate realm 'off by itself'; rather it designates the structure of the concrete with its conceptual clarification." Heidegger thus seeks to "disclose the ontological foundation which precedes any split between subject and object, and this foundation he finds in the phenomenon of the world itself" (*Existence* 32–33). Again, we are "always already" involved in the world, in the "simple thereness of our being." The world, then, is not understood as a pre-given presence, nor, least of all, is it equated with either the theological centered conception of creation or the secular representations of Being as substance, where we would understand the "cosmos" as simply "the whole of entities present" (*Poetry* 201). The world is indeed "phenomena," more specifically phenomena disclosed by language or, as Heidegger says, a "horizon illuminated" by language itself.

Especially in this context we can appreciate Heidegger's emphasis on fundamental ontology as it is associated with his purpose to return to that "elemental" sense of Being, less a philosophical concept and more the incipient sense of Being he maintained was embraced by the pre-Socratics. And, in this sense, Being is "earth." It is nature conceived primordially, free not just from the impediments of science and metaphysics but most significantly retaining for Heidegger the fluency of process, the "unfolding and emerging" power of nature, an "empowering power" that resides wholly within itself, called *physis* by the early Greeks (*Poetry* 42). Here, Being is not restrained by stipulations of "substance ontology," of things or entities derivative of the metaphysics of presence; nor, as earth or *physis,* can it easily become a mere label of metaphysics and in that way to be "forgotten." Moreover, as Heidegger's purpose to address the question of Being moves him back to the beginnings of Greek thought, this return, more than any other single factor, inevitably gives rise to the issue of language. Indeed, in Heidegger's interpretation, *physis* is explicitly the alliance of language and Being, a wholeness apparent to the pre-Socratics simply as "nature." As earth.

The issues involved here are no doubt uncommon, but even as Heidegger seems to tarry endlessly about the edges of inscrutability, never is he so practical and pointed as on this matter of *physis*. If, contrary to Heidegger, we begin with the proposition that language is our possession, in essence a system of signs we establish and use to represent reality, then there is no surer index of our separation from nature than language itself. If, in this view, language is primarily a means of "speaking out," expressing ourselves and exerting who we are, it is in that sense also a wedge that works to separate us from nature and define the essential aspects of our humanity in terms of that separation. Such a view may be useful and even valid in any number of respects, but as a consequence civilization becomes a series of barriers between ourselves and nature, as language is then reduced to an exclusively cultural occurrence, a means of not only distancing ourselves from nature but a tool far more basic, pliable, and effective than any intellectual or technological scheme to be turned against nature in order to bring it to account, to master it. However, if nature itself is the measure of all that matters, the "overpowering power" manifest in phenomena, then we are part of nature precisely because of language, and the latter achieves its elevated status in Heidegger's thought through its alliance with the power of nature as *physis*—precisely as *earth* in the later Heidegger. But there is in this transformation from *physis* to earth the sense that Being undergoes a most extreme mystification. Being as earth is necessarily practical, commonplace and comfortable as the place of our origin and dwelling, yet in the end it opens to receive us in our being-towards-death. So again, the paradox of the near and dear being what is most alien, the blatant "thereness" of our being. That generative force coming prior to "reality," prior even to humankind, prior still to the things related—a force, in effect, that creates the things related, to include both reality and humankind.

Heidegger's purpose in stressing this alliance of language and Being is, among other things, to redress the effects of an analytical mentality prevalent since Plato and Aristotle that fixes the beginning of all knowledge in the separation of subject from object, of our being from nature, if you will. He seeks to re-discover, to retrieve, a wholeness more aboriginal, preceding any such separation. The gist of the matter was introduced previously, but in this oneness of nature and culture Heidegger's idea of *physis* is definitive and, withal, manifest in relative simplicity and clarity. Being remains a mystery, the source of incessant questioning, but it is a linguistic phenomenon insofar as we are aware of it, insofar as we can know it at all. Under the rubric of *physis*, then, language is Being in its existential aspect as we encounter it in this world, as vital and immediate as the air we breathe, and as much a part of the process of nature as the good earth itself. In short, language is how, and where, the immanence of Being is manifest, where the "structure of the concrete" is most apparent and pronounced. Here our being is apparent in fundamental and very crucial acts of language, as we are gathered to the verbal environment of our literatures, mythologies, rituals, and ceremonies constituting the dynamic processes of *physis*, and to rhetoric worthy of the eminence of the "great *dynastes*." As Being, in all its uncanniness, is for Heidegger captured in the very utility of the commonplace *es gibt*—"there is" or, most literally, "it gives"—the apt rejoinder under the aegis of *physis* is *das Wort gibt*, "the word gives."[6]

Thus, the nature of *dasein* being "ontologically" situated in language is apparent. Here, in what is perhaps the most important likeness that can be drawn between Heidegger's idea of *dasein* and the pre-Socratic idea of the *psyche*, language is encountered in either case as *physis* itself. And it is always an encounter, for in this perspective we are first and foremost a recipient, a listener to language rather than its source, and language can thus be understood initially as a force impinging from the outside, a power that inevitably claims us insofar as we are open to it. And the implications run deeper still. As we achieve our participation in Being only through language, it is meet to say again, here in this Heideggerian context, that we are made of words. His analysis leads not only to the determination that our nature is nearer the *psyche* than the soul but that we always owe our being to the greater power of language, for we are essentially an extension of it, and, independent of that power, we are little else besides. Heidegger has claimed language is our master, our source of being, and here it is who we are. "Human being is *logos*," he says (*Introduction* 171). Or, perhaps even more aptly but in much the same vein, we are the "quintessence of dust with language," as Timothy Crusius say (23).

By virtue of this perspective we obtain that ascendancy of which I spoke earlier, though it is undeniably a very ancient and unusual means of deliverance. As Being comes to pass in language, the latter not only preempts the splendid idea of the soul espoused by Plato but anchors us instead in the peculiar reality that language does not merely express but, before all else, manifestly creates and establishes. The effect is that our redemption, if such is the proper word in this context, is never a deliverance from the quintessence of dust but always a reiteration of it, a fact that, insofar as we realize this fate rather than shun it, culminates in the tragic sense of life. As we neither own nor in the strictest sense even control language, being imperatively the quintessence of dust in any event, it truly may be that among all the creatures of the earth "there is nothing more wretchedly lamentable that human being," just as our condition is described in the *Iliad* (17.445–47; Krell 98). But, in another sense, as we ourselves are expressions of the power of language, of which the *Iliad* is a most superb example, there is a recompense, even a certain apotheosis to be had in this fate. Indeed, there is glory to be had in the *doxa* that language bestows. If our lives by virtue of language are moored eternally to earth, in language alone are they made tangible and immediate, are they ever fully lived and realized.

In the most essential sense, then, we owe our being, however lamentable or glorious, to the world that language has made. Here again, we do not come to the things of the world from the outside or by postulate. In the Heideggerian perspective we are "always already" involved in the world through language, being ontologically based in it, and therefore come to things always through language. Thus to state again a theme I have been struggling to develop, there is no "outside" to language, no reality beyond it, and to scrupulously follow the implications of this perspective, language is thereby the means of human presence in the world. Indeed, it is nothing less than human presence in the world. Or, more cogently, our very being is contingent on language, and we can no more move beyond language than beyond the reach of our own shadow, as George Steiner says. Affirming Merleau-Ponty's belief in this matter, he tells us the sole, truly external view of language is achieved only by "a total leap out of language, which is death" (*Babel* 111). Thus, language is Being in the only way it can be realized, in a narrative, the story that unfolds in the here and now between birth and death.

The Borderland: A Meditation on Cemeteries

Let me tell you, then, a story about language, and thus a story of life and death as Steiner would have it. Near where we live in rural Michigan, less than a quarter mile down the road, there is a small cemetery that beaks the regularity of corn fields and pasture land that stretches relentlessly all around. So, in the midst of all this, the cemetery ironically provides a very lively sense of human presence, and a sense of history and mythology, too. Some of the headstones mark the graves of early settlers, those who gave their names to roads, townships, streams and the like, names of nearby places, such as Dennison, Cone, and York. And there are veterans of the Civil War buried there, two of them of the First Michigan Cavalry Regiment and therefore men who quite possibly fought with Custer at Gettysburg. Stories could be told about them, and many have, if they indeed fought at Gettysburg.

My daughter, Cassie, has also taken a keen interest in the place, particularly in one set of headstones toward the back of the cemetery, near a tangled overgrowth of bushes and wait-a-minute vines running along the bank of a creek. The stones mark the graves of three children of a single family who died within weeks of one another in the early autumn of 1918. All of them are girls, one of them at her death about Cassie's age. And Cassie writes this stuff down. She writes everything down, so that much of the vital information fills her notebooks, names of the dead, dates of birth and death, epithets. As well she might, being what she has discovered about life at this place is such bad news, that we are constituted by death, and death comes for us all, even for little girls no older than herself it sometimes comes. But written over and over again, on page after page, taking up well over half of one of her notebooks are the names of the three girls.

Over and over again, that and little else besides, a morbid invocation or something equally curious from one so young. Occasionally there are a few false starts, maybe a sentence or two that seem to deal with the circumstances of the girls' lives and deaths, but then just disconnected phrases that lead nowhere. A profound case of writer's block to be sure, as these children lie stillborn on the page. Who are these kids? she asks. The question comes repeatedly. And it is easily enough answered by telling her stories—myths—that at death only the body dies, while the soul lives on in perpetual bliss, as it definitely does for these three young girls, now smiling down upon us. Or it is answered more simply still by telling her harsher myths, that who these kids are, or were, can be found only in their graves, in the bones, tufts of hair, and whatever other telltale remains we might find there. That they are dead and gone, pure and simple. Both are obvious and surely familiar answers, but whether we are wholly spirit or wholly dust they lead us in circles nevertheless, telling us essentially the same thing: that language limits, bears the mark of our mortality, that in either case it is interred with our bones. That in either case our being is beyond language, answers that Cassie finds unsatisfactory. Her notebooks tell her as much.

So it seems she is at least dimly aware that the person one is, the essence of one's being, is not exclusively a body or soul, neither an object, spirit, nor "being" of any conventional sort. It is instead an event, a happening, a story that unfolds in the course of a journey that is a life and death matter, as all Being—especially in its aspect as earth—is a narrative, whether individual or collective. Or, indeed, Being is language, and Cassie is too attuned to words to know otherwise. And now for the first time that she has ever been aware, though surely far from the last, words have failed her and, in a sense, have therefore failed the three girls whose names appear on the headstones. Insofar as these children are, or ever were, they are made of words, and their tragedy in those terrible autumn months of 1918 is untold and thereby no tragedy at all. Nothing at all. Maybe in time the language will be given to Cassie to bring them, and others like them, back to life, to join language and the quintessence of dust, if only to create their stories as they intersect with her own.

There is, of course, something about cemeteries that inspires utterance in any event, if less often gainful eloquence or words that work. If, as I say, they break the regularity of things, they not only disrupt routines of "everydayness" but do so as tokens of our mortality, instilling an awareness of life that comes only in an awareness of death, a mindfulness that Heidegger would say is the measure of authenticity. There is, therefore, something more, something quite curious in the mix. There is an alliance of language and death at work in this need for utterance as even Cassie was aware, as most assuredly was Plato himself. It prompted him to see both language and death as virtually inseparable, as accomplices that must be spurned for the sake of the soul and immortality. Some might see the alliance as unwholesome in any event, little more than an effort to distract us with soothing words from the burden of our mortality, which could in itself be the principal purpose of encomiums, eulogies, and funeral orations. However, what

I have in mind is not a particular type of discourse but the power of language as such—as, in fact, is sometimes seen in funeral orations—where the need for utterance is not based on an effort to elude death or our dread concerning it but to fully realize them both. Only in this way are we "released" into our own nature, as Heidegger says, allowing language to secure for us our being, "to grant an abode for mortals" (*Way* 129; *Poetry* 192), unless we should lose our being by being unaware, by letting it slip too far beneath the earth or rise too far above it. The Borderland.

Here, to be aware of language's power, of its union with Being, is to be stirred by the mystery of life and death, expressly by the tragedy that resides in their interface. It is to understand as well that this alliance of language and death is also one of language and life, of language bearing the mark not just of our mortality but also of our deliverance. Heidegger says we first become aware of the fact, quite ironically, when language fails us, when it no longer flows routinely and we are literally dumbstruck, brought to an impasse by language itself.[7] We seem not just unable to know what to say but to have utterly lost the means of saying it, as when, for instance, the plaintive question of "who are these kids?" is answered only in a hapless reiteration of names on a page. In such cases we begin to be aware that we do not own language, that we are not its masters; and then indeed, by and by, we come to appreciate that its power as Being is realized only in allowing ourselves to be claimed by language, by a listening that is also a belonging. In this way our nature is determined by language, as we are, then, less the knowing subject who shapes and uses language and more the subject shaped and used by language, immersed in it, released into its "way-making movement," as Heidegger says (*Way* 131). We are given over to that strange symmetry of life and death that is the "uncanniness" of Being. Such a process inspires the tragic sense of life, where, we might say, the recompense for the fate of Yorick awaiting us all is *Hamlet*, the likes of the play itself, enjoining us to appreciate that fate, even to be ennobled by that fate in the play's very lamentation of it. Whatever triumph, or glory, is to be achieved in the quintessence of dust with language, it abides here, in this fate.

These matters of life and death are most obviously, and tenaciously, joined to Being when Being is taken as *physis*, as earth itself. What is remarkable in Heidegger's analysis, and most worthwhile in this context, is joining *physis* to language in ways no less tenacious. A final example might help to clarify the idea, this one collected from a source I have used before, Garry Wills' *Lincoln at Gettysburg*, applied here to serve a somewhat similar purpose. Its focus is on yet another cemetery and need for utterance, on a near-flawless combination of time and place, and principally on a speaker borne of words that work as well as any of Shakespeare's.

The battle at Gettysburg took place in early July, 1863, and was as vicious as soldiers have ever fought. The Union was the clear winner, but various elements of the battle's aftermath offered nothing so definite, especially as we take into account those things included in what General Meade called the "debris of the battlefield." These consisted of shattered carcasses of horses and mules, wrecked caissons and canons, littered canteens, rifles, and various personal effects, and over eight thousand human bodies left to swell and decompose in the July heat. Disposal crews hastily buried as many bodies as possible, but they made a sorry job of it. The dead were dispatched into temporary graves so shallow as to be a mere blanket, Wills says, so that the rise and swell of bodies was perceptible even as they lay buried, the ground in places so yielding that it gave way completely under foot. In fact, as long as three weeks after the battle, "in many instances arms and legs and sometimes heads protruded from the ground and in several places hogs had actually rooted out the bodies and devoured them" (20–21). Over four months later, with the clearing and reburial of the soldiers in permanent graves not yet completed, and the odor of the battle's debris still in the air, Lincoln arrived to deliver his address at the dedication of the National Cemetery of Gettysburg. So the question persists, most conspicuously in this context, who are these people? And again the answer is simply given. They are the debris of the battlefield. The quintessence of dust. Men of flesh and bone and what comes of flesh and bone in violent contact with rifle fire, splintering metal, and shell shot; they are what becomes of the remains left in the heat for days and even weeks on the battlefield before being buried, and then sometimes rooted from their graves by hogs.

But in the Gettysburg Address they are redeemed, seemingly made to live again, as Lincoln sanctifies them, resurrects them in words to a life far more glorious. If mere words are a pittance to die for, given this time and place and Lincoln's words, these are not mere words. Nor is Lincoln's purpose at all unusual, but proper and predictable, for the Address is a funeral oration where by convention the dead are praised for their sacrifice. Moreover, as Wills says, this was the time of the Greek Revival in America, and Lincoln was drawing consciously or not on the traditions of ancient Greece, especially in this instance those of the funeral oratory of Pericles and Gorgias. He was therefore indebted to the Greeks not only for the basic style and content of his address but also for its élan, evident in that belief that words work in special ways to commemorate and memorialize, as they are drawn from a living tradition, a *doxa*, that they personify, foster, and enhance. So it is with the soldiers. They did not die in vain, Lincoln says, but from them we take "increased devotion to the cause for which they gave the last full measure of devotion." Through their sacrifice the nation has experienced "a new birth of freedom," and we thereby remain "dedicated to the proposition that all men are created equal." Delivered from the ignominy of the debris of the battlefield, these men are glorified in Lincoln's rhetoric, made cultural heroes in the classical sense, in ways no different from Achilles, Sarpedon, or Hector. They are portrayed as vital to a *volk* and its traditions, in this particular case establishing the distinctive and dominate features of a nation, of a government of, by, and for the people. Thus, they are appropriated by the words of the Address, given their place in a story that continues to unfold. By its authority they become part of a nation's *doxa*, its sacred narrative comprised indeed of those "mystic chords of memory" that Lincoln invoked in his First Inaugural, a belonging embracing us all through their sacrifice.

It therefore follows, as Wills says, that Lincoln's speech "hovers above the carnage," that in his rhetoric the battle is "lifted to a level of abstraction that purges it of its grosser matter." The "physical residue of the battle"—the debris—"is volatilized," he says, as "the many bloody and ignoble aspects of the battle" are "etherealized in the crucible of language" (37). Indeed, precisely because of these aspects of the battle, Gettysburg becomes hallowed ground, our native soil. Thus, from such words as "etherealized" and "abstraction" we must not assume that the carnage is the only certain reality, that it is *in deed* while Lincoln's Address is other than reality, something merely *in word* to shape façades to obscure or shroud reality. Though the distinction between words and reality is routine, it must be reconsidered to appreciate the paradox of language as *physis* and, through the joining of these, the ideal of *apate*. George Steiner, for instance, claims that language is "centrally fictive" and, as a result, "reality" is inevitably its enemy. Following Nietzsche's lead, he claims that language allows us to vanquish the more brutish realities of existence, those making life contradictory and cruel, even as language is coupled with the "organic trap" of death itself. In response to these afflictions, he says language sustains the "Life-Lie," "secreting" by its nature "the mythologies of hope, of fantasy, of self-deception" that allow us to persevere, even flourish, and by means of which we speak ourselves free of "total organic constraint" (*Babel* 226ff). We are, then, deceived by language, but in the peculiar relationship of language and reality that Steiner describes, reality itself is as much a deception as any "unreality" language presents in its place, as reality is what it is by virtue of language. Or, to put the matter more apparently within a Heideggerian perspective, we would say that *physis* joins the organic explicitly to language, drawing each into the wholeness that is *physis*. Through this notion of earth, nature reaches beyond material fact, to most expressly encompass language itself, which, as Being, is neither true nor false but simply *is* and, by virtue of which, we *are*. Here, then, as expressions of earth—of Being, *physis*, or nature—are the mythologies, sacred narratives, and deceptions that are simultaneously of nature and culture, the language constituting the *doxa* that links Yorick's fate with Hamlet's words, the carnage of Gettysburg, through Lincoln's words, with its glory. In a sense, to be free of organic constraint is another way of saying that we are made of words.

Nevertheless, the ontological difference prevails, though we retain a relationship to language as we do to Being, as we do in this context to nature itself. As language is the "overpowering power" in Heidegger's interpretation, it is most clearly identified with *physis*; but utterly as a process, a force at the heart of nature rather than the substance or mere epitome of things composing it. Heidegger is explicit. The issue here is not in flesh becoming earth but in language that is the earth. Language binds us to it, being of the nature of "the sea, earth, and animal" (*Introduction* 156), so much a part of *physis* that we hear its sound "rising like the earth" (*Way* 101). He thereby denies all metaphysical foundations for language, in particular denying language's basis in reason, in that cosmic lawfulness. Again, in its connection with *physis* language is the "overpowering power." As with Gorgias, who himself likens language to physical force, we witness the annihilation of *logos* as anything other than language; and consequently we attend to the order, and disorder, that it reveals. In this sense, then, language is "free," an autonomous external power bound neither to reason nor external reality, and only to human being to the extent that it is through language that Being is at all manifest. In this sense we are also creatures of language, "authentic" only as we are "released into our nature" by language. Here, in order to genuinely experience an "encounter" with language, to be "authentic," we must be open to the powers of nature, to what language has to say. It is a circumstance, says Heidegger, where we must, paradoxically, let language speak from within itself, to let it have its say. "What it says wells up from the formerly spoken and so far still unspoken Saying which pervades the design of language." This is the nature of *doxa*, and in our openness we listen, "*letting something be said to us*" (*Way* 124, his italics). In fact, on the basis of this perspective, Lincoln's accomplishment in the Gettysburg Address is this openness, a listening that is also a belonging to those "mystic chords" that precedes utterance and thereby comprises outright the sheerest acceptance and acknowledgment of one's place in language. All in all, a very fitting instance of Heidegger's belief that authentic speech, the "natural" aspect of language as "Saying," is "sounded" through the mouths of mortals (*Way* 129).

Given these strange juxtapositions of nature and culture, *physis* and language, seeming to reduce all things to the same, who is to say the dead of Gettysburg, as the mere debris of the battlefield, have a greater claim to reality than the words expressed through Lincoln would otherwise make of them? If death is the "total leap" out of language, then language in its alliance with *physis* is what gives them life and sustains it. As Kent Gramm expresses the belief, applying it specifically to Gettysburg, the biblical parallels to the risen Christ are explicit (Luke 24.5). The dead of Gettysburg are "giants in the earth," he says. "Buttons, rotten scraps of cloth, perhaps some bones remain—but where are *they*?" And then answers by simply rephrasing the question, "Why seek we the living among the dead?"[8]

To whatever extreme we might extend the matter, the point remains that words proclaiming what and *that* they are have far more endurance than bodies moldering in shallow graves. Hence, we gather from texts as ancient as the *Iliad* the realization that a heighten sense of carnage often leads to a heightened sense of glory. What we also gather is that language is corporeal, just as Plato claimed it to be, and joining it to life is at the same time to join it to death. And that realization is as near at hand as the cemetery. As it has since the time of Pericles, the cemetery has served quite appropriately as the essential *topos* for language in these matters of life and death. Wills tells us, for instance, that in the nineteenth century the cemetery was the supreme "locus of liminality," for it marked "a borderland between life and death, time and eternity, past and future" (74). With all my talk of cemeteries I hardly dispute it, but see the influence of that particular place heightened to an overall awareness of our mortality that at times is sufficiently compelling to join language to *physis,* to the earth, as language assumes its authority as the "overpowering power." There are critical times, Heidegger says, of "encounters" with language that "transmute" our relation to it, where by virtue of our mortality we bear witness to Being that, we must understand, would not "be" but for our mortality. In its nature as *physis,* language thereby draws us into its concern; we are "appropriated" by language, as it "gathers mortals into the appropriateness of their nature and holds them there" (*Way* 128–29). Here, as mortals, we are "needed and used to speak language" (134). The process provides crucial insights into the nature of *kairos* as I am interpreting it, as Heidegger stands Plato on his head, claiming we are appropriated *by* language rather than we appropriating it. But, for now, also crucial to this perspective is that the process marks that locus, that place or situation, where "the essential relation between language and death flashes up before us" (107). As it would, say, in a cemetery, whether the realization is evoked by a funeral oration or not, whether within the environs of a cemetery or not.

As I have suggested, one such "encounter" with language is apparent in Lincoln's address, where language is authentic to the degree it is drawn at the nexus, that "borderland" between life and death, time and eternity, mortality and immortality that in this context is no clear borderland at all. Here, instead, it is a revealing of Being, in this specific instance of the prevailing *doxa* of a nation "conceived in liberty and dedicated to a proposition" as it is quickened and then extended, made by language every bit as tangible, and real, as any other facet or feature of reality that we might imagine. Indeed, yet another hero of Gettysburg, Joshua Lawrence Chamberlain, claimed that the glory achieved by the soldiers under his command lingers palpably on the battlefield, specifically on a hill at Gettysburg called Little Round Top. The dead have consecrated the ground, he says, but it is a "deathless field" nonetheless, as they remain "transfigured" and "of the earth they glorified" through their sacrifice. It is native soil, where the "mighty presence" of "great things suffered and done" persists to extend its embrace even to modern day visitors to this place (201–02). Without a trace of irony, then, Chamberlain speaks of the dead as the "radiant fellowship of the fallen," and as late as 1913, on a return trip to Little Round Top, he claimed to bear witness to the heroic souls whose glory came, most critically, as a result of "proud young valor that rose above the mortal, and then was mortal after all" (37).

Chamberlain was not a professional soldier but a professor of rhetoric who, undoubtedly, taught and practiced a rhetoric of a relatively vintage sort. Thus, we might surmise he would believe rhetoric emerges from *mythos* and inevitably responds to it, a transaction amid a set of cultural beliefs, a *doxa* comprising who we are and can become at our best, our most "glorious." But it is a warrior's code he invokes, as ancient as Homer's, where things are resolved in the uncanny interface of life and death, where always the stark realization that we are alive is compelled through its chilling kinship with death. In this context we witness *doxa* in its existential aspect, of critical events that reflect the cultural *mythos* and rebound to mold it, a sense of *doxa* not only accounting for the glory, the "proud valor," of which Chamberlain speaks, but *doxa* in the sense of *xa*, racial memory, of cultural memory as such. Here we are not concerned with the mere recollection of the past but with the past as presence, of events instilling reverence and awe from which we draw life in a process sufficiently intense to transfigure particular times and places, even to render the killing ground of Little Round Top a "deathless field." Such transfigurations come about only in terms of the peculiar connections of nature and culture, of *doxa* in that sense. Language thereby achieves its purpose in its embodiment of the tragic circumstances of our being, rising above the mortal only as it is earthbound, most profoundly mortal after all. It is a peculiar manner of redemption, Christian in at least one sense, that life is gained in sacrificing it. In this, glory is a matter of words alone, yet words entail its activation in flesh and blood, of men quite expressly appropriated by words.

What Chamberlain says might seem terribly dated to us now, and no doubt can be dismissed as vainglorious, even as an exaltation of aggression and violence. But his words are as Lincoln's in capturing the solidarity of *physis* and language, and, as Lincoln, he seemed to believe in them implicitly. They mark the uneasy transition from a conventional understanding of *doxa* to *doxa* as glory, of language's embodiment in very tangible realities, of it "living out" human beings in the stories it creates. It is not my purpose to espouse a martial *ethos* but explicate an idea of *kairos* rooted in Gorgias's "Funeral Oration" that is expressive of a situation where the drawing near of death provides a likely awareness of matters meant to be captured in a funeral oration. As those of Lincoln and Chamberlain, Gorgias's oration deals in the same antitheses of life and death, mortality and immortality, compelling for us all but in these cases most pronounced for soldiers due to the salience of their sacrifice and its place as the paragon of the *doxa* of a people. Here, opposites meet and merge, as mortality and immortality, word and deed, even human and nonhuman coalesce in some in-between or borderland place. As a cemetery, to be sure, but just as aptly, and quite simply, a "situation." And here opposites are matters of essence, not of style alone, as Gorgias tells us of the effect of *kairos*, that of glorifying men whose courage was God-given and divine but who possessed, as men, the destiny of death, in that "excellence which is divine and a mortality which is human" (82.B6; Sprague 48). Or, more directly in the translation used by Wills, "men who showed godlike valor but died the death of men" (275). The circle is then complete, as in this borderland the ultimate antithesis is resolved into the ultimate synthesis.

We might invoke the idea of *apate* to account for these curious transformations from gruesome death to glory, from human to divine and then back to human again, to claim that death and mortality are banished and their opposites invested by means of deception outright, absent of any and all mystifying nuances. However, as I have indicated, we are not dealing with opposites pure and simple, but always with various and often contrary manifestations of what is the same. As such, grim realities are never transcended, denied, or otherwise removed but are at times transfigured by language. In this account deception is rooted in phenomenology, a matter of *doxa* in the most ancient sense, in that play of appearances where the latter are never a facade but of the essence of Being, as both Heidegger and Gorgias tell us. The process is strictly non-Platonic, where deception is not only an issue of how Being reveals itself through language but where, through language's alliance with *physis*, there is no reality beyond language at all to which it can correspond, faithfully or otherwise. Language is indeed language and nothing else besides. We are, then, subject to what it makes manifest, and deception comes about not at our discretion alone, if at all, but is essentially of the nature of language itself. Here, indeed, language may deliver us from "organic constraint" but never from *physis,* as it moors us in the immanence of Being, in a mortality completing the arc, allowing us to partake in the boundlessness of Being only by virtue of our mortality.

So it is here, in his thesis that language makes use of mortals, that Heidegger claims our most essential relationship to language is not one of grasping or pragmatic use, nor one of speaking at all, but one of audition, of hearing. Again, "we are trying to listen to the voice of Being," he says, to "let language have its say" (*Way* 124). To the degree we succeed, language is "delivered over to its own freedom" (131), as we are, by the same process, delivered over to it in an act of appropriation. Thus, we listen, not to hear what we have already heard but to hear and respond to the voice of Being, allowing passage not just to far more gracious realities but to realities as compelling as those seemingly left behind. Such manifestations of language provide a very clear indication that *logos* is free, a realization of the Sophistic *logos* embodying what I have called the "Gorgian Affirmation." Here, language is not only as near and persistent as our own shadow but, as I have stated, foremost among these manifestations of language is humankind itself.

That it is so, that language in effect creates humankind to obtain its expression, is both our dolor and deliverance, for language transfigures our mortality only as it affirms it. Again, the traditional sense of a prior, interior subjectivity is displaced or banished altogether. As even the idea of inside and outside, internal and external, is displaced, so is the traditional significance of the individual. Rather, to see these matters in terms of the unfolding process of *physis* is to see the individual as *dasein,* as an event rather than a thing, a being not just in time but *as* time in the course of its disclosure. As a narrative. A story. Here we might come to appreciate Heidegger's idea of *dasein* as the "irruption" of language, a notion that can be likened to the image of the gathering mass of a star, drawing to itself all that surrounds it, where it is its surroundings precisely, even as their "difference" is always present and apparent. And in the difference the sense of the uncanny prevails, that profound awareness of "Being-there" coming about in the relationship of Being and *dasein,* in precisely the "there" of Being and, accordingly, not in the particularity of separate things related. Into this "breach" or borderland, the "in-cident" or the "falling-between," Heidegger claims the powers of Being in their singularity with *physis* as language come forth and make their appearance (*Introduction 64*). Here the "irruption" of *dasein* is a facet of "de-cision," of Being manifest as language coming to us as "Saying." Language thereby "takes its origin," as Richardson says, "along with the irruption of There-Being, for in this irruption language is simply Being itself transformed into word" (293).

Thus, not only does Being achieve, through *dasein*, its means of expression in the world, but in this affinity—the "nighness" of *dasein*, language, and situation—*kairos* is the point of irruption eliciting at the right moment the telling effects of Being. So we might see it indeed as a "gathering," a process that is as well a way to begin to understand the Gorgian *kairos* as it is apparent in Heidegger, where language can be seen as an ingression inward, and further inward still, to create and shape our being at its source, and only then to blossom outward into life, into the surest warrant of reality.

Heidegger's Phenomenology of "Saying": Listening, Memory, and a Return to Xa

Memory believes before knowing remembers. Believes longer than recollects, longer than knowing ever wonders.

—William Faulkner

Gadamer says Heidegger's affinity for the Greeks was of such significance that it "distinguished him from all other phenomenologists from early on" (*Heidegger's* 142). In fact, one would be hard pressed to find any greater indication of this affinity than his phenomenology itself, especially as he joins it in such remarkable ways with his idea of language. In a statement mirroring his definition of Being, Heidegger says of language what we might expect: "language itself is—language and nothing else besides. Language is language" (*Poetry* 190). As he himself admits, the statement is tautological, but tautologies are hardly illicit in the context of his fundamental ontology as the "is" alone suffices without predication. Indeed, the intransitive "is" must take the stress in Heideggerian ontology because nothing stands behind the Being of things: there is only the "self-showing" of Being in phenomena. Therefore, Heidegger says, *"Ontology is possible only as phenomenology"* (*Being* 60, his italics). In a sense the statement serves as a recapitulation of the Gorgian tenet that "nothing exists," though more pertinent in this connection is the role language plays because of this phenomenology. In Heidegger, as Michael Hyde and Craig Smith tell us, "the disclosure of understanding is itself a disclosure of the human experience of language" (347). The point is crucial in not only establishing Heidegger's notion of Being in contrast to more traditional views but in developing further the kinship of Heideggerian and Gorgian phenomenology. In each the dominating power of language is evident as a matter of phenomenology alone, leading to distinctly similar perspectives on the nature of *kairos.*

For Heidegger, phenomenology is "appearing" in the light of language, of *logos* making manifest and letting us see something from the very thing which disclosure is about (*Being* 56). Once again, in this perspective appearance obtains its ontological as well as ontic status, as Heidegger contends that appearance is not subjective or imaginary but is inherent in Being as part of an "original unity." As such, language does not come after but goes in advance of what appears, serving as a means of disclosure, a way of bringing the Being of things into unconcealment. Language is then tantamount to creation, not precisely "causing" things but bringing them "to the fore," allowing them to appear, as Heidegger says. Hence, there are no "things-in-themselves" beyond language, or none that we can know, only a revealing and concealing, arrivals and departures prompted by language. In this manner Heidegger's conception of truth as *aletheia*—the revealing of that which is otherwise concealed—is a function of language alone. It stands in contrast to truth as *veritas,* where language is presumed to come after an independently existing reality and provide a more or less adequate correspondence to a to it. In this context we might grasp the importance of Heidegger's contention that "language speaks," that we do not first and foremost speak language but hear it, insofar as we listen. Language is thus the equivalent of Being necessarily, as we are "always already" involved in a web of cultural and physical imperatives, a verbal environment upon which his phenomenology is based, where things "show themselves" as language makes them manifest. And here, most evidently, "language speaks within *its* speaking, not within our own" (*Poetry* 190).

This phenomenology, in turn, supports Heidegger's various senses of *doxa,* to embrace not just the idea of "glory" but its more conventional meaning as "opinion." He tells us that *doxa* is "the regard in which a man stands," but that "in the broader sense" it is "the regard which every essent [thing] conceals and discloses in its appearance."

He explains by offering the example of a city, one presenting a distinctly "magnificent view." But he says the "aspect" that the city "has in itself"—and which it can offer only by virtue of having it in itself—is subject to the play of appearances inherent in Being, and thereby can be perceived in every case from differing perspectives, making it subject to constant change. Thus, in the most familiar sense, "the aspect is always one *we* take and make for ourselves." So, in Heidegger's interpretation we must be open to how the thing reveals itself. If we form an opinion about the thing without looking closely at it, or impose some perspective upon it through indolence or arrogance, the opinion we acquire as a result would have no support in the thing supposedly perceived. In either case we merely assume it to be "thus and thus" and our understanding then becomes a matter of mere opinion (*Introduction* 104). In effect, from the phenomenology that allows us to understand *doxa* as glory, we can also understand *doxa* to be essentially as Plato describes it, meaning "opinion" that has no relation to the nature of things in being either "imagined" or "subjective" (105).

Despite this emphasis on seeing, in all essential ways Heidegger's phenomenology concerns listening. In particular it concerns "Saying," the "voice of Being," and to further explore the link we might take a different example, not of a city but a village. I have identified *doxa* with the Vietnamese sense of *xa*, a word of complex, interwoven meanings, but meaning to us something as simple as "village." And it is very possible to see it as such, very simply or otherwise in water colors of muted blues and greens that meet in the mist of horizons on all sides, with foot paths and the dark red contours of earthen dikes that edge the rice paddies providing the only clear symmetry. In this mist those villagers would be hunched over, knee-deep in muddy water working the harvest, so routine and meager in their toil that even up close they seem to show no movement at all, but just to be part of the broad brushed landscape. In the dazing warmth of midday, and most of all toward evening when the colors deepened to shades of red and russet glistening across the paddies, all could be said to be blissful and beautiful all around. And there are those who have said as much. Indeed, those who are not Vietnamese and who never fought in that war might even say these villages absolutely exude a quaint and rustic charm, presenting a far more "magnificent" view than any city. But others would say that it takes one who was there in the war to understand that there is nothing in nature as unsettling as a Vietnamese village. In the swelter of the dry season it grows indolent and sun washed, menacingly upon the land, no different than the parched and twisted foliage in which it is ensconced. And when the rains come, it sags and settles, lies swollen amid stagnant bogs and thickets, a wellspring of the war's pestilence, giving off foul smells and a wet heat that rots things so fast they glow in the dark.

In either case, and all those in between, the "view" of the village is essentially a manifestation of what is concealed and what is revealed in *xa*, rather than what one "sees" as a matter of perspective alone. As such, the village is not one thing to the villagers, another to the visitor, and yet another to the soldier involved in that war. Nor is it first one thing and then, by and by, something else entirely, as we saw it then, and would see it now, decades later. It is, in fact, both things at once, many different things simultaneously, regardless of perspective, for here we are confronted with how Being reveals itself. We can therefore have little patience with those who wish to know some settled or certain reality of the Vietnamese village or to presume to know that reality apart from language. The only truth that we can know is fixed in the word to the point of disappearance, so that there can be no rising above the word, and there can be no essential distinction between language and reality, because in this scheme of things, just as it was in that mediation on cemeteries, reality is utterly a linguistic product in any event. The immediate issue, though, is to make a distinction between the "aspect a thing has in itself" and that which "we take and make for ourselves," a distinction very pertinent to the nature of the Vietnamese village and one that carries considerably more significance in this context than the way I posed the issue at the very outset, in that simple expedient of differing perspectives on campus traffic. But here, compelling as the issue may be, Heidegger's phenomenology takes a peculiar and particularly snarled turn as it relates specifically to the "Saying" of language.

To Heidegger, phenomena are manifest in events designated by the Greek term *legein*. The word means to talk, to hold discourse or to tell a tale, and comes from the same root as the word *logos*. But it also means, according to Heidegger, "letting be" or "letting lie forth," and to Heidegger the significance of *legein* comes in the congruence of the two meanings, as he claims the essence of telling comes to light in the mode of "letting appear." Being is thus manifest in the verbal environment, ultimately in what language expresses, "what it speaks and what it keeps silent." So here again, listening constitutes the basis of our authentic relation to language and thus to Being as such, and only in our observance of *legein* do we encounter things authentically, allowing something to be what it is rather than what we take and make of it by *our* language (*Thinking* 205–06).

Our sense of seeing, upon which we would inevitably rely, avails us little in this context. Heidegger, of course, warns us of the perils of "sight metaphysics," though he himself might seem to proceed precipitously, needlessly confusing the issue. In fact, he stipulates two aspects to Being, or *physis,* that seem to be not only based upon seeing, or some sight metaphor, but are in apparent conflict. Michael Zimmerman deals with this matter specifically in *Heidegger's Confrontation with Modernity,* claiming that this "puzzling duality" comes about in the differences Heidegger draws between *physis* as "self-emerging" and *physis* as "appearing." Zimmerman's point is well made and bears directly upon our discussion here, most pointedly on the two senses of *doxa* examined earlier. Insofar as Heidegger links *physis* to *phyein* ("to grow"), its meaning seems to be "self-emerging"; insofar as he links it to features of human presence, the aspect of "appearing" is emphasized. This distinction is not just another curious equivocation in Heidegger, for "self-emerging" is precisely that, seeming to reduce dramatically the significance of human presence if not to preclude it altogether, while *physis* as "appearing," manifest uniquely by our bearing witness to what emerges, is explicitly "dependent on human existence," as Zimmerman tells us (225).

That these two aspects of *physis* might thereby be incompatible is surely the case. Nonetheless, considered in light of the later Heidegger's emphasis on language, as I am considering things here, the distinction between "self-emerging" and "appearing" might not be so pronounced, even if obvious in other significant ways. Indeed, in this context we cannot so easily mark the difference between the two, which is where Zimmerman's discussion seems to lead in any event. It is not just that the distinction is based on the prior, comparable distinction between nature and culture but, given the preeminence of language in Heidegger's phenomenology—especially as we take it in terms of *legein,* as the concurrence of "letting be" and "to tell"—self-emerging and appearing are essentially different ways of designating the same event. Put the emphasis on listening and language, rather than seeing, and the distinction seems to disappear. To the degree we take them as separate, incompatible functions of *physis,* we also take language as something other than *physis,* as derivative of *physis* rather than of its essence. Among other reasons, Heidegger chooses to express Being as *physis*—and ultimately as earth—to stress its immanence as sheer process and power in the here and now; and what is necessarily critical in his interpretation of *physis* as language is that the here and now is not things and substances in their own right but those that emerge and appear solely by virtue of language's relatedness to Being. Only in this context would his curious thesis make sense that language alone speaks authentically and we, of our own devices, do not. Therefore, in this context it is not precisely that "appearing" is dependent on human existence but that the latter—human existence understood in terms of *dasein*—is itself a matter of emerging *and* appearing in the light of language by virtue of "Saying," by the "voice of Being." Thus, to say appearing is dependent on human existence is to say appearing is dependent on language. It is in this regard that I apply the idea of *xa*, for in this sense of *doxa* our use of language is so thoroughly merged with Saying that we are left with only what is reciprocal and mutual, a unity that precedes and persists beyond what is related and derivative.

This archaic conception of *doxa,* in the quite singular union of the physical and the cultural, might well constitute Heidegger's most essential and evident confrontation with modernity. In any event, this doxic image of language, achieved initially through a shared or common history, is inevitably the effect of dwelling in language. Heidegger, for instance, writes often of the superior nature of the Greek language and quite imperiously, it seems, says the same of the German language, claiming that it bears some kinship to ancient Greek that other modern languages lack. But as Francoise Dastur notes, Heidegger's contention pertains not to the quality of the Greek (and presumably the German) language over and above other languages but to the different relation that Heidegger says the Greeks bore to their language, "a relation of dwelling and an instrumental relation" ("Language" 363). And why, then, should their language not be deemed superior, given their emphasis on language as the "house of Being" that Heidegger claims it to be? In his phenomenology language is creative in its own right rather than merely a means to exert whatever powers we ourselves presume to possess apart from language. Again, in this peculiar perspective, we are creatures of *physis,* of language and reality as one and the same, and ultimately are creators of neither.

Moreover, as language brings things "to the fore," phenomenology thus becomes a "presential" ontology that in crucial respects can be likened to the Gorgian view of phenomenology explored in a previous chapter. Indeed, in Gorgias's treatise, "On Non-Being," the very distinction between Being and Non-being is undone by a conception of language whereby Being is disclosed in instances of phenomena attained through language, for in the Gorgian world view there is no correspondence between "objective" reality and language but purely an "irrational *logos*" holding sway in the necessary absence of any such correspondence. Here again, language is not a subsequent expression applied to what appears but the means through which things are brought to the fore, to "unconcealment." Thus, the phenomenologies of both Gorgias and Heidegger attest to the fundamentally linguistic character of experience, and do so in a context that makes it possible to relate Gorgias's notion of the irrational *logos* to significant aspects of Heidegger's theory of language constituting the source of "authentic" discourse.

In these phenomenologies lie the connections between listening and Saying, and in those to the Gorgian *kairos* itself. In *Being and Time* Heidegger defines language functioning in its ontological dynamic as *Rede,* specifically language that is "authentic" in that it maintains its relationship with Being. "Saying"—*Sagen*—is then not just a related version of *Rede* but its very source. I have discussed the former in terms of the myths, rituals, and customs of a people, drawn from their poetic language, but here a more abstract development is in order. Heidegger tells us that language is experienced not only in asserting and "speaking out"—as *sprechen*—but as Saying—a "pure draft" or "calling from Being"—that is experienced as a "peal of stillness of not anything human" (*Poetry* 207). Therefore, as Paul Ricoeur says in explanation of Heidegger's position, "*Saying* designates the existential constitution and speaking its worldly aspect which falls into the empirical" ("Task" 155). Language, then, insofar as it is authentic, occurs in hearing and listening, heeding and being silent and attentive to the "Saying" of language present in the call of language. L. M. Vail terms the "call" a "mysterious injunction" (167) through which the "primordial" word "speaks" to us and through us, so that "mortal speech is founded on a listening response" (178). William Richardson designates the call as the "hail" of the "aboriginal" *logos,* "a response to which takes the form of human language" (578). Putting an even finer edge on the matter, Heidegger himself says he derived the notion of "Saying" from the Heraclitean *logos*, so that, in "Saying," the plea of Heraclitus—"to heed not me but the Word"—is given its Heideggerian significance (*Early* 59–78). In another context, Heidegger says of "Saying" that it "releases human nature into its own," but only in order that we, in our openness, "may encounter and answer Saying" whereby in the course of our "appropriation" by language, "the encountering saying of mortals is answering" (*Way* 129).

Thus, mortal speech—the human *logos* manifest in asserting and speaking out—is founded on our response to Saying and comes to pass in the relationship prevailing between the two. Here the authenticity of mortal speech, *Rede* as such, depends upon its "accord" (*ent-sprechen*) with Saying, the "voice of Being," rather than some scheme of enduring reality apart from language. Indeed, in this particular context our authenticity is explicitly joined to that of language in ways vital to our understanding of the Gorgian *kairos*. We are "made appropriate for Saying," Heidegger says. Human being is then "released into that needfulness out of which man is used for bringing soundless Saying to the sound of language" (*Way* 129). Therefore, most crucially, Saying is fundamental to the verbal environment of *physis*, to our dwelling in language, as mortal speech alone, without accord, falls under the onus of *Gerede*, the "idle chatter" of the "public realm." Once again, the Heideggerian position is that "it is language alone which speaks authentically" (134). Hearing, or listening to the voice of Being, thus accomplishes for Heidegger "the primary and authentic way in which Dasein is open for its ownmost-potentially-for-Being—as in hearing the voice of the friend whom every Dasein carries with it" (*Being* 206). That language, in our authenticity, therefore appropriates and uses us is one of the more compelling if commonplace implications of *legein*, of "telling as letting lie" (*Thinking* 205–06). But to complete the idea of *legein*, we return to *xa*, to see how the "telling as letting lie" of the village comes about or, in fact, miscarries completely.

In all of this we must take note that there is some wisdom in the perceptions of even the most heedless infantryman, for in countless ways his life and the lives of those with him depended upon seeing the village in its most sinister aspect. And if these perceptions were flawed, and the ignominy of the war attaches to him as a result, it attaches in equal measure to those who sent him there and at whose behest he did his work. In either case there are features of that war that shape it altogether in terms of the steadfast denial of *legein,* where "telling as letting lie" was displaced by aspects of the village, and the country as a whole, "we take and make for ourselves." And we are left, in our failure to listen, with *doxa* only in *that* sense. Indeed, there is a hoary tale from the war that marks the process specifically. An artillery officer, explaining why the village of Ben Tre was razed by the immense fire power at our disposal in that war, claimed that Ben Tre was held by the Viet Cong and that we had to free it, to save it from Communist rule, so it was forthwith leveled by artillery fire. He said it became necessary to destroy the village in order to save it.

If it is an all too familiar tale, it provides a clear instance of the arrogance Heidegger indicts, that presumption to use language to fashion and control Being or, in this case, that part of it constituting Ben Tre. To say that this language is "inauthentic" is to seriously understate the matter, but more grievously it is inauthentic in that it preempts Saying, thereby abrogating *xa,* the "letting lie" of the village. No doubt we wish to see it in the fixed and no uncertain terms of "reality," or at least without "puzzling dualities." But, as I say, in this interpretation language itself is the relation between subject and object, being prior to each and thereby determining the "reality" of each, in this way abating the differences between "self-emerging" and "appearing." And it is more. Language bears the stamp of *physis,* joining the authenticity of language with our own, and to see language or ourselves as otherwise is to betray that "needfulness" out of which we are "used for bringing soundless Saying to the sound of language."

As the artillery officer uses language to take and make of Ben Tre what he wishes, he presumes in his taking and making the justification to destroy it, as the village is then his creation to do with as he wills. This peculiar use of language, a glaring usurpation of power, is by far the most insidious aspect of "idle talk," as it destroys our "dwelling" in language as surely as any village. Language used in this way is both common and current, though it has a pedigree every bit as ancient as the pre-Socratic *logos* itself. It is witchcraft pure and simple that the artillery officer practices, and like any witch he does his work with telling, often horrifying effect.[9] *Legein,* then, can be understood as a counter-spell to witchcraft. As the congruence of "to lay and to tell," it not only shapes the relatedness of self-emerging and appearing but is crucial to the authenticity of language, ultimately defining our place and purpose in our dwelling in language. Thus, there is wisdom also in Heidegger's words, whatever ignominy might otherwise attach to *his* name in this regard. "What lies before us is primary," he tells us, "especially when it lies there *before* all the laying and setting that are *man's* work, when it lies there prior to all that man lays out, lays down, or lays in ruin" (*Thinking* 205, his italics).

This concept of "Saying" brings us yet nearer to an understanding of *kairos,* as it corresponds in significant ways to Gorgias's "irrational *logos.*" In each instance we "let something be said to us" by granting language its mastery. In ways the phenomenon can be understood in the sense of those voices heard by Homeric heroes, as stock phrases or epithets borne by language that, in effect, speak to us *as* language. But a more direct means to the issue is gained by a further comparison of Gorgias's idea of the *psyche* and Heidegger's *dasein*. Each presents similar features of human nature that come to light through similar perspectives on the nature of language. Nowhere in the Gorgian fragments is there a focused discussion of the nature of our being, yet the attributes of the *psyche* can be surmised, as Charles Segal does so well, from what Gorgias says of language. Nor, at long last, does Heidegger offer a definition of the nature of our being. Joan Stambaugh, for instance, says one of Heidegger's most basic insights is that "we do not know what man is" (12). But here we need not surmise. Heidegger claims that we are what we are only as we are devoted to the call of language, used by language for the speaking of language. Here we need not see the similarity of *dasein* and the *psyche* as one of kind or substance, reducing each to body, spirit, élan vital or some such. Rather, as I have stated, their similarity is one of function, of the vitality of unfolding events. From the beginning Heidegger claimed *dasein* to be the relationship to Being—the "*da,*" the "there" of Being. As such, *dasein* is not only the place of Being's occurrence in the world but, in what becomes more and more apparent in Heidegger's later works, *dasein* is the relationship to Being occurring essentially *as* language. It is then manifest as language in ways similar to Gorgias's conception of the *psyche*. But, even as such, still "we do not know what man is." In ways to be explored in the following chapter, the *psyche* and *dasein* are made of words and then they are not, the same as language and yet other than language, in the playing out of the phenomena of one's words and one's self as the relationship to Being is realized in the dynamics of the ontological difference.

Nevertheless, as each seems contingent on neither a soul nor substance independent of language, they are subject to a mortality of flesh and blood, a physicality that is present only as it is contained in language. Here the nature of each is akin, and not so ironically, to the image given by some Christian mystics, surely by the likes of St. John of the Cross who remarks that the virtuous soul is transparent, a window through which God is shinning and is otherwise empty, devoid of the stains of ego, so that nothing is present in that soul but the light of God. Indeed, as should be apparent by now, to equate the nature of God with language is hardly extraordinary. Consummately He *is* the Word. But here, of course, the emphasis is not on transcendence but the immanence of Being—absolutely on the very earth in this context—and, as Heidegger says, the soul seeks the earth rather than fleeing from it, in order to "poetically build and dwell upon it" (*Way* 163). In this context both *dasein* and the *psyche* not only obtain their identification and most essential function in language but do so in ontological views of language opposed to logocentric perspectives. Through each, we necessarily embrace that conception of language that Theodore Kisiel chooses to describe by way of the Book of Acts (17.28) in the phrase I have invoked repeatedly, informing us that language is the "indigenous field" in which we "live, move, and have our being" (91). Here the immanent God, who is the Word, becomes a most likely metaphor for Being, that "indigenous field" as precisely earth itself.

We might say, then, the likeness of "Saying" and the "irrational *logos*" is based on the similarity or functional equivalence of *dasein* and the *psyche*. "Hearkening"—listening—is the term Heidegger uses to designate our most proper bearing to the Saying of language (*Being* 207). He further specifies its nature, its "call," by speaking of it in terms of an encounter, of "undergoing an experience with language" that brings to mind the basic feature of Gorgias's notion of the irrational *logos,* especially of the "near physical impact" that Segal claims language brings to bear in the Gorgian conception. Heidegger says:

> To undergo an experience with something—be it a thing, a person, or a god—means that this something befalls us, strikes us, comes over us, throws us around, and transforms us. When we talk of "undergoing" an experience, we mean specifically that the experience is not of our own making; to undergo here means that we endure it, suffer it, receive it as it strikes us and submit to it. It is this something itself that

comes about, comes to pass, happens.... To undergo an experience with language means, then, to let ourselves be properly solicited by the claim of language by entering into and submitting to it. (*Way* 57)

Thus, not only does Heidegger's idea of language involve similarities to the violent nature of the Gorgian *logos*, but the notion of "Saying" establishes language as the critical means of "gathering" us with Being in the event of appropriation. Insofar as we are "solicited" by the claim of language, we are "solicited" by Being as well. Given the phenomenological perspectives of both Gorgias and Heidegger, language is identified with Being in either case, as in each Being is not so much revealed but, *as* language, is revealing. Here it is important to realize that words are not mere terms, "like buckets and kegs from which we scoop a content," as Heidegger says, but they are "well springs that are found and dug up in the telling" (*Thinking* 130). In fact, he says words involve "memory" (*Andecken*)—indeed, comprise it—as memory embraces in this sense not matters of individual recall or recollection alone but the power of language manifest in Saying, in "constant concentrated abiding" with what is past, present, and to come in the "oneness" of our "present being" (140). Or, that is to say, what comes in the "fullness" of time. Heidegger claims this process has its being by virtue of "*thanc*," the word having its roots in Old English and meaning "original memory." Thus, in "undergoing an experience with language," memory is placed not in the self but in the word and can then be conceived as Saying in its purely temporal aspect. Here the object resides essentially in the word and, as *xa* so vividly reveals, cannot be legitimately commandeered and owned in a concept.

There are the workings of *kairos* in this sense of memory, as it moves with powers that in ways determine our being even before we are, thus being prior not only to individual recollection but at times utterly subsuming it in this far larger wholeness. If knowing is to appropriate, to seize and master, "original memory" is to be appropriated, and by that same power believing is to be held in belief, to be seized by it. So it is with *xa*, even at times for those who were not Vietnamese yet were "solicited" by its claim nevertheless, appropriated by it. As Gorgias indicates in his "Funeral Oration," war puts us at the mercy of certain basic passions—frustrations, hatreds, even exhilarations—that all cultures would seem to suppress, yet war often makes cultural virtues of these same passions, giving rise to situations that clearly return us to the more primeval effects of our relationship with earth and *physis*, to an "aboriginal," pre-Socratic, or Gorgian *logos* out of which these passions seem to emerge.

Vietnam was a place of such situations. And a place astonishingly resistant to what we would make of it by our political biases, stylish technology, and military ordnance. But our effort there was so massive, so bizarre and unavailing in its effect, that making sense of the place was utterly a matter of *our* adjustment to it, very aptly a matter of listening within a situation where being attuned to an "aboriginal" *logos* was hardly outrageous. An encounter with language awaited us there, truly in the degree Heidegger described, as something that "comes over us" and "throws us around," transforms us in the experience of hearing and heeding. Language and language alone confirms old realities and creates new ones as well, possessing at certain, critical times an impact that is physical in effect, to the end that language is often violence itself. If "glory" is hardly the proper word for the experience, and "heroic souls" even more improper yet for those enduring it, there was for some this awareness gained, complete and unqualified in the realization that the most essential wisdom lies in the word. Especially in this word *xa*.

The sacred relationships it designates, particularly those with the earth itself, draw all those concerned, even would-be outsiders, into the all-embracing circle of *physis*. Here the word not only expresses the event but comprises it. Rather than being some concept, epitome, or abstraction two or three places removed from reality, it is reality itself, a corporeal process, alive and of the flesh. Those who were our adversaries in that war knew this instinctively. More than knew it, they embodied it and brought it to bear, the substance of the word then seeming to rise from the earth like those wet, heavy fogs over there, but to have its effect so severe and vicious to quite literally spill blood and shatter bones in some strange and horrific encounter with language indeed. In very visceral ways, then, some came away from the experience with an awareness based on that listening that was a belonging to *xa* itself. In that way it is an awareness that did not belong only to those opposing us, nor least of all one that came from phrase books or Vietnamese language school but from *xa* manifest as sacred ground in the dire mix of these matters of life and death. Here the word expresses an event that is on-going and self-evident, all this time quickening the experience of that time and place in blood memories that through the passion of that sacrifice mark the paradox of *xa* being as much ours as theirs. If it hardly results in some grandeur or glorification, there is for some this awareness still, this voice of the outsider within—perhaps an "irrational *logos*," as Gorgias says; or the "peal of not anything human," as Heidegger says, that is nevertheless the "voice of a friend" that compels participation in an event that dominates and ultimately determines the person undergoing the experience. The experience has nothing to do with armies, theirs or ours, nor with human standards of justice, of right and wrong, as Gorgias himself has said. Nor even with war in any essential sense. It deals instead with *kairos* as the realization of the word in relation to the whole of Being, of its truth lying not in our saying but in the listening and belonging to the Saying that *kairos* bears.[10]

And if only some had come to such an understanding, and if they were not tragic heroes nor at all glorious, their experience was nevertheless authentic in that they came to understand the village in a way they did not take and make for themselves, as not the sort of thing they destroyed in order to save. Maybe this understanding did not come in an instant but panned out eventually, in some "fullness" of time. And maybe it was not precisely as that of the hero of Proust's *Remembrance of Things Past*, who takes the rice cakes and in that "whirling medley of radiant hues" is transported back in deep remembrance to his village of Cambrey. But if the Vietnamese village is far different than Cambrey, the "original memory" of *xa* is something very much like deep remembrance. The experience of each is based on a solicitation, and each culminates in *kairos* as well, what Proust describes as a spontaneous accord between the being that feels and the object felt. And in the case of *xa*, surely as we interpret it in Heidegger's sense of memory, the experience is embedded in the word, a force indeed rising from the earth through the passion of some sacrifice, that blood investment in native soil. Heidegger's phenomenology is thereby apparent not only in how things reveal themselves by virtue of *logos* but even more crucially how we ourselves are revealed by means of this peculiar emphasis on the hearing of *logos*. Thus, the experience of *xa* lies neither in things nor even ourselves but transpires in between, in that borderland appropriating each to the word in those rare moments of spontaneous accord that is *kairos*, the moment of authentic language. We then perceive the latter as "the central point where 'I' and the world meet," as Gadamer says, where indeed they "manifest their original unity" (*Truth* 431).

The nature of this unity is, of course, critical to understanding a point I endeavored to make previously, through N. Scott Momaday among others, that there comes a point when we are in "reality" made of words. This conception embraces the most essential and complete effect of Saying, but its significance is apparent in more conspicuous and practical ways. Heidegger asserts poetry is Saying's basic means of expression, not because it stands farthest removed from *Gerede* but because, by Heidegger's definition, it both draws upon and determines the traditions, beliefs, and rituals of a people in the magnificent mix of language and Being that is *doxa*. Here the nature of knowledge is itself transformed in Heidegger's analysis, manifest in "original memory" or "deep remembrance" embedded in the poetic word. This way of knowledge is *thanc* indeed, being to Heidegger not just "original memory" but meaning simultaneously "to think" and "to thank." As "logical-representational" thinking is a "reduction and an impoverishment" of *thanc*, the latter prevails in the "steadfast intimate concentration upon the things that essentially speak to us in every thoughtful meditation." Thus, to be solicited by the claim of language, to encounter it in the attitude of *thanc*—allowing it manifestly to "speak to us in every thoughtful meditation"—quite clearly puts the emphasis on listening rather than expressing. It is a "devotion," Heidegger says, "held in listening" (*Thinking* 140–41).

Once again we seem to be into regions so thickly obscured by Heidegger's idiom, his mystical musings, that we can scarcely see his phenomenology of Saying as quite basic, very close at hand. Yet it remains so distant from the main currents of Western thought that Heidegger's philosophy, especially in its emphasis on language, has often been compared to those of the Far East, to the Japanese in particular.[11] But many of these connections can be found more clearly and conveniently in American Indian beliefs and practices, definitely as they are expressed by N. Scott Momaday. His works not only deal with these beliefs but seem to evoke many of the attitudes, attributes, and perspectives of the pre-Socratics in the process, in that sense seeming to serve as a way of passage from their world to ours, indeed from Heidegger's world to ours. Momaday thereby serves my purposes here. And if, as the case may be, we learn more in the end of Momaday through Heidegger than we do of Heidegger through Momaday, that is no great harm. In either event, it is worth the effort, because we learn more of language and listening, and ultimately of the Gorgian *kairos*, through both than we would from each separately.

LISTENING AND LANGUAGE: AN AMERICAN INDIAN COMPLEMENT

Acts of sacrifice make sacred the earth. Language and the sacred are indivisible. The earth and all its appearances and expressions exist in names and stories and prayers and spells.

—Momaday, *Man Made of Words*

I once taught on an Indian reservation in North Dakota and was myself taken to school there to learn some lessons in cultural vitality and endurance from American Indians. It was back during the time of the nation's bicentennial, and agreeably enough the anniversary of the Custer battle as well, so that event became the focus of one of our class discussions. And then in the midst of it, a student who hadn't said a word all semester, a guy by the name of Carry Moccasin, abruptly announced, "I was there."[12] He was a real "long hair," so traditional he seemed to spook even the other Indians in class, but I was so pleased he finally had something to say that I was well into my account saying I was there too, trying to engage him further in the discussion by telling him of my visit to the Little Big Horn the previous fall. "No, you don't understand," he said. "I was there. I rode with Chief Gall. We killed Custer and those others."

I waited for the rest, some point or follow up—a thesis of some sort—but nothing came. It just didn't seem to be that kind of a story. Now, I have always been of the opinion that "reality" is a vastly overrated commodity in any event, but this was utterly beyond the pale. Call it illusion or the persistence of the oral tradition, too many medicine men and too many stories passed on and believed from the time of Custer and well before, this was the mentality of racial memory at work, even though an extraordinarily eerie demonstration of it.

I cannot say whether Carry Moccasin was sincere or putting me on, but that confusion, too, would be in keeping with the work of racial memory. But there was an aptness to the experience, an integrity typical in an environment where ideas of the sort I have been discussing seem to have a way of being distinctly matters of common sense and practice. A similar ease and directness seems to be the case with

Momaday's paradigm of racial memory, and I use it here to flesh out some of the more mysterious characteristics that Heidegger attributes to both *doxa* and *kairos*. I wish to reiterate this connection and others of the same sort, pressing the analysis even further, to a point where Saying, listening, and language—their interactions and correlations—are brought into clearer relief by focusing on certain aspects of Momaday's thought. Especially in his novel, *House Made of Dawn,* the unity of language and *physis* is rendered, if not with the same precision, with much greater vitality than Heidegger is ever able to muster. But for both Momaday and Heidegger the ultimate source of "Saying" lies in this unity, and in Momaday it is quickened, given flesh through the stories, rituals, and ceremonies understood by any given community of speakers of a language. And if the stories differ significantly from culture to culture, there is this underlying *logos,* these shared meanings within a community of discourse, providing in the coupling of hearing and belonging the basis of not only our ability to communicate but more essentially the surest expression and verification of our being.

Again we are concerned primarily with the cultural rather than abstract, discursive particulars of language, and to understand the nature of language in this context we must understand culture as embracing, as necessary and characteristic elements of its constitution, features of the natural world. In accordance with this idea of *logos,* as I have stressed all along, there is no distinction between what we are by culture and what we are by nature, for both are essentially the same in this perspective. Especially in this context of American Indian thought the nature of language is that of the earth itself, the earth in full measure, where the natural and cultural coincide. The notion not only parallels the pre-Socratic conception of *doxa,* but nowhere, I believe, is it more ably illustrated in modern thought than by Momaday and his idea of racial memory. In Momaday, language does not only represent external reality but in certain ways is given precedence, such that there is no external reality except in terms of a "primordial" spirituality that embraces our connection with nature, a spirituality that, true to the ancient idea of *logos,* language creates and then invests with meaning and order insofar as we are open to it, insofar as we listen. And insofar as language does not wholly create the world, and never one precisely to our measure, it creates the only world that we can know, which is expressly the mark of this "aboriginal" or "primordial" *logos.*

The idea implies any number of things, all of them accommodated in some fashion in Heraclitus's thesis that all things are one. That we are, if only in some very ultimate sense, one with nature, and language is the means of manifesting this oneness, shaping a relationship so complete that nature, as much as we ourselves, is also made of words. So it is the idea of listening inherent in the relationship that interests me, for attendant on this idea of *logos* is the necessary corollary that nature "speaks." And when language is given precedence over human being, as it is here, the only proper attitude of human being is to pay heed to what nature says, to hearken in deep, reverential silence so that listening might transpire. The result is a peculiarly American Indian awareness of "Saying," where Saying is neither arcane nor indirect, nor simply the work of metaphor. Listen to Walking Buffalo, a Stoney Creek Indian quoted by Vine Deloria in his book, *God is Red*. "Did you know trees talk?" he asks.

> "Well, they do. They talk to each other, and they'll talk to you if you listen. Trouble is, white people don't listen. They never learned to listen to Indians, so I don't suppose they'll listen to other voices of nature. But I have learned a lot from trees; sometimes about the weather, sometimes about animals, sometimes about the Great Spirit." (104)

It is an amazing statement but makes apparent two senses of *logos* distinguishable in terms of the different emphases of seeing and hearing. Plato's philosophy, and in large part Western philosophy as such, is under the dominion of what Heidegger claims to be "sight metaphysics." The eye is the "natural inlet" to the soul, Plato says in the *Phaedrus* (255d), and we behold the Forms, the fundamental features of True Reality in a vision that is eternal, and utterly nonverbal. In this view language is not only corporeal, tainted by time and flesh, but is corrupt by virtue of the very process of seeing. Essential to the latter is the division of the world into the one who sees and the thing seen; the subject-object separation persists, as seeing in all cases distances and separates. Reality is then "the stasis of an *eidos*," as Kisiel says, "a fixed and determinate visage which stands before us, the viewers, as an intelligible entity" ("Translator's Introduction" xxv). Here, as spectators, we come with the bias that what is said is what we put there, liable to our ruses or deficiencies, but what is seen we assume is there by the nature of things, by the starkest turn of reality, utterly true and unassailable. Thus, from the perspective of "sight metaphysics," *logos*, as the correspondence between the world and our way of knowing it, is consistent, unconditional, and the same for all, necessarily a matter of the light of reason.

But there is no such implication in the idea of *logos* as language. Here the emphasis is not just on listening but on a perspective other than that of the detached observer or seer. In our disposition of openness, as we "let something be said to us," we are involved in a process we participate in bringing about. Listening in this sense means not just listening "to a sound of a word, as the expression of a speaker," Heidegger says, but it prompts participation and belonging insofar as it is successful and genuine. Again, Heidegger's play on words is particularly apt, as he invokes the similar sense and sound of *gehört* and *gehören* (*hören/gehören*) to make the point. "We have heard [*gehört*] when we *belong to* [*gehören*] the matter addressed" (*Early* 66). The eye appropriates what it sees, but words do not grasp, do not "take things in" as does seeing. They evoke. Or, more aptly yet, in heedful listening we ourselves are appropriated by language in certain situations, caught in its web and made part of the community it creates, as sound surrounds, encompasses, and assimilates rather than separates.

Hence, to extend a point previously made, in this perspective *doxa* is not only a collection of beliefs that we hold as we would mere opinion, but a power bestowed in the view that beliefs come prior to us, holding us, determining who we are. The site or focal point of the self is obscured or absent altogether in this perspective, and here the curious turnabout of believer and belief is apparent. For instance, when Black Elk, the Oglala Sioux medicine man of John Neihardt's classic, *Black Elk Speaks,* tells us in rather specific detail of the strange happenings in his vision, it is as if his recollection is not crucial but that the events of the memory, seemingly of their own account and in place of the person of Black Elk himself, constitute the deciding factor. "I did not have to remember these things," he says. "They remembered themselves after all these years" (44–45).

The phenomenon may not be identical to *thanc* but is truly of its nature in this withdrawal of the self as the source of our beliefs and memories, of them having us rather than the other way around, perhaps being akin in this respect to the way our dreams control us rather than we our dreams. But here there is no perspective of "waking consciousness" that we can know, one that would allow us to leave our dreams, or our language, and come to know what "really is." That, of course, was Plato's purpose, and my response has been that we continue to dream nevertheless, for in the snares of language even the awakened state is but another dream still. Language, then, holds dominion in this view. As either our dwelling place or prison house, it discredits Platonic notions of reality. This seems especially apparent in Momaday's thought, as the oral tradition displaces the idea of language having its source exclusively in a speaker, a "father" of *logos* or in anything other than language as a living, vibrant organism in its own right. However mysterious the point may seem, it is of the utmost practicality still. When, as a natural fact, trees talk, the emphasis is naturally on listening rather than seeing, an emphasis thereby passing from us as knowing subjects to this "voice" of Being, with the result that the distinction between "what is" and "what is said"—as that between "self-emerging" and "appearing"—becomes particularly problematic. In fact to see these as separate is to participate in a fundamentally illicit undertaking that denies the essential nature of each, as they subsist only in their relationship to one another. Thus, we listen, for racial memory is not only the substance of the semantic environment of the here and now, but we listen because in this peculiar earthly realm what is said is most veritably what is.

Such was the case with *xa*, as the issue is not whether our relationship with *physis* is ancestral or acquired but that, in war or otherwise, all that we can know of "reality" is temporal and mortal, of the earth and language. In this sense we might grasp the significance of Momaday's belief that "acts of sacrifice make sacred the earth." As "language and the sacred are indivisible," the earth therefore exists "in names and stories and prayers and spells." This idea of the sacred, this felicitous joining of *physis* and culture to constitute, in effect, a pre-Socratic sense of *doxa*, seems an enduring feature of American Indian thought. In very essential ways the sacred is expressly this coming together of "Saying" and "speaking" that can be used to illustrate as well as any war story or funeral oration the nature of the Gorgian *kairos*. Indeed, such a process lies at the heart of the Navajo cosmos, where the unfamiliar is inherent in the familiar, the miraculous comprising what seems otherwise most routine, even monotonous.

I have spoken of "aho," the word meaning "mist" or "breath of life" to the Navajo. Simply, the word is rooted in the belief that breathing itself is a sacred act. Accordingly, the Navajo perceive the universe itself as the breath of life—as *Nilch'i*, the "Holy Wind." *Nilch'i* is Being *as* physical process, most conspicuously the power giving life and movement to beings. More essentially it is the "aboriginal" *logos*, and when sacredly heeded it imparts wisdom made articulate in "aho," of breath given form and substance in human speech, in stories, songs, and ceremonies. And the idea is not utterly out of place in Western culture. Queen Gertrude herself says as much, claiming in as wise an insight as any to be found in *Hamlet*, that "words are made of breath and breath of life" (4.1.198–99). Of course, Heidegger invokes it as well, alluding to Herder's belief that "on a bit of moving air depends everything human that men on earth have ever . . . done and will ever do," and that, for the poet at least, breath means "the word and the nature of language" (*Poetry* 139–40). Thus, for the Navajo, as much for Heidegger, thinking itself is this peculiar variation of *thanc*, determined by *Nilch'i* as joined with speaking at those moments we are open to *Nilch'i*, in a listening that obscures differences between internal and external, thinking and speaking, words and deeds. It is a most magnificent expression of the power of *physis*. Here, for instance, is the coupling so vital in the formation of the hero, that nobility of action and words in a *logos* where they are one in *kairos*. Here, as well, we might more fully appreciate the experience Gorgias assigns to the warriors of his "Funeral Oration," especially as we shall see it informed by Heidegger's paradigm of that "voice of a friend every *dasein* carries with it." And here also to the Navaho the experience is not embedded in the calamity of war but is as natural as breathing in and breathing out. In the ambiance of *Nilch'i*, the voice comes from within but at the same time from beyond and above, as indeed a call of "not anything human." The experience occurs in lieu of mandates of the metaphysics of presence, nor, seen in this perspective, would it seemed to be based on any reality other than words, nor upon some reality which words would seem to depend. It seems, then, to capture in certain opportune moments, by virtue of *Nilch'i*, the peculiar merging of Saying and speaking that in this perspective is the nature of *kairos*.

In all of this the aboriginal *logos* embodies language explicitly, a view given full substance in Monday's novel, *House Made of Dawn*. The story concerns a man named Abel, a young Pueblo Indian estranged from the traditions of his own community yet unable to adhere to anything new in white culture. In fact, with the exception of the book's prologue and closing passages, Abel's presence throughout the novel is marked only by an ill-defined *pathos*. In one sense he is precisely the wooden Indian one character in the novel suggests him to be; in another, perhaps more clinical sense, he seems afflicted with a spiritual malaise so intense as to be schizophrenic, his personality utterly smashed as he suffers one misadventure after another. Momaday identifies the source of Abel's angst by saying, as we might expect, that Abel is "inarticulate," no longer attuned to the "old rhythms of the tongue" (56).

In the end Abel's voice is restored. Central to his deliverance is his grandfather, Francisco. Though physically ailing himself, Francisco embodies the racial memory of his people, and his healing presence prompts Abel's return to his tribal roots. However, along the way to his spiritual deliverance, Abel encounters a most amazing character, a preacher called John "Big Bluff" Tosamah, "The Priest of the Sun."

The name can only be ironic, for the Priest of the Sun conducts his services in the cold, damp basement of a warehouse where the sun never shines, indeed where it is present only as a reddish-yellow cutout, a huge cardboard oval fixed to the wall. Tosamah is described as "shaggy and awful-looking," "lithe as a cat," and having the voice of a "great dog" (85). In this and every other way, he bears the earmarks of coyote, a trickster figure in Native American mythologies of the Southwest. He therefore shatters decorum, is a bit treacherous, and sometimes plays the fool or dolt, so we had best be wary of what he says, though we ignore or disbelieve him only at our peril. So this figure, "Big Bluff" Tosamah, the Priest of the Sun, offers a sermon in Abel's presence, taking his text from the Gospel of Saint John. In it, he talks about the Word and, by extensions smooth and apparent, about rhetoric:

> "*In principio erat Verbum.* . . . In the beginning was the word. . . . Now what do you suppose old John meant by that? That cat was a preacher, and, well, you know how it is with preachers; he had something big on his mind. On my, it was big; it was the truth, and it was heavy, and old John hurried to set it down. And in his hurry he said too much. . . . It was the truth, all right, but it was more than the truth. The Truth was overgrown with fat, and the fat was John's God, and God stood between John and the Truth. . . . In the beginning was the Word . . . Brothers and sisters, that was the truth, the whole of it, the essential and eternal Truth, the bone and muscle of the

Truth. But he went on, Old John, because he was a preacher. The perfect vision faded from his mind. . . . He couldn't let the Truth alone. He couldn't see that he had come to the end of the Truth, and he went on. He tried to make it bigger and better than it was, but instead he only demeaned and encumbered it. He made it soft and big with fat. He was a preacher, and he made a complex sentence of the Truth, two sentences, three, a paragraph. He made a sermon and theology of the truth. He imposed his idea of God upon the everlasting Truth. (85)

In tone and content this is the nature of the better part of Tosamah's sermon. In his inimitable way, Tosamah assails Saint John for his stone-deaf insensibility, for "Old John" had no reverence for the "sudden and profound" impact of the Word. Being a "white man," he talked about the word, but never experienced it. He talked about it, talked through it and around it, and went on to construct an elaborate scheme about the Word, in the process reducing the truth of the Word to all that is irrelevant and remote (85–87). And so Tosamah rambles on and on in reiteration of a theme, in his verbosity committing the very fault he accuses Saint John and all white men of committing. But again, we had best be wary, and very discerning, for Tosamah is a trickster. Running through his sermon is the message that truth, all that we can know of it, lies in language, and in this and other matters, Tosamah is no doubt Momaday's mouthpiece.[13] Truth is verbal, and to say there is something beyond language that it symbolizes—say, some realm of ideal forms or any world preceding or coming after our own to which language does not have access—is to burden and obscure the truth. As Tosamah says, it is the fat of Saint John's God standing between John and the Truth that renders life opaque to those experiences that enrich life, giving it meaning and nobility.[14]

If we are indeed afflicted in Western culture with a view of nature that projects it as totally other than ourselves, as indifferent or even hostile to concerns of the human spirit, Tosamah's sermon challenges this position, having not just a positive side to it but a way of redemption that follows the rebuke of the white man's language. His focus is on words, not as a means of representation, but as a source of creation, for nothing is pre-existent or prior to words in Momaday's presentation of the American Indian world view. "A word has power in and of itself," he says. "It comes from nothing into sound and meaning; it gives origin to all things. By means of words can a man deal with the world on equal terms. And the word is sacred" (*Rainy* 42). This is also Tosamah's view, and though his guile and bluster may be bothersome, he offers a remarkably cogent statement of it in spite of himself. He thus figures significantly in Abel's salvation, though given his disposition it is never certain whether it is for good or ill.

However, this is clearly not the case with Abel's grandfather, whose influence on Abel is benevolent and generous throughout. Indeed, insofar as we are concerned with the functioning of racial memory, Momaday's greatest triumph in *House Made of Dawn* is the relationship he draws between Abel and Francisco. Throughout the major portion of the novel, they hardly speak to one another, and seem never to understand each other when they do. But they are linked by a recurring image that begins as nothing more than a fragment from Francisco's memory of his youth, an image of a ceremonial race "for good hunting and harvests" as it is first introduced in the book (11). But later the image gathers intensity, begins to crop up in Abel's consciousness, as perhaps a vestige of a previous mentality or premonition of what is to come. In the novel's most essential and expressive scene, its presentation comes in a twilight realm of consciousness, where silence prevails utterly, and in the ensuing dialogue that transpires in this silence the effect of racial memory is pervasive and compelling.

The image is rife with the *mythos* of the sacred tradition: of men, "whole and indispensable," running with "great dignity and calm," who are the measure of perspective and meaning in the universe, who by their very presence give expression to creation itself. In them there is no distinction between what is individual and what is communal. They thus run "simply in recognition and with respect," and "deep in the channel . . . in the way of least resistance, no resistance" (96). In the course of this memory of the ceremonial race, near the novel's end, as Francisco lies dying, Abel waits upon him, and upon nature to take its course. And in the seventh dawn of that long vigil, Francisco's words become coherent to Abel, but coherent only because Abel is in a sense transfigured by them. His words are muted, but sacred and reverently spoken, enticed by six separate dream states or memories drawn from his youth. As Francisco speaks, Abel dozes, but still he "could hear the faintest edge of his grandfather's voice on the deep and distant breathing out of sight, going on and on toward dawn" (197).

If the dream states of Francisco are not lucid nor, least of all, lucidly understood by Abel, they nevertheless comprise a dialogue between the two, though one prevailing more in silence than sound. Like the merging of our being with the land, individual with communal consciousness, there is little distinction to be made between what is in Abel and what is before him, between what is in Abel and what is in Francisco. At that propitious moment their bond is so perfectly reciprocal that what is related is wholly subordinate to the relationship itself, a bond that is here manifest in the wisdom of Francisco shared with Abel in racial memory. Nothing happens to Abel, nor even within him, but all that is critical comes to pass between Abel and Francisco and, through Francisco, between Abel and the racial memory—the "original memory"—of his Indian roots and heritage. "To significations words accrue," Heidegger says (*Being* 204), and such is the case here. That is the nature of the aboriginal *logos*. On the basis of words a communion transpires between Abel and Francisco, one that surely parallels Heidegger's paradigm of *thanc*. He provides a more complete explanation, saying that *thanc* not only discloses the things that speak to us "in every thoughtful mediation" but that it means the "mind, the heart, the heart's core, that innermost essence of man which reaches outward most fully and to the outermost limits, and so decisively that rightly considered, the idea of an inner and outer world does not arise" (*Thinking* 140, 144). It is as if the nature of the spoken word is realized, authentically, only to the degree it does not originate with the speaker but reaches out through him to claim the hearer. It "lays hold" of the hearer, "even makes the hearer into a speaker, if perhaps only a soundless one."[15] And such is precisely the case here. Abel listens, and in silence—"sound's sanctuary," Momaday says—Abel finds his voice if only to hear it through Francisco. He hears and he belongs. Abel's voice is restored. Through Francisco he re-enters the racial memory of his people, their *doxa*, and in this integration, racial memory becomes his salvation.

And Momaday returns to the image of the race, at once both ceremonial and sacrificial, a flawless expression of *thanc*. Abel and Francisco hear the runners at dawn, "a hundred men, two hundred, three, not fast, but running easily and forever," so that the sound of their going is as one. "Listen," he says, "it's the race of the dead, and it happens here" (186). And, imperatively, it is not the power of the runners on display but an expression of the power of nature, as it is manifest as *physis*. Momaday explains, saying elsewhere, "to watch those runners is to know that they draw with every step some elementary power which resides at the core of the earth and which, for all our civilized ways, is lost upon us who have lost the art of going in the flow of things."[16]

Perspectives such as this affirm a sense of dialogue where listening, a "constant concentrated abiding," is given precedence. A dialogue not between two sentient centers that was Plato's wont to call our souls, nor yet one established with a text or some such, but with the earth—that idea of nature understood as *physis*—so that both we and nature in the most essential sense of *doxa* are appropriated by words, made of words. It is fundamentally a mythic form of perception, whereby particulars are woven into the whole, into the perfect oneness of an undivided sphere. In this perspective the activity of thinking is itself a manifestation of *thanc*, resting more on involvement and participation than a dialectic in the strictest sense of detachment, analysis, or any other scheme of dividing the whole into constituent parts for the purpose of study and understanding. Not only is the idea of knowledge transformed as a result, but the realm of mystery is extended to embrace what is otherwise prosaic, indifferent, or antagonistic. Here the task of thinking is not "to know" in the accepted sense of "grasping" or "mastering," as we would "grasp" a concept or "master" a subject matter. On the contrary, thinking is an ontological matter, manifest in the way we encounter the natural world and to the degree we express that encounter in gratitude and celebration of Being, as an experience apparent in reverence, awe, and respect.

To experience Being—*physis* or earth—in this manner can be understood essentially as a matter of sacrifice, a realization of *thanc* by which we give in return for what is given. As Joseph Campbell tells us, sacrifices are not gifts, bribes, or dues paid to placate the powers that be, but are "fresh enactments, here and now, of the god's own sacrifice in the beginning, through which he, she, or it became incarnate in the world process" (181). Campbell is not speaking here of the God of philosophers and theologians—the immortal, omniscient, and omnipotent One—as that God cannot be sacrificed nor even sacrifice Himself. He is speaking instead of the dying gods of mythtellers and believers, those who must be mortal to be gods at all. In their sacrifice and death, they are affirmed in their "belongingness" to Being, being of the flesh, indeed of the earth to "become incarnate in the world process." In this view our being is not diminished by our mortality but exalted, as that is the nature of our relation to Being, in *our being* that "nearness" to Being, to what Heidegger calls the "unobtrusive governess occurring essentially as language" ("Letter" 212). Thus, we abide, "held in listening" to "what lies before us"; we "go in the flow of things." We "tell as letting lie." Or we "follow our bliss." And the natural world, rather than rearing in primitive hostility, circles back and embraces our being, redundant with the life embracing us, as we thus discover its references within ourselves insofar as we are attuned to the voice of nature, to the language of the earth. Insofar as we listen.

We hardly exaggerate the matter to say that such an attitude turns the world of rhetoric on its ear. And rhetoric is the issue here, at least rhetoric of the type that engaged the Sophists. As we have seen, from the time of Plato we have been diligent to make critical distinctions between language and nature, words and things, between "what is said" and "what is." But the point that Momaday makes in response is compelling in its simplicity, and clearly similar to that made by Heidegger. What is at the heart of racial memory is the idea that the relationships we bear to one another, to "outside" reality—to the earth as such—is altogether manifest in language. Within the oral tradition "what is said" is thereby merged with "what is," allowing events to "remember" themselves in "names and stories and prayers and spells." Because memory believes before knowing remembers. Maybe this was the case with Carry Moccasin. I would not say otherwise, though it is the manner of his believing that is most engaging. The idea of the story is not as distant nor diminished as we might think, surely not that story of Custer among the Standing Rock Lakota. Indeed, it would seem Carry Moccasin had authentic existence only insofar as he participated in the story, that it was to him the surest expression and verification of his being. Even at that, given the besieged state of Lakota culture, his existence was thereby less precarious in so many respects than that of his better adjusted, more "sensible" classmates.

The paradigm is old, far older than even the Greek rhapsodist, with the focus always on the listener, where the storyteller is himself first and foremost a listener. Yet the story is the thing, the "Saying" it manifests. It is the place language becomes most conscious of itself, and the language user and listener least so. Here, listeners hear themselves into the story, becoming part of it, as did Abel, and perhaps Carry Moccasin as well. Or, more aptly yet, in heedful listening they are by nature appropriated by language, caught in its web and made part of the community it creates. As I say, sound encompasses rather than separates, and we do not master but submit to it, undergoing an encounter with language that at times transforms us. In this view language and it alone is *logos,* and in the resulting wholeness all things—to include the "truth"—are reduced to functions of language.

This is generally Heidegger's conception, as in crucial respects it seems also to be Gorgias's. In his famous trilemma and other tracts on rhetoric and philosophy, he divests us of the natural world only to the degree we assume it exists apart from language. More pointedly, language as the great *dynastes,* in effect, is equated with the natural world, the latter appearing altogether as phenomena generated by language. Doubtless, Gorgias proposed a rhetoric that aimed to annul or dilute the sense of self-will in the listener, but in pursuing the view to the utmost we realize the rhetor achieves this goal only as he himself defers to the power of language, gives himself over to it, as in this rhetoric the deceiver is himself most deceived. Once again, in the Gorgian perspective language is "free," not primarily a faculty belonging to humankind but a process that engages us, "happens" to us, through our reciprocal bond with *physis* conceived ultimately *as* language. Or, to put it otherwise, "nature speaks." We "dwell" in language. Accordingly, every genuine act of language is an act of faith, as we must first listen to language before we speak, let ourselves be claimed by it, for only as we are claimed by language, by its "Saying," do we speak authentically, even as we deceive and are deceived in doing so. We then let language itself speak, language that comes of its own accord in our openness to it, as it did for Abel, establishing coherence, substance, and meaning in the fabulous mix of culture and nature that constitutes our dwelling in language, our racial memory or *doxa.* We would then live in a "House Made of Dawn," and given Momaday's rendering of the cipher, language is indeed, as Heidegger so often claimed, the "House of Being."

Time, Situation, and the Blink of an Eye

Thus, it follows, as Otto Pöggeler tells us, that in Heidegger's phenomenology "the occurrence of truth is essentially an occurrence of language." But things undergo interesting transformations in his thinking, especially things like "truth." To Heidegger truth has little to do with a precise, methodical correspondence to "reality" but is *aletheia*, the power of revealing and concealing, the play of appearing and perceiving vested in language. In this sense it is necessarily joined to *doxa*, "bound up with the historical speaking of a people and its basic words and principles" (*Heidegger's* 237). If the conception seems atavistic, it is because it truly is, marked by a return to a "primitive" mentality where the prelogical, mysterious powers of language are predominant. However, in another sense, the conception is merely an acknowledgment of powers that have always resided in language in obvious and practical ways, if very often condemned and discounted with great success over the centuries. Whatever the case, Heidegger's emphasis on the "Saying" of language inverts ordinary notions of language, *verba* seeming to be the only sure sanction of *res*, by virtue of *physis* grounding us at all times in language. Heidegger therefore claims that language as Saying moves in a realm beyond a mere system of signs, possessing the speaker and going in advance of his performance as a speaker (*Piety* 86). Here we are concerned with previously covered ground, but the perspective is critical for it is the basis of the idea of *kairos* I am attempting to explicate. Still, the question yet to be addressed concerning the Gorgian *kairos* is the same: What are the circumstances under which language "speaks," so that we may hear the "voice of Being"? What is the process, more exactly, that delivers "Saying" (*Sagen*) to "speaking" (*sprechen*) that takes the form of human language?

Heidegger claims that to heed the call of language, its Saying, requires a "resoluteness" on the part of *dasein*, an openness to Being and thus to language itself. He tells us this openness can lead to encounters with language, to consequential events that take shape in the here and now, and are thereby apparent only in the context of time. As such, they are "situational," coming about in the tension between concentration and distraction, "resoluteness" and "average everydayness." As these events are manifest temporally, they take their measure from Being in *that* sense, in contrast to metaphysics, to the idea of Being as presence. Here it is important to recall that in Plato's conception of time—or Aristotle's, for that matter—the present exists as a mere boundary between past and future, and in that sense has no temporal existence at all. Or, as would seem to be the precise Platonic emphasis, the present persists perpetually, as always being present and changeless, as *ousia* or eternal Being itself that, by extension, is the basis of our consciousness, ego, or subjectivity; that it is, in essence, our immortal soul. But to Heidegger Being is never outside of time, and in seeking to retrieve it from the "oblivion" to which he says philosophy consigns it, he claims that "the central problematic of all ontology is rooted in the phenomenon of time, if rightly seen and rightly explained" (*Being* 40). On the metaphysical plane time is represented as a parameter, he says. It is one dimensional and measured out in numbers, outside the course of human events. It is "leveled" off into an infinite succession of "nows" closed off from each other, without any internal link—"unapproachable" and without the nature of "nighness"—bereft of "open face-to-face encounters" that can "neither bring about nor measure nearness" (*Way* 104).

In this conception time is "inauthentic," a mere "fiction" that Heidegger associates with *das Man*, with "everyman" (*Being* 379ff). It is an "average everydayness" whereby *dasein* "becomes blind to its possibilities, and tranquilizes itself with that which is merely 'actual'" (239). Average everydayness, the "leveled" time of eternity, is "nonsituational" as well, abstracted from experience and merely derivative of authentic time. Indeed, the latter is itself the source of *dasein*'s authenticity, as time is understood in its most fundamental sense as a phenomenon "organic" to *dasein*, giving it substance, comprising it utterly in the rhythm of consequential events. Thus, Heidegger says, "time is not eternity" but finite and "fashions itself as a human, historical being-there." In the strictest sense, he claims we cannot say "there *was* a time when man was not. At all *times* man was and is and will be, because time produces itself only insofar as man is" (*Introduction* 84). Time, then, is immanent—as are Being and language in the same respect—manifest in the phenomena of the here and now. In Heidegger's ontological analysis, time is thereby not just inseparable from the finitude of *dasein*, but "primordial temporality"—the original unity of past, present, and future—structures *dasein*'s existence. He says as long as *dasein* exists, "it is never past, but always is as already having *been*, in the sense of 'I *am*-as-having-been'" (*Being* 376). Such is plainly the experience of Momaday's hero in *House Made of Dawn* and Proust's in *Remembrance of Things Past*. And, by this same perspective, the future is not the abstract, conceptualized "not yet," but it, too, is organic to *dasein* as the issue of anticipation. Heidegger thus fashions time—past, present, and future—as a process that unfolds in terms of the integration of these three "ecstasies," or directions, of temporality.[17]

A crucial feature of this view is that time not only offers a structure of occurrence but that the future is in one sense the most critical of these ecstasies of temporality. It alone provides the absolute warrant of *dasein's* finitude, its "Being-towards-death" (*Being* 378), and therefore constitutes the "primary meaning of existentiality" (376). More importantly, the future itself is manifest in *dasein's* "anticipatory resoluteness," in the latter's connection with Being and by that measure the resulting degree of *dasein's* authenticity and inauthenticity. Here, rather than possessing any particular place or position in its own right, the future rebounds to yield *dasein's* "primordial temporality," investing significance in the other ecstasies comprising *dasein,* the past and, given my concerns here, the peculiar nature of the present. As Heidegger says, *dasein* "projects" itself, exists "ahead of itself," capable of recalling the present from distraction of "everydayness" and reaching—"projecting"—into the future to realize the various possibilities afforded by its "own-most-possibility-for-Being." Thus, in the end, whether *dasein* projects itself or projects some sense of an authentic present makes no difference in this context, for what we customarily surmise to be a self, a soul, or human ego is but a process in this perspective, a narrative or story, as only time and language would mandate. Seen in this light, time is not only an integration of "moments" shaped by critical events, but a process allowing us to perceive *dasein* as a focus of concern and care, a form of the self persisting as an all "embracing" present, extending not just into the future as possibility but into the past as legacy. In contrast to the empty expanse of eternity, that linear sequence of indistinguishable now points, its specific temporality, and authenticity, is apparent in the "fullness of time," as the was, is, and will-be are realized in the instant, the *Augenblick*. At that instant the present is held in resoluteness, a resoluteness not necessarily of the self but of the situation in which the self is implied, a resoluteness that abides in the tension of seemingly discordant conditions of the stability of "things that are" on the one hand, and Being itself as unfolding and unremitting process on the other. Heidegger says:

> To the anticipation which goes with resoluteness, there belongs a Present in accordance with which a resolution discloses the Situation. In resoluteness, the Present is not only brought back from distraction with objects of one's closest concern, but it gets held in the future and in having been. That *Present* which is held in authentic temporality and which

thus is *authentic itself,* we call the *"moment of vision"* [*Augenblick*]. (*Being* 387)

In the *Augenblick*—the "blink" or "twinkling of an eye"—we are given precisely the Heideggerian rendering of *kairos*. The *Augenblick* is *kairos*, the way the term appears in the Diels and Kranz translations of Gorgias's "Funeral Oration," and the way Heidegger himself sees the issue, as he explicitly equates the two terms in his *Basic Problems of Phenomenology* (288). More importantly, in Heidegger's handling *kairos* is a most un-Platonic notion of the instant, fixing as it does the mark of authenticity utterly in the temporal realm. In his analysis, as *dasein* is grounded in the temporal realm, only in the latter is there the tension or polarity in understanding Being, where, as L. M. Vail says, there exist alterations or variations between an "everyday" or pedestrian understanding of Being and the "visionary moment," the *Augenblick,* that ultimately yields an authentic engagement with Being (12). Here, Heidegger's explanation is apt, particularly cogent to the matters at hand. As given above, *kairos* retrieves the present from "distraction," thereby establishing it as "authentic" by vesting it in the "to be" and the "having been." Indeed, in this explanation *kairos* does not come "in the present" at all but determines it, as the present is not an independent, pre-existing entity but, insofar as it is authentic, owes its "presence" utterly to *kairos*. In this sense of *kairos*, in the *Augenblick*, we are thus given the measure of authentic temporality. By its power the temporal ecstases of past, present, and future are mutually implied and "generative," so that *dasein* is not just grounded in temporality but by virtue of *kairos* authentically grounded in Being as well.

In Heidegger's later works these often threadbare reckonings concerning Being are repeatedly rounded out in terms of language. Especially as he comes to use *"physis"* and "earth" in place of Being, designating a process more clearly corporeal in nature, the association of Being and language is explicit. Here, Pöggeler's comments linking truth and language with the "historical speaking of a people" are a most apparent indication of the later Heidegger's expression of Being. As time is fashioned as a "human historical Being-there," it is fashioned itself by a culture's sacred narratives, myths, and rituals. In this connection of language, Being, and time, the idea of the *Augenblick* is most closely akin to the Gorgian *kairos*, especially as we look to two additional features critical to its makeup, those of "decision" and "situation." Each of these helps to explain the phenomenon of "authentic" human language coming to pass as the creative liaison between the three "ecstases" of temporality, where the *Augenblick* is itself decisive, disclosing the situation as the locale of "truth," of its happening in the event of language.

In dealing with the idea of decision and its relation to *kairos* we might return, more profitably, to Paul Tillich. In his work the metaphysics of presence gives way to aspects of knowing based not on the separation of subject and object, but upon their "intimacy"—not on "distance from but nearness to life, a community of knowing and the known," as he puts it (*Interpretation* 148). Indeed, Tillich's contention that knowledge has its ultimate source in "decisions" rather than logical analyses or empirical examinations gives us, I believe, an ontological conception of the matter and another significant insight into the nature of the Gorgian *kairos*. "All knowledge," he says, "even the most exact, the most subject to methodical technique contains fundamental interpretations rooted neither in formal evidence, nor in material probability, but in original views, in basic decisions" (143). What is most critical is that these "original views," these "basic decisions," are not those of the self-contained ego, the isolated subject detached and distanced from an object who "objectively" examines it, but they stem from a "vitalist relationship" or mutual implication that obtains between the two (148)—decisions grounded, let us say, in their "dialogue" that marks the Gorgian *kairos* initially as the vitalist relationship itself. Such an analysis seems to comport with statements offered many times over by Mario Untersteiner and others that the "decision" is not that of the subject involved but is willed by *kairos* (181).

As a matter of course, Heidegger's explanation of the nature of decision is more complicated than Tillich's, entangled throughout with his imperatives concerning language, but it thereby affords a more fitting access to the idea of the Gorgian *kairos*. The sense of a "vitalist relationship" is maintained, but here, conspicuously in the context of the *Augenblick*, it is seen as both a "separating out" and a "bringing together," a dialectic wholly of language played out as the ontological difference. Heidegger sees our language, the human *logos*, as the result of the "*Unter-Scheid*," the "difference" or "scission" (*Scheid*) between (*Unter*) Being and things, whose approach or convergence is manifest in "*Ent-scheidung*," the "de-cision" prevailing in the tension between them. So again this emphasis on the borderland. In this view Being and *dasein*, "world" and "thing" in the diction of this specific context—or, just as aptly, "Saying" and "speaking" (the human *logos*)—"traverse a middle" and "penetrate one another," and in the middle "they are one." But, as Heidegger insists, it is not a "fusion" but an "intimacy" that prevails, as can only be possible as each "divides itself cleanly and remains separate" (*Poetry* 202).

The conception is grounded, as all things Heideggerian, in the ontological difference. Without the difference there is no *dasein*, for then its definition would be merged in the complete wholeness of Being, which to every indication is death itself. Nor would there be Being without *dasein*, for the latter not only bears witness to Being but is in fact its sole means of manifestation. In the absence of either there can be no coming to pass of Being in language, as language provides the way of our participation in Being, and this participation is the measure of life itself (Richardson 579). Here, in the difference of Being and *dasein*, yet by virtue of their intimacy, the role of decision is strategic, as it indicates at once both "scission" and "de-cision," a separation that is also a coming together again. What results is a dialectic of a peculiar Heideggerian sort, with an emphasis in both ends and means wholly on language. Indeed, the relationship that obtains between Being and *dasein* occurs in the paradoxical sense of what Heidegger calls a "unifying scission," where "Saying" becomes a constitutive element of *dasein* in order for it to be an element of Being at all. The power of Being is never annulled nor diminished in this process but is unveiled and, more significantly, in "de-cision" it appropriates us to itself in *Ereignis*, in the process of its own coming to pass in language.[18]

Decision, then, does not mean human judgment or choice but initially the "scission" inherent in the separation of Being and *dasein*, leading back to the ontological difference itself (*Introduction* 110). And, just as fundamentally, it is also through "de-cision" that we experience, at propitious moments, the bringing together of "Saying" and "speaking," so human language authentically comes to presence in the moment of our response to Saying, to the "call" of Being. Thus, language, insofar as it is authentic, does not have human activity alone as its source but reposes in our relationship to language's origins as "Saying," ultimately constituting the "dwelling place for the life of man" (*Poetry* 191–92), that sense of *doxa* marked by the "great *dynastes*" of language. Most critically, however, in these circumstances of decision, of this "unifying scission," we indeed "undergo an experience with language" as, say, did Abel in that closing scene of *House Made of Dawn*, where language quite naturally "brings itself to language." Heidegger says that "one would think that this happens anyway, any time one speaks." But he explains that this is usually not the case, in that *Gerede*, "everyday language," is the rule:

> At whatever time and in whatever way we speak a language, language itself never has the floor. Any number of things are given voice in speaking, above all what we are speaking about: a set of facts, an occurrence, a question, a matter of concern. Only because in everyday speaking language does not bring itself to language but holds back, are we able simply to go ahead and speak a language, and so to deal with something and negotiate something by speaking. (*Way* 59)

Interestingly, Heidegger places "languages" that transcend cultural boundaries in this context, and is therefore especially critical of the idioms of modern technology and even of analytic and philosophical examinations of languages, condemning what he calls the "technicalization of all languages into the sole operative instrument of interplanetary information" (*Way* 58). His condemnation is apparently based in his belief that these languages are bereft of Saying, lacking the moral substance provided by a culture's customs, rituals, and traditions. More essentially, however, in this passage Heidegger is referring to "everyday speaking," to the process of "passing the word along" as part of day to day commerce, where we are obliged, if not in some sense obligated, to consider ourselves as the ultimate source of language. Most often we proceed in blissful ignorance of our role in this matter, as William Richardson says, for *dasein* is limited and language is therefore "profoundly negatived" by *dasein*'s finitude. As a result, *dasein* can accomplish only a "finite containment" of language because the latter, as Saying, persists without difference from Being itself (293). Consequently, it seems that we have invented language, that it lies at our beck and call, when in fact we discover ourselves only in and through language and thus, once again, it is language "that first brings man about, brings him into existence" (*Poetry* 192). Thus, what characterizes the event of de-cision is being claimed by language, prompting an awareness as a function of language that in various contexts Heidegger calls the "irruption" of *dasein*. De-cision then comes in the revelation of Saying, and comes only infrequently, at critical moments, marking language as something not exclusively human nor, once again, even human at all, and that ironically marks ourselves, as *dasein*, as something not exclusively "human." It is a very enticing idea, was addressed previously and will be again in due course, but for now we must realize that such a circumstance bears categorically upon the nature of the Gorgian *kairos*.

In this perspective language is truly the master and, as I have suggested, is necessarily joined to a circular logic, its only referents being other words in ever expanding tautologies. The moment is what it is because, by breaking the routine of what is regular and commonplace, it makes us aware; and by breaking the routine and making us aware, it achieves, straight away, its status as the moment. But in the de-cision of Saying and speaking, where we are mastered by language, we meet with and endure the experience of *kairos*, being taken by it, and in the instant indeed "undergo an experience with language." Awareness is here a function of the instant alone, and ironically it comes most readily when we are least in control, when we are altogether at the indulgence of "Saying." Gerald Bruns characterizes the effect of this encounter as an "epistemological crisis," one that ironically leaves us, just as Heidegger claims, "dumbstruck and dispossessed" (100). It is an apt choice of words indeed, as "crisis" is the very issue of the experience, of this turmoil and dispossession. In this context experience does not mean gaining knowledge or mastering a subject matter, or "being in the know," but as Bruns says, "being experienced" or "being open to what happens" (100). Even more essentially, it means "being experienced" in the reflexive sense, where the action is directed back upon the erstwhile doer of the action, where we are the thing or event that is experienced. The crisis precipitates not only reappraisals of accepted notions of the authenticity of the ego-subject, but all taken for granted imperatives of the human soul, leaving us dispossessed even of the seemingly most reliable assurances of the nature of the self. A renewed attitude toward language is the effect, for we are here dispossessed in particular of illusions that we are the masters of language. Dumbstruck in that sense, most definitely.

This, too, bears upon the nature of the Gorgian *kairos*. We would then see the "situation" not as predetermined or in any way complete in its own right but as the "scene" of de-cision itself, by its nature confirming a peculiar concurrence with the person involved, something of the sort I have previously suggested, in the sphere of the mystery of Being of which Marcel speaks, "where the distinction between what is in me and what is before me loses its meaning and its initial validity" (I.260). In any event, to isolate the speaker from the situation and thereby see the two as distinct is to resolve the event of *kairos* along lines of objectification, with the speaker as the subject and the situation as an object he confronts and manages, resulting in an emphasis on the speaker as the source of what to say and how to say it. Doing this is to frame the issue exclusively in terms of the Platonic *kairos*, specifically as the speaker's adaptation of language to what is fitting in time, place, and circumstance. And that, as I have claimed, is an interpretation at odds with the idea of the Gorgian *kairos* I am presenting here. In the latter any sense of the separateness of the speaker and the situation is derivative, following well in the wake of their de-cision, for there is initially a meditative correspondence between the two, surely of the sort we see between Abel and Francisco and more profoundly in the warriors and their situation depicted in Gorgias's "Funeral Oration." In the dialogue that results, they draw near, tend to touch in a radical closure of the distance between them that *kairos* prompts, so that both the speaker and the situation, like Saying and speaking, can be fully appreciated only in terms of their relationship to one another.

Such an interpretation is, of course, implied in the very phenomenon of *dasein*. Heidegger says the situation is disclosed as the *"da"* of *dasein*, the "there" in "which the existent is there" (*Being* 346). Here, again, the situation consists in its relationship with *dasein*, a relationship grounded in *dasein*'s awareness of the simple "that it is," providing a clearing by virtue of de-cision, a refuge from the obscurity and distraction of everydayness. In this light we might conceive of the situation in ways very similar to *dasein* itself, as an "in-between area," William Richardson says, a space of intimacy or "mediation" between Being and *dasein* that gives rise to a "luminous clarity" where truth occurs (421). The situation is thus apparent in terms of the event of *kairos*, never as a circumstance separate from *kairos* or separate from *dasein*, and always as a function of *dasein*'s being in the midst of things, of its "Being-in-the-world." The situation is that place where *kairos* occurs and exerts its effects, where we hear the call of language and are claimed by it in our "hearkening attunement" that is, at once, a listening and belonging (*Early* 67).

I realize that the "call" remains, at long last, as mysterious as ever, an unfortunate turn of events in that the call, the "Saying" of language, is essential to understanding this interpretation of the Gorgian *kairos*. Here the issue centers on just how literally to take Heidegger, as seems to be the case invariably. But to conceive of the call as either literal or metaphorical is of little help, because that distinction is itself based upon a practice of objectification fully in conflict with Heidegger's theory of language. There are, for instance, examples of the sort offered by Julian Jaynes, who argues in his extravagant interpretation of the pre-Socratic mentality that the heroes of Homeric legend were bereft of consciousness, and that in its place the "voices of the Gods" were present in the right hemisphere of the brain and the "call" was heard in the left. Even as traditional ideas of consciousness are deposed in his analysis, Jaynes' commitment to the idea of self-presence informs the whole of it. In fact, however creative or bizarre his account may be, in certain strategic aspects it is strikingly akin to Plato's doctrine of the soul. A sort of inner sanctum of the mind prevails, a place beyond measures of time and space, inviolate and set off against external reality. Such a literalism can come about only by way of a thorough enchantment with the pre-Socratic mystery, yet here is an explanation proposed by way of a conception of language that posits it as something wholly linear, as a thing that flows from a source, through a medium, to a receptor, even as these agencies are all situated in the mind. So, for all its allure, Jaynes' explanation of the pre-Socratic mentality is based on a relatively conventional epistemology, a refashioning of Plato's dialectic where separate components of the soul literally carry on a conversation.

For his part Heidegger is no less enchanted with the pre-Socratic mystery than Jaynes, but instead of representing language conceptually, as essentially a means of communication, he affirms its ontological status, its nature as Being. Again, as he tells us, we are concerned here not with phonetic data, information gathering, vocal chords or various other physical or psychological accouterments but, quite explicitly, with language in its essence as *physis*. He therefore emphasizes its most visceral aspects, saying we hear the sound of language "rising like the earth," possessing an "earthyness" that he claims is "held with the harmony that attunes the regions of the world's structure, playing them in chorus" (*Way* 101). We might occupy ourselves looking for hidden meanings in this, ways to clear the snarls, but the effort would be for naught. The issue, as ever with Heidegger, is not what lies behind the words, what they represent or even their facility as a means of communication, but their singularity with Being, that in this context mark words as the catalyst of *Ereignis,* of a process of appropriation. And so the notion of the call relies on the view that language is not authentic, nor even language in its most essential sense, when used as critique, observation, definition, or even explanation; but is so only as it comes prior to these uses, rooted then in an ontology that in all critical circumstances mandates the proposition that language uses us. Thus, in this account *logos* is free, language its own master, and in Heidegger's equating its "earthyness" with a certain physical prowess, the account parallels in significant respects the Gorgian view of language as the great *dynastes.*

I have indicated that this idea of *kairos* is apparent in a passage from *Being and Time,* where Heidegger says, "hearing constitutes the primary and authentic way in which Dasein is open for its own most potentiality-for-Being—as in hearing the voice of a friend whom every Dasein carries with it" (206). The passage provides, in Jacques Derrida's analysis at least, one of the more vital connections of Heidegger's preoccupation with *dasein* in *Being and Time* with his subsequent emphasis on language in his later works. And it provides as well, in the paradigm of "the voice" *dasein* carries with it, an apt equivalent of the Gorgian *kairos.* Heidegger identifies this voice as the "call of conscience," but in his formulation the call is not an inner voice, at least not initially, nor does not come from someone else in the world and often comes against our expectations and even against our will. It "comes *from* me and yet *from beyond and over me*" (320, his italics). If this explanation is as elusive as what it means to explain, perhaps the complexities are expressive of Heidegger's intent concerning this specific matter of the call, indicative of what Derrida calls a "pre-platonic hearing of *logos*" ("Heidegger's" 173). Whether the appropriate phrase is "pre-platonic," "pre-Socratic," or "Gorgian"—as I have chosen variously to designate the phenomenon—it definitely provides a most obliging means to our understanding of *dasein,* that it is neither self nor subject, man nor woman. As it must, *dasein* carries the voice within itself, yet *dasein* is, at the same time, always "beside itself," for there is no sense of self-presence inherent in it, as it subsists only in terms of its relation to Being. In particular, as I have stipulated, *dasein* is an agent of the semantic forces enveloping it. It is the "irruption" of these forces, so that in the phenomenon of the call, in *kairos, dasein* is affirmed in *doxa,* the verbal environment that constitutes its being.

The notion of the call, of Saying, is thus more complex than some mere cipher for "inner voice," being intricate enough to warrant Heidegger's oblique, even convoluted descriptions of it. Here, in the call of language, not only are traditional notions of subjective awareness challenged, but specifically in the case of *dasein* there is no interior subjectivity from which the voice might originate. Or, given what we have seen of it, the idea of internal and external does not enter into these deliberations, for that distinction is ultimately resolved into something more essential and vital than each. The proper perspective in this matter, then, lies in a wholeness that precedes any analysis we might care to apply, in an inclusiveness that hearing prompts, and excluding in particular the source of the call in the singularity of the self, as an "inner voice" of the soul or some such. I have spoken of this wholeness in terms expressed by Heidegger and Heraclitus, especially the latter's contention that it is wise to say that all things are one. In fact, it is through this wholeness that we can understand further the connections of *dasein*, language, and the peculiar nature of the ontological difference. In this perspective the call of language is apparent as a matter of a belonging integral to the call, achieved not by self will or rule but as de-cision, an implication of the event of *dasein* itself. It is a belonging by nature and culture to *doxa*, as that is the voice, in effect, that *dasein* hears and to which it belongs.

Here the call is a consequence—in a way, a recompense—of the ontological difference, of *dasein*'s temporal nature. The difference always abides but always a greater proximity is at hand, as that very difference, by virtue of the "unifying scission," gives rise to the sense of wholeness in the first place. As the call originates "from within me" while at the same time "from beyond and over me," the "voice of the friend" calls *dasein* back from its separation from this wholeness of Being, from its "fallen state" of being dispersed "into inauthenticity," as Dastur says ("Language" 360). It has the character of a call because *dasein* is removed from Being, but insofar as the call is heeded a greater sense of identity between the call and *dasein* is achieved, between what calls and what is called. Again, the measure of *dasein* is never self-presence but its temporality, a measure of the ontological difference that ensues inevitably in our distance from Being. Thus, Dastur says, the call has its source in the "radical finitude" of *dasein*, as *dasein* is summoned to comprehend Being (361) that can indeed be comprehended only by virtue of *dasein*'s "finitude." And as the summons in all cases is resolved into language, into "discourses," as Heidegger says, they remain "discourses in the uncanny mode of keeping silent" (*Being* 322). Indeed, this emphasis on hearing the call of language, rather than initially on speaking, epitomizes the Gorgian *kairos*.

This motif of "keeping silent" to hear the "voice of a friend" is to perceive *kairos* as it is played out in the extraordinary exchange between Abel and Francisco in the closing pages of *House Made of Dawn*. The scene's vitality is vested in its silence, as Abel listens, silent to hear Francisco's words though they verge on delirium. But *doxa* is here, present in Momaday's sense of racial memory, that province of words and communal memory where the separateness of Francisco and Abel is overcome, redeemed in the vital relationship between the two subsisting wholly as language. In Momaday's presentation Francisco's memory, evident in the image of the runners at dawn, becomes Abel's, and Abel thereby "enters into the old rhythms of the tongue." Or, in Heidegger's idiom, Francisco embodies Saying and Abel the human response, who in his openness to Saying has been "brought into his own" by language, "appropriated" by it, so that he experiences the words, embodies them, and does not clutter up the moment with idle chatter. It is "a matter of letting language speak from within language, of itself, saying its nature" (*Way* 58). Here, listening is a belonging in the most profound sense, of being held by our beliefs in the particular grace that *kairos* bestows. In racial memory, living itself is a remembering, a listening and belonging to *doxa* that encompasses both past and future in the focus of the present moment. In his listening and belonging Abel personifies this *doxa*. In that moment he *is* the word.

Thus, it is hardly ironic that a similar emphasis is apparent in the Christian tradition where the paragon of a dying god is fundamental. In this respect the Christ figure is as we are, a being whose authenticity can be achieved only in being subject to death. I have mentioned the motif of halos as signs of glory, of *doxa*, but especially revealing in this context are the various, mostly medieval depictions of Christ, of his image adorned by a halo divided by into quadrants. It is the perfect circle of Being intersected by the cross, the symbol of his sacrifice and death, and a most significant allusion to the ontological difference as the mystery of this god's incarnation as the son of man. This persona of Christ—in the paradox of his oneness with Being a measure purely of his separation from it—is not, as I have said, the god of the philosophers and theologians, whose science precludes such things, but the god of storytellers and listeners. Like the Greek hero, the glory he achieves in dying both manifests and enriches the *doxa* from which it is drawn, from which he himself is drawn. He is then the bridge or passage to the Christian *doxa* that he literally embodies, a passage marked by his mortality and death in a sacrifice so profound as to overcome each. In this sense he is the way, the *kairos,* just as he is so often depicted in Christian mythology. As much as Abel, he is made of words. He *is* the word, its categorical archetype in the Christian tradition.

In these instances there is not only an absence of clear distinctions between internal and external, word and deed, but purely a knowledge gathered, powerful and imposing, rising to the level of Being in language's embodiment in the blink of an eye. *Kairos*, then, is a disclosure of authentic being realized only in that moment we heed and faithfully respond to call of Being. And it is the moment simply for the reason that the call of language makes it so, shattering the decorum and regularities of everydayness in its connection to language and temporality, to a mortality transfigured that is earth itself. Of all that goes into it, this is the most essential attribute of *kairos* in the Gorgian sense, surely that seeming best to reflect the fate of the Athenian warriors eulogized by Gorgias. So it fits that his purpose is to glorify these men, as in this case the key to the Gorgian *kairos* lies not initially, nor even essentially, in a speaker's use of language but in an openness, an act of listening that leads to the apotheosis related in his "Oration," where the Saying of language—the call heeded by the warriors in *kairos*—is inherent in the *doxa* it both reflects and engenders. Here, in particular, *doxa* is manifest as glory, in the acclaim of their countrymen, as a "longing" for these warriors that has not died but paradoxically "lives on immortal, among bodies not immortal, though they do not live" (82.B6; Sprague 49).

They were warriors then in the Homeric sense, evincing a prowess not exactly their own but rising as an expression of the power of *physis*, of earth. The power is then manifestly an expression of language itself, of a persuasive *logos* borne of *kairos*. As *logos* is joined to these life and death matters, it is not only corporeally apparent but on that basis transfigures, bearing the apotheosis that is the glory of the warriors, resolved in the only irrefutable affirmation of their mortality possible. These warriors thereby "achieved an excellence which is divine" only because they possessed by virtue of their human nature "the destiny of death." Here, by way of the Gorgian *kairos*, not only are the divine and the human absolutely dependent on one another, but in this perspective there can be no separation of these men from their experience and, as Gorgias would seem to suggest, in the dynamics of *doxa* no separation between themselves and others of the *polis*. If their fate were simply inherent in the human condition, in their sacrifice it is brought to the fore, made obvious and compelling by a situation shaped by the drawing near of death. Enveloped as they were in that immense sensibility, the spell of "everydayness" was broken. They were thus overtaken by an "ontological" awareness that moves beyond propriety, the mere "rigidity of the law," and embodies instead the persuasive force of an irrational *logos*—a universal divine law, Gorgias says—released in the instant of decision, in a *kairos* that has as its purpose "the right thing at the right moment."

This sense of apotheosis is fundamental to Lincoln's Address, to Chamberlain's conviction concerning the "heroic souls" defending Little Round Top. It is manifest in Sarpedon, Hector, and others of the *Iliad*. And if war seems to make the experience of *kairos* all the more conspicuous, its realization is as likely to transpire otherwise, even in what may seem to be the most humble and wayside circumstances. As it is realized in the Christ myth, played out in some fashion in the lives of all people everywhere. Or in the experience of Abel and Francisco. Even that of Carry Moccasin. *Kairos* transfigures not just these warriors, saviors, and Indians, fictional or otherwise, but all who participate in the *doxa*, embracing those who tell the story as much as those of whom the story is told. It embraces most especially those who listen, as the word is then given flesh, their flesh, in their listening and belonging.

In this sense the Gorgian *kairos*, and Gorgian rhetoric as such, can be understood essentially as a matter of listening, in terms of the latter removed from any metaphysical basis in logocentric views of language and placed wholly in phenomenal, temporal world of words, in that borderland of "divine power" immanent as earth. *Kairos* can then be seen as the Heideggerian *Augenblick*, as the de-cision of Saying and our speaking, of hearing language and responding to it, and thus to become in that instant, in the blink of an eye, the word itself. By this peculiar emphasis on *verba,* the full rhetorical implications of *kairos* become apparent. To put it more directly, the issue centers on the word "appropriate," whether it signals our felicitous use of language, to make it suitable to fit our purposes and the occasion; or whether it itself expresses the action and relegates us to its means of expression. In this latter case, "appropriate" is "to be appropriated," as in this view of language apprehension itself is not a faculty belonging to man but, as Heidegger says, "the happening that has man" (*Introduction* 141).

Of course, these ideas regarding language can be dismissed, and have been repeatedly, as the results of excursions far too mystical and removed from the practical, day to day concerns of rhetoric. Yet they remain practical for many of the same reasons they are mysterious. Indeed, mystery is very often the way we used to think about things before we learned to think otherwise, as this interpretation of the Gorgian *kairos* surely exemplifies to some a way of thinking that has long since passed into antiquity. Or it could be that we have lost the ability to think about things that really matter, having long since forgotten, as Heidegger contends, the question of Being.[19]

Once again, what we know is far less than who we are, even as we fall prey to the arrogance of believing that we make our own knowledge, that reality itself exists at the behest of our powers of reasoning. Being is inherently the mystery, and in Heidegger's joining it to language in forms such as *thanc*, the mystery is acknowledged, even revered. He establishes a number of such connections that were philosophy's purpose to elude all along. As the quest for philosophic truth has been shaped traditionally by what is eternal and non-verbal, the most significant barriers to this goal were quite naturally discovered in the folds of language and time. At least such seems to be the case in Plato's account, where language and time not only stand in contrast to Truth and Reality, but are precisely the evils from which philosophy would deliver us. The soul itself is the essential manifestation of what Plato deems to be eternal, a steady state of serenity and dispassion, and expressly in that sense is a reflection of ordinary time and language, what Heidegger condemns as *dasein* absorbed in its everydayness, the "They-self" (*Being* 149). Of course, the Platonic attitude is one to put Sophistic rhetoric in a most unfavorable light, to say nothing more of what its condemnation of language and time does specifically to the Gorgian *kairos*.

And then in remarkable contrast Heidegger identifies Being in its immanence with both language and time. Being is manifest in Saying, he says, in the root unfolding of language that allows things to be, bringing them to the fore by the force of a phenomenology joining language with Being, as Being in its immanence is itself joined with the earth. In this perspective *kairos* does not necessarily pertain to the stupendous, astonishing, or dire but is essentially an encounter with language, with Saying, and thereby with the unfamiliar or uncanny in breaking routines of everydayness, the spells of tedium. It is mysterious in that sense, as we are taken outside of ourselves, brought to the realization that we are used by forces beyond our control, and that there are no easily gained truths, perhaps for us no certain truths at all that are of consequence. If so, nothing is so practical. And how could it be otherwise, given the exigencies of the here and now, given indeed our mortal fate and all that we are, the very mysteries Plato would explain away or have us transcend? In this alliance of language and Being that Heidegger portrays—as does Gorgias, in effect—the issue concerning the authenticity of language is resolved at long last as a reverence for language as for life itself, in the belief that language is a gift never belonging to us but allowed by some special grace that ultimately determines the beings that we are.

Thus, quite often it seems—even to the most reasonable of people, perhaps especially to them—as if the governing principle of the universe is the lack of governance. In fact, it could be that the most significant lesson to be drawn from this ontological perspective on language, that all things are in some sense made of words, is that all things, that reality as such, are subject to all the contradictions and ambiguities of language, the result of some strange but glorious mix, a cosmos suffuse with chaos, at times presumptively chaos itself. In this perspective there is no greater sense of Being beyond the here and now to which we can be delivered, or even come to know. We are cave dwellers perpetually, peering into the shadows, seeing always as through a glass darkly. Nothing has a definite shape, clear and constant, and we are confronted with the appearances of Being, most aptly its changing and shifting ruses making it strictly apropos to the Gorgian *kairos*. In this perspective the uncanny, the "irrational *logos*," has its place within the rational *taxis*, is integral to it. Insofar as we are made of words, the uncanny is not only all around us but most essentially within us. It is the "nonhuman," as we are foreigners to ourselves, possessed by otherness.

In this view *kairos* is a threshold that opens to the uncanny, marking the transition to novelties of chance and change, blurring differences between truth and deception, right and wrong. Essentially, the Gorgian *kairos* is the channel of the transformation of *doxa* into para-

doxa, and in this sense *kairos* is most evidently rooted in the distant past of myth, legend, and lore. Here we encounter a tradition as imposing as anything Plato and the philosophers have proposed in its place. Indeed, its presence is literally manifest by nothing less than a personification of Being as ancient and enduring as our reckonings with the things that are, a creature of epic proportions that we have come to call the trickster.

6 Paradox and the Power of the Possible: *Kairos* as the Mark of the Trickster

Eshu, do not undo me,
Do not falsify the words of my mouth
Do not misguide the movements of my feet.
You who translate yesterday's words into novel utterances,
Do not undo me.

—Oriki Esu[1]

Here I begin where I began before, to say that there are times indeed when Gorgias's sense of *kairos* seems too remote, altogether too arcane to recover, while Heidegger's thought, if sufficiently archaic itself in crucial respects to give shape to the Gorgian *kairos,* is for that very reason no less mystifying. Nevertheless, the issue seems to rest with the perspectives of each on the nature of language, and for each these are expressed in decidedly ontological terms. The likenesses are surely apparent as each deals in matters of language as much as philosophy, especially as we see in philosophy the enormous influence of Plato, where Being is unchanging, transcendent, and eternal. In contrast, eternity is never the issue to Gorgias and Heidegger, as in their analyses things temporally abide, shaped more by possibility than actuality, and where, of particular significance in realms of rhetoric, distinctions between language and Being are virtually erased. Here, in contrast to Plato, the only possible sense of Being comes in the course of process and change, and we then see its work in the flux of cultural beliefs, in the unfamiliar within the familiar, in shifting contours of *doxa* yielding endless displays of para-*doxa*. As Being is made flush with the earth and coupled with language, *logos* is free and goes its own way, displacing conventional notion of reality and reason. So my emphasis in this chapter is on this mystery of Being and language, as perhaps it has been all along, as it necessarily must be given the perspectives of both Gorgias and Heidegger on the nature and power of language.

And if this perspective seems mixed with a repudiation of reason, perhaps it only seems that way. Rhetoric itself has been construed traditionally as a *techne*, as an art or science providing a methodical means of using language to exert one's will to achieve particular ends and purposes. But more essentially rhetoric not only deals with language but is necessarily subordinate to it, and, as I have argued, in Heidegger's view the power of language, as Being itself, is mysterious, unfathomable by reason alone in any number of respects. In any event, it is hardly unusual to claim his views on Being and language are in various ways contrary to the mandates of reason, certainly as they are apparent in many conventional *technai* of rhetoric. In terms of Gerald Bruns's remarks specifically, I have discussed the irregular or unconventional turn of Heidegger's thought. Indeed, Bruns goes on to say that repetition and confusion are constituent features of the "dark, comic world" that Heidegger constructs (174 *passim*). The phrase seems particularly apt, but Bruns, whose views on this matter are definitely worthy of respect, does not offer the judgment to accuse or condemn. As I say, there is as much to be learned and realized in the dark of Plato's cave, that wisdom of catacombs and kivas, as can ever be gained in the blinding glare of his reality. And, in Heidegger, there is truly no sense without shadow, and to know his world, as Gadamer suggests, it is essential to know the night. And it is not Heidegger's "dark" world alone that is the focus of our concerns but always the more inclusive matter of Being itself, in which issues of darkness necessarily loom, for they partake so obviously of the mystery of Being. There is in Heidegger's analysis an enduring sense of the "uncanny" to Being, a wholeness that embraces, simultaneously and in equal measure, reason and contradiction, lucidity and ambiguity, *doxa* and para-*doxa*. In this sense Being encompasses more than just reason, and surely more than what is only dark and comic.

Indeed, the mystery that there is something rather than nothing, conceived in the strictest analytic sense, is no mystery at all. The rationale runs as follows: If there is nothing, then we *are not*, and thereby utterly incapable of being mystified by the fact of our nothingness. But the fact that we *are* annuls the mystery, as life is then a necessary condition for all things, categorically excluding nothingness. Thus, nothingness, as Being itself, is "absolute or not at all." This is Parmenides' argument specifically (Diels 28B 8.7), and generally seems to be Plato's as well (*The Sophist* 239c–241e), as each endeavors to refute those who would have us believe that "what is not has some sort of Being" (240c). Yet life remains mysterious all the same, surely a profound and passionate thing as it is lived in the shadow of death, of nothingness. All rationales to the contrary, the mystery begins in embracing all that seems to run counter to reason and then, circling back in the great wholeness of Being, embraces reason as well. In some fashion or another all cultures have come to some manageable accord with this mystery of Being, doing so in any number of creative responses, even if in this enlightened age it is to ignore the mystery altogether. But certain responses remain far more ancient than reason itself, in stories, myths, and rituals. Though most of these may seem now as those whited sepulchres filled with dead men's bones, at least one persists, seemingly by necessity given to continual rebirth, and that is the myth of the trickster, whose purpose is most appropriately, as Sheila Moon says, "to keep the beauty of realization from being static" (98).

The myth implies this much and more, to embrace most crucially matters of *physis*, language, and rhetoric. The trickster quite literally brings us down to earth to encounter here, rather than in some transcendental rendezvous, the nearness of Being. Indeed, in significant ways earth is realized here in its peculiar Heideggerian emphasis. Thus, it is neither a planet nor even nature, though it is, as Reginald Lilly says, "some particular earth," of a people or works of art, "a place of rootedness capable of proffering, given an epoch and world, a nonhistorical possibility" (xvi). This is the connection I have attempted to develop from the start, of *physis* with *doxa* and each with language itself, and in this chapter I attempt to develop it more accessibly in terms of the mythological figure of the trickster. He is most explicitly *logos* as language, forever defying its meaning solely as reason. In his person he harbors language's ambiguities, contradictions, and incongruities, his antics allowing us to see language in that perspective. George Steiner says the genius of language lies in its "dialectic of alternity," its aptness for "disguise" and "false garb" being "a counter-factuality" that is "overwhelmingly positive and creative" (*Babel* 226). Such attributes are as much the mark of the trickster as of language, serving as a way to understand the dynamics of the Gorgian *kairos* not just in contrast to the Platonic *kairos* but as a means to understand the connection of Being and language that is the very nature of the Gorgian *kairos*.

Thus, in a sense Heidegger bears some kinship to the trickster, and not just in aspects of his "dark, comic world," but more obviously in the fact that no thinker, ancient or modern, spells out alternatives to the main currents of Western thought as ably and completely. If, as a result, his thought is laden with convoluted ways of expression, that is as it must be, though as we examine some of the more familiar characteristics of the trickster, we might thereby escape what is most cryptic in Heidegger's jargon while still retaining in important respects what is essentially his. And here, in these remaining pages, we see the trickster not just as a mythological figure attached to any particular culture but as an archetype or way of thought that serves as a paradigm to understand further Heidegger, Sophistry, and the Gorgian *kairos*. Thus, while much of this chapter comes as a summary of ideas already in place, this approach provides a way of appreciating the "arcane" as neither cryptic nor static but as the "beauty of realization" given flesh, quickened in the figure of the trickster.

A Primer on the Trickster: The Pythagoreans and "The Terrible Secret of the Irrational"

So, to begin this fleshing out, I open with a story. The Pythagorean Brotherhood believed in keeping their affairs well ordered and regular. They were suspicious of outsiders and eccentric in various other ways, but as all cultists they were very close-knit, bound together in this particular case by the incalculably valuable wisdom that the world is comprehensible through rational means. They believed that all things could be reduced to relations of numbers, that indeed all things *were* numbers. This wisdom was fixed in the purest simplicity, in a *logos* certifying that the world was so constituted that it would necessarily give up its mysteries if we but applied the rational methods of discovery that they embraced. Surely the most notable of these methods was the Pythagorean theorem itself, for with just a little agility and discipline on the practitioner's part it would promptly dispense some of the most remarkable of nature's truths. But there was a flaw. When the sides of a square were taken as single integers, the dissecting diagonal could only be expressed as the square root of two, an irrational number. That their beloved theorem would yield such an anomaly in a matter so basic was not just a cause for grave concern but a scandal to be contained at all costs. You see, the rationality of the world was dreadfully serious business to the brothers, and discretion in such matters was a responsibility they took deeply to heart. Fortunately, it was just the type of anomaly that their secretive community was well equipped to handle. In fact, legend has it that one of their number, a relatively witless malcontent by the name of Hippasus, was taken out to sea and drowned for revealing to outsiders this "terrible secret of the irrational."[2]

There is a lesson to be learned here, one that all of us in some measure have learned remarkably well. As much as the Pythagoreans, we, too, often take great umbrage when things seem to run counter to reason, as if that involves, in itself, a great miscarriage of justice. We thereby presume to construct a reserve or retreat, some immense ambiance of non-contradiction and clarity, if not perfect truth itself, that secures our management of the mystery of Being. And the matter is wholly peremptory, involving more than *logos* regarded as reason alone but *logos* as the warrant of the rationality of the world. From the time of the Pythagoreans and well before, *logos* designated the lawfulness of the flux and was therefore the accord necessary to all knowledge, that perfect fit between external reality and ourselves that would accommodate differences, whether ontological or epistemological, so that the known could be known by virtue of its susceptibility to our way of knowing it. Initially, then, *logos* meant language, a very obvious result in that language provides the primal means of our knowing anything. That it somewhat belatedly comes to mean reason, even to the exclusion of language, is the result of the increasing refinement and sophistication that philosophy brought to the relationship of the knower and the known. Plato was a rationalist and idealist, presenting a particular philosophy among many others, but the philosopher's faith is principally his legacy, this belief that reason is not only our surest hedge against the dark, corrupting influences of ambiguity and contradiction but that the realm of perfect order and clarity that reason yields is the only real one.

So it is, Heidegger claims, *nihil est sine ratione*—"Nothing is without reason"—is the principle we accept as unerring. He says the phrase expresses our faith that there is a direct route to reality to be achieved by means of reason, a reality that is solitary, steadfast, and unsullied by human presence, perspective, or language. Obvious corollaries are that ambiguities are to be avoided, that deception is a departure from the unvarnished truths that reason yields, that both ambiguity and deception are the results of the failure of language to properly foster and apply the rule of reason. In this ethereal realm of changeless reality, something either is or is not, and to say that it is simultaneously both is to commit the quintessential error, a contradiction. Indeed, the latter is of such weight that it has the authority of Latin inscription in its own right: *Esse non potest, quod implicat contradictionem.* To wit, "Whatever implies a contradiction cannot be" (*Principle* 17). It is, of course, one of the first principles of reason that the Pythagoreans learned and endeavored to apply regarding their theorem. And, as a final, most fitting corollary, all problems that reason seeks to correct—ambiguities, contradictions, and deceptions—are considered the result of the moral or mental defect of the person posing them, or they are the result of the irrational, the legacy of language's claim to *logos,* that in this context of philosophy and Pythagorean tribulations amount to essentially the same thing.

It is therefore important to recall that neither reason nor the reality it discloses provides our most immediate and enduring awareness of the world. That is accomplished by language itself, and, if the "truth" be told, the world it discloses is replete with ambiguity, contradiction, and deception. Language not only reflects this world but in most essential ways creates it. Indeed, according to Heidegger, language *is* this world—Being itself—for here we dwell in the experience of language insofar as we dwell in the world at all, a ready made world where language is our master. Thus, this world is itself a "nonhuman occurrence," Heidegger says, that "behooves" our being. We are "thrown" into it, discovering or finding ourselves in the midst of things with no clear indications of beginnings or ends. Here, things are given presence in the flux of the revealing and concealing power of language, and reason is just another of the tricks of language, a figure of speech, say, no different than anaphora or antithesis. Here, as well, instances of the "nonhuman," the "uncanny," are of great significance. In these instances reason fails us, and in attempting to apply it we are at a loss for a satisfactory account of "what is," left simply to experience the wonder and often shattering impact of this place.

This is the world that Heidegger seeks to bring to our awareness, and once we give it some consideration such a world is neither illusory nor at all archaic. It has always been with us and hitherto acknowledged in various and interesting ways. As it was by Heraclitus,

Heidegger's mentor in many of these matter. Jean Beaufret tells us of the peculiar, "ever-changing God of Heraclitus," who is all strife and struggle and yet is harmony as well, "the original unity of all oppositions." In fact, such a harmony is always implicit in the interface of these opposing forces, a curious rendition of ying and yang at the heart of pre-Socratic thought (72–73). Indeed, this is the world of the trickster, the magician and lord of misrule, the shape-shifter who confounds our anticipations of the way things should be and anchors us instead in an earthly realm where truth and deception, clarity and ambiguity, sense and nonsense are complementary rather than essentially opposing forces. In equal measure these would-be opposites are inextricably woven into the process of life and growth of this world, into that mystery of Being impervious to claims that they are "contradictory."

The trickster, then, is the terrible secret of the irrational in the flesh, exposed and realized. Even embraced. In myth, story, and ritual he is reason's antithesis, having achieved in his long history a status as nefarious as reason's is reputable. He is the exemplary manifestation of *logos* as language, for he is, in very crucial ways, language itself. He proceeds by hint and indirection, ambiguity and deception, being both hero and buffoon, giver and taker, as often the one tricked as the one who tricks, embracing in this manner anomalies central to our being. As opposed to the transcendental realm where reason rules, the trickster lives his life flush with the earth, in the mess and muddle, appearing very often as carnal and amoral as a result. Indeed, he is not just encumbered by physical needs but driven by them, seeming at times constituted by them. He would reduce us to "mere bellies," as Lewis Hyde says, though there is another organ of the anatomy, the male anatomy at any rate, that in many trickster tales looms at least as large. Consequently, his rapacious hunger for food is not only matched but generally exceeded by that for sex. In one of the tales from the Winnebago Cycle, for instance, trickster is burdened with a phallus of such enormous proportions that he carries it with him in a sack. Not only is it so heavy and ungainly that he must sling the sack over his back to bear it, but it is so demanding and perverse as to act continually of its own volition, as when it strikes out on its own and discovers and rapes a chief's daughter. Sure enough, Plato warned us of the danger, specifically chastising the phallus in *Timaeus,* claiming "the organ of generation" to be so rebellious and self-willed that it seeks to gain absolute mastery, "like an animal disobedient to reason."[3] In fact, Plato surely understates the matter, as in the Winnebago tale, and many others similar to it, the size and weight of trickster's penis is the complete measure of his being. It is who he is. In any event, it can be claimed that in this particular disposition the trickster's influence is farthest removed from the purpose Plato assigned to our immortal soul, as the seat of our reason, wisdom, and dealings with the transcendental, our solitary warrant of spiritual significance.

The trickster is thereby appropriately dubbed "the master of intercourse" (Pelton 122). But he is doubly so, as the context here embraces the primordial and earthly, both the physical and erstwhile symbolic. Thus, beyond the bawdy sexual emphasis, intercourse is as likely a matter of words alone, sometimes in a most august sense, as the trickster is often portrayed as the messenger of the gods. As is the trickster Hermes, the creator of language and master of deceit, who specifically in that role serves as an intermediary between humankind and the powers of Being, nature, or *physis*. In this capacity the trickster deals in banter and allure, noble or otherwise, in an intercourse of words that bring couplings and connections that can indeed, especially in his identification with *physis*, be seen simply as another expression of his sexual prowess. Thus, through it all, the most telling characteristic of the trickster is his verbal prowess, equally the result of his identification with earth, as he wields skillfully and with great aplomb the power of language. He is the consummate storyteller, a word user and abuser, an out and out smooth talker who transforms the world by words. In some sense he is the source of Isocrates' inspiration, being the very embodiment of *logos* as the creative force of civilization lifting us above brutish resolutions of conflict; in another sense he simply has a way with words because he has no other way. So he stands in opposition to the guileless, often clueless hero with swelling muscles and a club, as Hermes is opposed to Hercules, Odysseus to Achilles, or as the scheming and fast talking Br'er Rabbit is opposed to the brute, dumb force of Br'er Bear who confronts all adversaries with a single, simple mindset, and the expression, "I'm going to knock his head clean off!"[4] The sense of language personified by the trickster in this regard, though owing little to reason and its sundry dialectics or dictates, would seem to serve facilely as a counterpoint to violence. Seemingly, this transition to more idyllic resolutions of conflicts is rhetoric's crowning achievement, marking its very essence.

But there is far more to the issue. In his personification of language, the trickster is a shape-shifter, purely an apparition, seeming at times to be a mere feature of the landscape, and then at other times to be an animal, a monster, or best friend, always capable of assuming the profuse and varied appearances that Being can project. Yet he is no less real for his many guises, as real in them as otherwise, for they *are* his reality. One of the more famous instances of this shape-shifting occurs in Yoruba mythology, in a story featuring the trickster figure Eshu. It has been given currency in rhetorical theory by Henry Louis Gates, and it tells of two very close friends sworn to everlasting goodwill. They were farmers, working their adjoining fields as Eshu came traveling the boundary between them. He wore a cap in two colors, one side white and the other black, the seam between the colors of his cap as sundering as the boundary he trod. In some versions of the story Eshu even engages the two men separately in idle talk before proceeding on his way, and in other versions he rides a horse shaded similarly to his cap, one side white and the other black.

But once he leaves, the two friends—seemingly so inseparable to see all things reasonably and thus the same way—fall into a quarrel over the color of the cap worn by the passerby. One sees it as white, the other as black; and their dispute continues vehemently, so intense that in some accounts it becomes physically violent and has to be resolved by adjudication, a proceeding at which Eshu appears, only to offer the excuse that "to sow dissension is my great delight." Such a justification reveals, as does the entire story, the ambiguities inherent in knowledge, as Eshu does great violence to old habits of perception and understanding. He tells the farmers that in their pledge of eternal friendship they failed to reckon with his authority, and that those who failed to put Eshu first in all matters have only themselves to blame when things go wrong.[5] In an odd turnabout he comes not to bring peace but brandishing a sword, sometimes inspiring outright violence and always agonizing reappraisals. The irony becomes the trickster, as it does rhetoric. Even as he might be the messenger of the gods, the gods are forever fickle. Even as he might be the means of peaceful resolutions of conflicts, before all else he is language, harboring all its equivocations and strife. Lewis Hyde says that Eshu stands between the two worlds of the friends, "or between two of anything trying to communicate, not like some high-fidelity sender-receiver, but more like the atmosphere itself, shifting, cloudy, full of static and the smoke of human fires" (126).

This is a most satisfactory image, serving to frame the connection of Being and the trickster's personification of language, of earth precisely. Nature likes to hide, says Heraclitus, but here the meaning of the fragment is not simply that we discern things darkly but that sometimes our expectations of the way they should be are dashed utterly, and, as the two friends in the Eshu tale, we undergo an experience that is unsettling and agitating, sometimes violent. In the trickster's role as shape-shifter, we can comprehend Being as indeed a verbal environment that is unstable for the very reason of being verbal, of events fostered in the flux of appearances, of different things at different times brought to the fore by language. And in his identification with language the trickster stirs the mix, always at odds with complacencies and well laid plans. He ignores established rules, shatters taboos and proprieties, always prodding the intellect and spurning the rigid and sterile dogmas that stave off awareness and insight, sowing dissension and forcing new realizations. If he is for these reasons the benefactor of humankind, in any number of ways he is also our foil, often the Devil himself, inflicting upon us the mysteries of suffering and death. And insofar as he is the taker, he is also the giver. So in other guises the trickster is the Christ, redeeming what has been taken, providing humankind deliverance from suffering and death by way of the mystery of life itself. He therefore stands forever at points alpha and omega, at all thresholds, dwelling perpetually in the borderlands. He is the archetypal "earth-rim-roamer," marking the gateways between conflicting worlds and world views.[6]

In this general view, the trickster most particularly comprises in his connection with earth the wholeness of the mysteries of life and death. In his person each is realized only by virtue of the other, each being ultimately the same in their earthly disposition of the here and now. Yet in all his guises the trickster is the contradiction that comprises our lives, making this world, as surely it is, half the Devil's and half God's own. Thus, unlike the world that reason projects, there is always implied in the presence of the trickster not just contradiction, ambiguity, and paradox but accordingly a sense of incompleteness, of us never knowing or possessing the full account, because through the trickster's influence all things are real only insofar as they are protean and pliable, matters of possibility rather than actuality. This is truly the nature of the terrible secret of the irrational, as every effort to reform the irrationality of the world simply confirms it further. And here, most specifically, our relationship to whatever we deem to be reality comes prior to "reality" itself, establishing not just reality as incomplete but also ourselves, as each is the issue of possibility. This is the work of the trickster in his most essential sense, his most crucial guise as the mystery of Being itself, where his every ruse is linked to language and forthwith, in peculiar ways, to rhetoric.

The Trickster and Language as Listening: The Place of "The Quiet Power of the Possible"

Even as the relationship of the trickster to language and rhetoric is apparent in the power language possesses over physical force, once that analysis is pressed, the far more obvious case, and the one more in keeping with the creative ambiguity of the trickster, is that language has this power by virtue of its likeness to physical force. As language not only functions on the level of physical reality but in no small part determines it, the measure of truth and reality in such a circumstance is possibility, never actuality. Accordingly, deception in this context is not essentially a departure from truth but simply the effect of the nature of language, as words shape and ultimately create reality, where the power of speech can be compared indeed to "the effect of drugs on the bodily state." Hence, without embellishment or stretch of the imagination one might claim "persuasion by speech is equivalent to abduction by force." Not only is the influence of Gorgias clearly apparent in all of this, but that of Heidegger, too, in his assertion that Being remains mysterious, manifest to *dasein* as the "simple nearness of an unobtrusive governance" occurring essentially as language itself ("Letter" 212); or that it is, with no contradiction implied, manifest in the experience we undergo with language, that "befalls us, strikes us, comes over us, overwhelms and transforms us" (*Way* 57).

So I return to a familiar theme, that language has an "elucitory" or "evocative" function, a means of revealing and shaping reality to the possibilities inherent in language. In fact, Henry Johnstone sees Heidegger's philosophy in its entirety as serving this function. He claims that Heidegger "conceives philosophy as basically rhetorical" in that his philosophy appropriates rhetoric's evocative function to reorient hearers in favor of a more "authentic" life (68). And as John Poulakos has illustrated in a series of articles, the relation of Heidegger's philosophy to rhetoric is apparent in Heidegger's "radical reflection on the nature of language" ("Possible" 216), a perspective on language's "evocative" function that Poulakos claims Heidegger shares with the Sophists in general and Gorgias in particular. Poulakos focuses specifically on the dynamics of Gorgias's "Helen" to illustrate this idea of language, of it going in advance of appearance, where the ill-fame of Helen is concealed and, through the power of *logos,* she instead appears in the light of innocence.[7] And there hardly exists a more apt example than Helen of Troy to illustrate the point. She is how she appears in accordance with the power of language. Her true nature is equivocal and forever in dispute, and in her various, conflicting guises, the woman herself bears the mark of the trickster. If the power of *logos* condemns her to ignominy, it also redeems her, even praises and glorifies her, as language possesses the power to prove opposites and all possibilities in between.[8] Such power is the measure of myth, as reality is so conspicuously the issue of *doxa* in this context. As it is surely the measure of rhetoric for the same reason, being the case especially as the connections with Heidegger are pressed. His thesis of Being's capacity "to show and hide itself," Poulakos says, "and the potentiality of language to reveal and conceal, introduce a set of questions that identify the basic problematic of all rhetorical practice" (220). What informs Poulakos' perspective in particular is the belief that rhetoric has its place in Heidegger's philosophy in terms of this crucial facet of Being as the "quiet power of the possible." Heidegger says:

> When I speak of the "quiet power of the possible," I do not mean the possible of a merely represented *possibilitas,* nor *potentia* as *essentia* of an *actus* of *existentia;* rather I mean Being itself, which in its favoring presides over thinking and hence over the essence of humanity, and that means over its relation to Being. To enable something here means to preserve it in its essence, to maintain it in its element. ("Letter" 196–97)

Poulakos joins Heidegger's idea of possibility to rhetoric by claiming that rhetoric is a practical discipline through which we procure a "region of possibility," a terrain where "the possible can be actualized" ("The Possible" 221). As Heidegger explains it above, the "quiet power of the possible" is inevitably the result of his conception of Being as immanence, of *physis* as the power of creation abiding as the earth itself. Therefore, it is a process of "enabling," of preserving or maintaining something "in its element," of letting Being be itself. By that pattern Heidegger says that *dasein,* as the Greeks experienced it, is a being whose authenticity consists not in self expression but in discovering the world in "letting beings [things] come to appear in Being as they are" ("Letter" 198). In this interpretation of authenticity, as we have discussed it in terms of *xa* and otherwise, the emphasis shifts abruptly from self expression to the discovery of Being through our openness to it, thereby allowing Being to be expressed through us. Here, in particular, the legacy of the trickster is apparent in this very formidable power manifest in continually shifting appearances, in the revealing and concealing of things that may seem now and then to be actualized but persist always as pure process and possibility.

Thus, much as the trickster, as Eshu or even Helen, Being "likes to hide" and is revealed only in the appearances or guises it assumes or presents. This perspective, perhaps, marks a departure from Poulakos's view of Heidegger and rhetoric, surely a departure from most conventional views of rhetoric, in that here the "quiet power of the possible" is utterly a function of Being, of language, and not a matter of our privilege in the use of language. As this power is manifest only through Being, it cannot be seized and still remain the power of the possible. Again, the essential point is that *dasein,* in its authenticity, is the relation to Being, and its thinking and language can be authentic only as each is achieved in the lure of Being, in response to the call of Being. Thus, asserting or expressing ourselves, speaking as such, is merely derivative, secondary if not altogether alien to this power of Being, as again Heidegger places the emphasis on "hearing and listening, heeding and being silent and attentive" (198), for these are the criteria of belonging, of "enabling" Being, preserving it "in its essence," maintaining it "in its element." To exert our independence from this power, or to presume to control it, precludes listening and belonging, a miscue reflecting that of the farmers in the Eshu story. In particular Heidegger condemns the separateness that this subjectivity entails, as it prompts beliefs that we can transcend material creation, the eventualities of time and place. To Heidegger, such beliefs are bound firmly to metaphysics, and we thereby fail not only to maintain Being in its element, but, more grievously, in its place we perceive only reflections of ourselves, utterly self-obsessed and distorted ones at that.

Accordingly, Heidegger claims the "question of Being" is forgotten, and in the aftermath, to "procure validity," our thinking is restricted to issues of epistemology alone, and we then seek to redeem the loss by invoking various versions of *techne* that are mere "instruments" of education, "classroom matters" called philosophies (197). More directly, he says, what is forgotten is the nature of language, for language is the "house of Being" and only therein do we dwell authentically. The consequent "downfall of language" is the result of language coming under "the domination of modern metaphysics of subjectivity." Here, as we strive to secure validity, "language surrenders itself to our mere willing and trafficking as an instrument of domination of beings [things]" (199). This is the misfortune afflicting modern humankind, he says, as it was, indeed, the misfortune of the Pythagoreans confronting the "terrible secret of the irrational," surely that of Hippasus; as it was the misfortune of the quarrelsome friends in the Eshu trickster tale, who relied on themselves and "failed to reckon with Eshu" in their separateness from one another and from Being. In any event, the "quiet power of the possible" is preempted or obscured as we project all things and events as actualities and thus encounter them in a "calculative business-like way" or, scientifically and philosophically, by way of "explanations and proofs" (199). Thus, given this interpretation, it is critical to understand that a Heideggerian—or Sophistic—rhetoric must take into consideration the peculiar nature of the "quiet power of the possible."

Heidegger's concern, then, surely rests with the authenticity of *dasein,* but in his design that concern remains incidental to the authenticity of language. As language is severed from its basis in Being, it is groundless, becomes mere "idle talk" (*Gerede*) depriving *dasein* itself of any viable connection with Being. Though idle talk is language as it is most commonly used, and Heidegger therefore means to offer no moral judgment by the term, it is still language removed from its "primordial belongingness" to Being, language that has "fallen out of its element" and has thereby lost its evocative function, discouraging "any new inquiry and disputations" and simply "suppressing thought and holding back" (*Being* 212–14; "Letter" 198). As such, "idle talk" is basically situationless; or, as Heidegger says, it presents a situation that is utterly "general," annulling any sense of temporal exigency, of those distinct and matchless particularities that constitute the situation itself. *Dasein* then falls prey to the "dictatorship of the public realm" and, as we have seen, is absorbed into the generalities and banalities of the everyday "They Self" (*Man-selbst*), where life is reduced to uniform compliance, to following "the route of *gossiping* and *passing the word along*" (*Being* 212).

In contrast, language based in its "belongingness" to Being is authentic, and what is ontologically constitutive of authentic language is resolved in the impetus of situation, the "there" of language's disclosure, as Heidegger says. Authentic language thereby cleaves to the thoughts, feelings, and aspirations giving rise to it; that is to say, to these dispositions not as things exclusively our own but as gathered via the animus of *dasein,* of "Being there" as the place of the unfolding manifestation of Being as language that, in these dispositions, "situates" us *as* language. Here, most definitely, the distinction between nature and culture, as we see them now aligned with Being and language, is undone in the wholeness of *doxa.* Indeed, the difference between authentic language and idle talk—between language bound to and ultimately constituting the situation and language that is situationless—is precisely the difference between the two senses of *doxa* previously discussed and brings each into sharper relief regarding the phenomenon of the Gorgian *kairos.*

Plato's condemnation of *doxa* as mere opinion has its parallel in Heidegger's representation of "idle talk," but as we have seen, among other things, *doxa* means "glory" to Heidegger. Here it is pure process, a play of appearances disclosed in the mutability and uncertainty of the verbal environment in which we dwell. In this context *doxa* itself comprises "the power of the possible" in that it rises, no doubt very "quietly," in the relationship between *dasein* and Being, as that relationship "enables" humanity, preserves and maintains it "in its essence." In contrast to Plato, there is no clear separation but a dynamism rising from the very "nearness" of the two in the borderland that is the event of the ontological difference. We can again invoke the commonplace of the dancer and the dance to explain the matter, as they are what they are only in their relationship to one another; and what they might be separately is secondary, derivative, and ultimately false or inauthentic. Accordingly, the things related have no integrity in themselves, apart from one another. Yet, as the ontological difference prevails, we are concerned with a relationship still, never an identification.

Ironically, then, it is the difference itself that accounts for idle talk, for in the absence of a perfect unity with Being, the relationship with

Being is most often corrupted, made inauthentic, aggravating the difference and in a process of deterioration bringing about a subject to object relationship. As *dasein* maintains itself in idle talk, it "floats unattached," says Heidegger, separated from the world, from those in it, from Being itself. And, as the relationship to Being, *dasein* is then separated from itself.[9] What is vital, on the other hand, is *dasein*'s persistence as "Being-there," its essential "belongingness" to Being, evident as well in the ontological difference. This sense of the situation, as the place where *dasein* and Being converge in the manifestation of authentic language, provides a good indication of *doxa* as that verbal environment in which we dwell, ultimately authenticating our being and thereby providing a connection of this sense of *doxa* with the Gorgian *kairos*.

As I say, much of this comes by way of a review of arguments previously made, but here, as they are offered in terms of the trickster, a more specific, even spirited portrayal of the Gorgian *kairos* is in the offing. It is therefore important to realize that the idea of the trickster presumes an animate universe, a cosmos—or, just as aptly, a chaos—profuse with life. The trickster thereby personifies not just a dynamic presence we encounter in the world but the world itself *as* dynamic presence, a world that we do not merely encounter but one that most emphatically encounters us in immediate and very direct ways. Thus, not only is the trickster's presence manifest in ways similar to Heidegger's rendering of Being, but what is most essential in each is a dynamism that can overwhelm the sense of a personal or individual consciousness; and therefore our perspectives, individual or otherwise, are necessarily hedged in tentativeness, uncertainty, and doubt. The issue then becomes one of constantly shifting perspectives, or more properly, in light of Being as dynamic presence, it shifts utterly from disconnected, individual perspectives to Being as such, to how Being reveals itself in appearances. Here the trickster—as the shape-shifter, as Being itself—is the force of insight, compelling fresh perspectives as he undermines our customary dealings with the world, confounding rational discourse and, at times, seeming to confound the validity of reason entirely. Thus, we encounter "outside reality" not as *geganstände*, a collection of objects we "stand against," but encounter it through our awareness of a force or power, a presence in form, substance, and spirit far greater than our own by virtue of our relationship to it. Curiously, in this relationship we most empathically encounter ourselves.

So, as we have seen, traditional notions of subject and object are displaced or annulled altogether, but specifically in this context, as Robert Pelton says, the trickster is a reflection of the world experienced as an active subject in its own right. Not as a Jungian archetype or some such, but as an active subject seeming to function outside ourselves as the perceiving subject, upon which our role as active subject is itself ironically contingent. In a sense, then, every action *is* a reaction, to the extent that that distinction is even valid in this context. Pelton applies Lienhardt's use of the term *Passio* to explain the issue, as the word carries the meanings of both separation from "divinity," thus accentuating the sense of individuality, while at the same time sanctioning the nullification of individuality by affirming the nature of experience as "being acted upon" (257). We then become "active reciprocals" in all crucial events, experiencing life in ways beyond our control and even our comprehension in that to act at all we are necessarily, and simultaneously, acted upon. Under the influence of the trickster our experience is indeed shaped by the ontological difference, that in this context we can understand as obscuring binaries and opposites, serving as a threshold of coming together yet keeping separate, as does the trickster himself. He is, then, the outsider within, the essential affirmation of *logos* as language rather than reason, so easefully bearing the power and validity of "contra-diction." As Pelton says, he offers what is "radically ambiguous, essentially anomalous, inescapably multivalent—facing both in and out, linking above and below, animal-like and godlike, social cog and individual solitude, shaped and shaping, part of all that is but as a subject knowing its own apartness" (258).

In this sense the trickster is, as well, the personification of a "primordial" state of affairs that would seem to be particularly to Heidegger's liking, a personification that to all intents is indicative of the revelatory power of earth itself. In either case, Being is simply as it is given—or as it gives, ambiguously and unpredictably—rather than a derivative commodity, remade and packaged at our bidding to fit our philosophies. In just such a context, Heidegger offers his view of an "authentic" relationship with Being that is rendered first and foremost in listening rather than speaking. Here we play the trickster's game, to say it makes no difference whether perspectives are individual or a function of how Being "reveals" itself, as that difference is necessarily artificial, subsequent to our relation to Being. But insofar as Being "gives," our openness to Being is a matter of listening as the source and nature of our relationship to it. In speaking we identify with the ego, a particular thing that is realized as subject in expressing itself through language, thereby creating, through language, greater distance and distinction between itself and what environs it; whereas in listening we are by virtue of language a process that participates in Being. In this way listening not only possesses a nature equal to that of speaking but is the more sublime in being joined to something outside the ego that is crucial and real, yet at the same time something to which we must be joined to be ourselves at all.

The shift in emphasis is apparent in subtle but always extremely important ways. In the instance of speaking, especially as it is applied in traditional theories of rhetoric, we devise *technai*, schemes and strategies to gain advantage in speech, using words to extend our independence and mastery; however, in listening we are used by language, even embodying it to the degree that in Heidegger's thought we owe our "authentic" existence to this relationship we have with Being by virtue of language. Gerald Burns makes a vital point concerning the distinction, saying, "to approach language from the standpoint of speaking is already to have assumed the theoretical framework of the monological subject" (21). Hence, a rhetorical *techne*, however underdeveloped it may be otherwise, is in place from the very outset when the emphasis is placed on self-expression, on speaking out as a means to exert our power through language. Listening, on the other hand, dissipates language as structure and mere means of communication. It not only opens us to the ambiguity, confusion, and "smoke of human fires" that can be the equivalent of violence itself, but in rare moments of grace can lead also to that "active reciprocation" that is language in its most essential sense.

Considering all that the above puts to issue, we can appreciate Plato's efforts to create a greater integrity of the self, his purpose in endowing us with an immortal soul. But very often essential anomalies cannot be denied, even in Plato's deliberations. No doubt his condemnations of the "willfulness" of the phallus are merely an extension of his admonition to care for the soul, but at the same time they are so burlesque as to alert us to the presence of the trickster even here, right in the pages of Plato. As I have said, there are those who claim the trickster is "mere belly" and his schemes and intrigues reveal all of us mortals to be the same. Yet, whether the trickster reduces us to mere belly, or whatever physical feature or organ, we can surmise that in no small part Plato's purpose in creating the soul was to grant us a nobility that was utterly spiritual, free from the foibles of the flesh, to thereby enable us to reach a truth and sense of our being that were decidedly "belly free."[10] The *Phaedrus* itself can be seen as a most conspicuous example of this purpose. Plato's sublimation of physical love to achieve in dialectical ascent the union with the divine would not be countenanced by the trickster nor, more pointedly, would Plato's corresponding effort to sever the corporality of language from the "spiritual" essence of our being. In fact, to present the issue from the trickster's side, we might say that we are neither physical nor spiritual but both simultaneously, as distinctions of this sort do not hold in this context. Or, more aptly, we might say that the trickster divests us of a soul altogether, reducing us to a series of moods depending on the various situations in which we might be involved, so that we are, much to his fancy, essentially manifestations or various attitudes apparent in nothing so much as the primitive idea of the *psyche*. In this way there is nothing of the self, ego, or individual independent of some situation or another. It is abrogated by the work of the trickster as "active subject," where the individual and the situation are always interdependent, together comprising the ingredients necessary for our being. We are, then, material manifestations in this integration with *physis*, this integration with earth and words. Again, the relationship is primary. We come into being by virtue of this verbal environment, our listening both its genesis and means of enduring vitality.

Indeed, in the beginning was the word, but well prior to Saint John, the trickster was there to tell us as much.[11] In his world the issue is never vested in what we say, to him or whomever, but in what he says to us and in our resolve to listen, to be open to it. He deceives, tricks us as a matter of course, but we refuse to heed his word only at our peril. In story after story, we bear witness to the dire fate of those who refuse to listen to the trickster, who are too busy speaking to listen, or who dismiss what they hear as absurd. And here, especially, there are crucial similarities to things Heidegger says regarding Being, for to realize this primal sense of *logos* we must be open to Being, to something outside ourselves that is still profoundly within, to listen and most often to be silent in our listening to consummate our relationship with it.

In this vein Heidegger reforms the idea of thinking itself, in the process recapitulating in many respects his theory of language. Insofar as we assume thinking to be essentially a matter of reason, *logos* in *that* sense, we most likely take what lies before us as inert and separate, yielding to our deliberations and efforts to impose our will upon it. By reason's mandate we assume the ability to rid ourselves of prejudicial perspectives, of "mere" perspectives altogether, where we not only perceive things in the same or similar ways but by virtue of reason often assume our perception is not a perspective at all but most definitely truth and reality as such. Amid such beliefs the trickster lies in wait to dispute and defeat, and to him it seems to make little difference even as we adopt rehabilitated models of reasoning that provide practical, probable, pragmatic ways to truth. He confounds them all, the models, *technai*, and strategies that in particular sever subject from object, for we must reckon with him in all cases as an "active subject." He is, then, the external element of our involvement in specific contexts or situations; yet, in *being* the relationship, he is internal as well, always the outsider within. He is the process where our relationship to Being is fundamental and constitutive, where "thinking" is utterly a matter of this relationship rather than an activity separate and detached. As such, the trickster can be seen not only as a force that prompts thinking but as its constant and crucial ingredient. Through his aegis—or Being's—thinking is not something that we do exclusively or independently, involved only with the workings of the mind. Nor is thinking in this context a matter of ratiocination alone. Nor is it to Heidegger necessarily conceptual or systematic at all, shorn as it is in essence of any exclusively epistemological basis in reason or representation.

Rather, in the manner of language, thinking beckons, coming as a call from Being—or, in fact, the trickster—that "enjoins our nature to think." Indeed, no less than language, thinking belongs to Being (*Thinking* 240). It is language's equivalent, or double, and therefore not something that goes on in our heads but a relationship between us and what we can understand as a "nearness" outside ourselves, yet within, of Being itself as that "unobtrusive governance" occurring essentially as language itself. We would then experience or "undergo" the call of thought in essentially the same manner as we do the call of language. Rather than being a wholly internalized process that is then exerted outwards to direct and dominate, thinking is the effect of the relationship to Being that is precisely the "quiet power of the possible." Thus, in Heidegger, the call of thought is "to be attentive to things as they are," unfettered by the conceit of thinking as grasping or mastering, and in that way, as J. Glenn Gray says, "to let them be as they are, and think them and ourselves together" ("Introduction" xiv).

This sense of the quiet power of the possible brings us back to those connections Heidegger draws from the word *thanc*, especially as we see them in this emphasis on listening. Again, Heidegger translates *thanc* to mean both "to think" and "to give thanks," and so to him it means "piety," thinking as a "concentrated abiding with something" in "unrelenting retention" that is "original memory." Thus, *thanc* returns us to the idea of *doxa*, being what memory "draws up" and "recovers again and again," so that what is within reaches out "most fully" to what is without and "the idea of an inner and outer does not arise" (*Thinking* 140, 144). This perspective can indeed be considered irrational, surely one to subvert the conventional ways in which we have come to know the world. However, one of the more most engaging aspects of this coupling of listening with thinking is that it deposes the soul or self, evoking in its place the idea of *dasein* or *psyche* precisely as an openness to Being as the quiet power of the possible. Of course, thinking is the crucial measure of our being in any event, the *sine qua non* of both the soul and *dasein*. But here their difference rests upon these different interpretations of thinking. As we have seen, the concept of the soul is in a sense the most obvious mark of our separateness from Being, apparent in Heidegger's contention that the further our consciousness is "heightened," the further we are excluded from the world. Yet for this reason it is the very standard of objectivity, and thereby the more intense and successful the effort to care for our souls, the more we are then removed from the imperatives of listening and the "piety" of thinking. However, in the spell of the latter, where the distinction between inside and outside is abated, thinking abides in reverence that is essentially the quiet power of the possible. And though the ontological difference remains inviolate, and in vital ways we are always alone, the state to which we aspire is apparent in the old precept that we do not hold Being in our hearts but are all held in the heart of Being. Our resolve in this aspiring, our openness in "listening," is the quiet power of the possible in its most vital aspect, as it is the "active subject" only by virtue of the "piety" of our thinking.

It should then come as no surprise that Heidegger's explication of thinking, patterned as it is after his sense of Being and language, is rooted in a pre-Platonic sense of the Greek dialectic. As Gadamer says, Greek thinking "did not seek to base the objectivity of knowledge on subjectivity" but "always saw itself as an element of [B]eing itself." Thus, the *logos* expressed in the dialectic "was not a movement performed by thought, but the movement of the object itself that thought experiences" (*Truth* 418). This account surely bears comparison to Plato's dialectic, though the essential difference, one that I have emphasized throughout, is that Plato fashioned the dialectic to his metaphysical view, thereby raising it to a "new reflective level," as Gadamer says, "one which seeks to free itself entirely from the power of language" (422).

However, in terms of Heidegger's idea of the piety of thinking, *doxa* is the manifestation of Being in its most tangible aspect, providing the measure, Plato to the contrary, of *what is*.[12] Clearly this idea of thinking removes us from the "true knowledge" of Plato's dialectic, to a way of knowing not just apparent in the linguistic nature of human experience but more specifically in Heidegger's belief that language is "the house of Being." And, as we have seen, more specifically yet in von Humboldt's belief that language is not ours but "we spin it out of our own being and ensnare ourselves in it," as it thereby "draws a magic circle round the people to which it belongs"[13]

And to which *they* belong in their listening. In this a rhetoric is implied, and if one dispossessed of a "monological subject," a rhetoric still, implicit in the ontological dimension that the Heideggerian emphasis on listening and thinking entails. Accordingly, the sense of our "incompleteness" remains, and if we are then conceived as a series of moods evoked by language, by that same measure we can achieve nobility as well, even glory, in the *doxa* we embody. Indeed, in this incompleteness we are enabled to realize the possibility of ourselves or, more precisely, to realize ourselves as possibility. In contrast to conceiving our nature as eternal, a stable and enduring spiritual presence, possibility is then, much to the trickster's purpose, inherently a physical process, especially as we see it expressed rhetorically in the power manifest as language and *doxa*. Yet everything comes full circle in the wholeness of Being the trickster personifies. He is above all else a cultural phenomenon, but only as he also comprises in his robust physicality a necessary connection to nature, to the earth. In this context, surely, Heidegger's idea of language is eminently viable, embracing elements of both nature and culture, Being and the verbal environment as one, where *doxa* is thereby constituted, as Charles Guignon has told us, in the "Saying" of a people, the stories, rituals, and myths presenting their understanding of reality. Here, where language and thinking are in these ways indistinguishable, their union marks a rhetoric of listening, a piety of thinking where we reckon with Eshu by means of a *techne* based on the deceptive powers of language. Here, as well, we might understand the Gorgian *kairos* as an instance of this *techne*, a most likely access to *physis* in terms of this quiet power of the possible. As Heidegger develops it, this power is expressed through us by virtue of this *techne*, as we thereby achieve our identity in its sheerest ontological aspect, our authenticity in the "blood consciousness" that comes about in being claimed by language.

THE TRICKSTER AND RHETORIC: DECEPTION AS *TECHNE*

Bruce Gronbeck made the case years ago that Gorgias's rhetoric can be considered a *"techne* of *apate,"* a rhetoric apparent as a "physical phenomenon" bringing together the physical world and the physical body in order to move the latter in practical ways (33). It was a remarkably good insight and remains so, especially in that *techne* and *physis* are generally taken as opposites, the first as something we do and the second as something done on its own, of nature simply taking its course. But as *physis* itself is taken as an active presence, there is a peculiar interaction of the two, peculiar enough in this perspective to blur distinctions between them. Thus, as was the case with a number of things examined thus far, this discussion of *techne* proceeds in ways different from customary perspectives and, much as those other things, begins and ends in the ontological dimension of language. Here the trickster serves as the personification of this *techne*, not only because he deceives but because we are not dealing primarily with a means or methodology but a lively presence having an integrity and even an acuity of its own, an awareness in ways separate from ours that serves as the threshold or gateway to this ontological dimension.

No doubt the power of *physis* comes most keenly to mind in its outlandish aspect, in tumult and spectacle revealing nothing so much as the character of the trickster. Here indeed he is preeminently the shape shifter, often dashing our lofty aspirations and always our expectations of how things ought to be. He moors us instead in the material world of *physis*, where Being is ceaselessly Becoming in its immance as earth, having none of the stability and little of the splendor that Plato gives to it. The trickster's purpose is therefore critical, even sacred, if hardly ever solemn. He embroils us in Being's immanence, in the flux of the here and now that cannot be ignored or neglected. It is on this level, in any event, that Heidegger encourages us "to let Being be" rather than "forgetting" it or assuming to control it. To let it be enables us to listen to the voice of Being and from that, he claims, arises our nature as thinkers. He says we think authentically only in heeding and responding to Being, and here we understand that to mean we must heed the *logos* speaking through us, even as in this perspective that means, in turn, to hear and heed the trickster's voice through our own. It is not just coincidental that the trickster Hermes gave language to humankind, or that the Yoruba believe Eshu did the same, so that along with language comes the imperatives of the trickster, those ambiguities and contradictions of language and Being to which we are heir. In this discord language itself is the persistent verification of the trickster's presence.

And we are centered in language, in no small part placed there in story and myth by the trickster. He leads us astray and so often into trouble that why should he not be equated, all the more profoundly, with the rebelliousness of the "organ of generation" that Plato condemned for reason's sake? *Logos* is free, so we become entangled in the trickster's pestering and pranks, at times utterly entranced by his language, by language as such, as his presence confirms that language is the consummate measure of our being. Indeed, as Robert Pelton says, "the trickster reveals the human world itself to be language" (270). He is "hermeneutics in action, creating language out of his own body like a spider spinning its web, spelling out the ways of intercourse with upright penis, probing ceaselessly all opacity for hidden designs, and forever rejecting every form of muteness" (243). Thus, whatever the ways of intercourse, given the trickster's presence there is no place for nonverbal, transcendental realms of the soul; there is earth alone, the verbal environment of the here and now where it is not we who play with words, as Heidegger says, but where "the nature of language plays with us" (*Thinking* 118). The priority of the relation is reiterated, between us and Being, and this, too, is in keeping with the character of the trickster, because the trickster is both us and what is outside us and beyond. Whether we spin language out of our own being, as von Humboldt says, or the trickster out of his, as Pelton says, it is all the same. In any case reality is neither exclusively within nor without but is the event of their relationship, appearing or withdrawing from awareness, sequentially or concomitantly, as the nature of language "plays with us."

Here, especially in light of the trickster's presence, this power of language is not only intrinsic to *doxa* but both the source and nature of a "Sophistic" *techne* of rhetoric. Gorgias tells us that language is the great *dynastes*, that *doxa* is the medium in which the rhetor works, and, by implication, that deception is the principal attribute of the rhetorical *techne*. As the trickster reveals this earthly realm essentially as a linguistic creation, deception is not only a rhetorical tactic we employ but a necessary consequence of the power of language. This *techne* is hardly conventional, certainly not a method external to language that we imposed upon it. Rather, here the power of language is primary, residing within language as *physis* itself, manifest in the power to reveal and conceal "reality." Accordingly, *techne* is an ontological rather than epistemological concern. By means of this "*techne*" we tap the quiet power of the possible, drawing upon it in the sheerest sense of "letting Being be," in that way indeed "reckoning with Eshu." This *techne* is itself hermeneutics in action, though it comes not at our instigation but in our relationship with Being, with language *as* language, in a process where we must first be deceived ourselves in order to deceive. In such circumstances we are at once both the trickster and the tricked, the deceiver and the deceived, and we are all the wiser because of the deception, just as Gorgias said so long ago.

Being—or *physis*—rises of itself, "flourishes on its own, in no way compelled," though as *dasein* we are essential to the process, "Being there" to provide the means of its expression, allowing Being "to be." In Heidegger's assessment we therefore fail grievously in arrogating to ourselves the power to use and manipulate Being, thus affirming that sense of *techne* he condemns as "human knowledge without qualification," the "kind of knowledge that guides and grounds confrontation with and mastery over beings [things]" (*Nietzsche* 1:81). In this sense *techne* is aligned with metaphysical thinking, and language is a tool we apply with the knowledge or skills given in this *techne*. Gerald Bruns goes so far as to say this conception is representative of the "old rhetorical idea of artistry as mastery, where mastery of materials of discourse develops into the means of social and political control." Perhaps Bruns goes too far, but quite to the purpose of the subject, he concludes that the "conquest of language means conquest of the world."[14] In any event, this disposition to use language for our aggrandizement alone is the sort of thing chastised by the trickster in more humble cultures, those less "civilized" than our own.

This will to master, specifically manifest in our usurpation of language from its source in Being, is not only the sense of *techne* that Heidegger condemns but is also what the presence of the trickster is meant to preclude, and to punish should it appear too prominently, this presumption to bend Being to human will. Yet even here, the trickster's authority is exerted not just among "primitives" but is as omnipresent and insistent as the mystery of Being itself. He is in all places at all times and, as I have suggested, in times as current as the Vietnam war. He was sure enough there, in-country and undaunted by the most prodigious effects of our technology to rid ourselves of his presence, even as these very effects made his presence all the more apparent. As he seems to be in the midst of our more recent efforts to bring democracy to other parts of the world. The violence, the unspeakable horrors that so often follows in his wake, can scarcely be considered a game or amusement. But the trickster is who he is in spite of himself, even in this day and age when we have been so inclined to ignore or dismiss his presence altogether. Perhaps the horrors thereby come in retribution for our arrogance and neglect, seeking as we might through a *techne* to impose our vision of Being on Being itself. But whether the horrors come from the trickster or are brought on by ourselves, they are self-inflicted in either event.

Therefore, "to let Being be" is never a matter of simply quelling our hubris but, once again, of achieving a "piety" of thinking that is above all else a listening, a paying heed "in constant abiding." As I say, doing so comprises the basis of our authenticity in conceiving language as *legein*, as it also comprises a *techne* in this congruence of "to let lie" and "to tell." But whether we succeed or fail in this piety of thinking—though especially as we fail—Being becomes evident regardless in the antics of the trickster. He embodies an apparent otherness that prompts various insights and perspectives that we often assume to be ours alone, incessantly revealing and concealing reality in all its ambiguities and contradictions, forever qualifying and disqualifying our claims to mastery and truth. As the "active subject" the trickster is behind these matters, his work especially apparent in the shift from views of reality we assume to possess to how reality reveals itself, taking the prerogative of decision from us and placing it with our relationship to Being.

Thus, even as we pay heed "in constant abiding," perhaps as we do so in particular, we then seem to be dispossessed of a workable *techne* by the very dynamics of *physis*—its "spontaneous unfolding," whether conceived in aspects of "self-emerging" or "appearing"—usurping the human role. I have raised these issues before and do so again as they concern in significant ways Heidegger's view of the nature of human knowledge, a "*techne*" gathered as it is from precepts so ancient to provide in crucial respects a proper analogue to the ways of the trickster. Yet, as before, as seemingly in all other things in Heidegger, this view leads straight away to his emphasis on language. In fact, in the "difference" between self-emerging and appearing there is an affirmation of language as primary and fundamental, especially as we perceive language as the expression of *physis*, ultimately as a matter of *doxa*. We ourselves emerge and appear in language, Heidegger says, such that in this most vital event there can be no "outside" to language. Here, language speaks, not us, and insofar as we appreciate the idea that *physis* is the "voice of Being," we see the difference between self-emerging and appearing as an essential implication of the ontological difference, a "unifying scission" in its own right. In the creative tension between the two persists the power that is language, that difference between Saying and the human response, *Sagen* and *sprechen*, giving focus to the peculiar sense of *techne* that Heidegger espouses, indeed giving focus to the peculiar sense of the Gorgian *kairos* as a *techne*.

Again we recollect, having seen that Heidegger affirms the superior nature of the Greek language, his opinion based not on any inherent quality possessed by the Greek language but by the different relation he claims the Greeks bore to their language, "a relation of dwelling" first and foremost. According to Francoise Dastur, this dwelling was essentially an openness to the "strangeness" of language, an "uncanniness" recapitulating that of our own being and therefore possessing a vitality we can appreciate but neither possess nor master ("Language" 363). In this emphasis the Greek language is indeed superior, as then it is itself creative, possessing powers in its own right rather than being merely the instrument to express whatever powers we ourselves presume to retain apart from language by means of a *techne*. Here the difference between "self-emerging" and "appearing" is precisely the creative tension intrinsic to the trickster's presence, between us and an otherness that we might, but for the presence of the trickster, designate as "objective" and take as the staple of a *techne*. The tension is initially established in the vitality the trickster invests in all things apart from us, a tension distinct and significant because, by virtue of the trickster, "outside reality" is not shaped to our schemes and efforts but is beyond the reach of customary notions of *technai*. It is hinged instead to something that in critical ways is independent of human existence but, like human existence, an active subject in its own right, the peculiar presence of the "nonhuman" residing within as much as without.

Thus, as I have discussed, in this perspective things are neither inert nor stable, neither inside nor out, as the trickster upsets customary boundaries. Distinctions of this sort cannot be made in any essential sense, as in the sway of the trickster all things are instead complementary aspects of the mystery, this outsider within. It is his nature, as it is the nature of language itself as Heidegger depicts it. So it is, this creative tension between inside and outside, ultimately between the human and nonhuman, *is* language to Heidegger, and, as language, it "comes to presence" in the ontological difference, what he calls the "relation of all relations."[15] In a similar sense language is the trickster, at once the human and nonhuman, being as Eshu treading the borderline, embodying the problematic distinction between self-emerging and appearing, that ambiguity between inside and out. In this light *techne* has its effect in terms of this paradox, of Saying and the human response, of dwelling in language where we are claimed by Being only as Being is there by virtue of our presence, the relation of the two disclosing as meaningless the issue of whether language is ours or Being's alone.

So, in essence, self-emerging and appearing are one in a *techne* that obtains its meaning in the conviction that language is a natural force, a power of such magnitude that we are, as a result, simultaneously the used and user of language, the deceived as much as the deceiver. This is earth in its most compelling significance, *physis* as the verbal environment in which we dwell, returning us again to that archaic idea of *doxa* as both a natural and cultural force in the revealing and concealing power of language. Otherwise we are as the two farmers in the Eshu tale, each perceiving the black and white hat as either black or white alone. But as the ontological difference dispels the notion of some stable reality outside ourselves, focusing our attention instead on how Being reveals itself—its transformations, vicissitudes, even capriciousness—the appearing of Being is one with its self-emerging, its "reality" equally true and false at different times or even simultaneously.[16] It is then an awareness that joins the color of Eshu's cap to our perceptions of it, an authenticity achieved in the concurrence of appearing and self-emerging. Thus, the trickster is both independent of and dependent upon human experience, active subject both within and without, self-emerging and appearing, in this manner as much a reflection of us as we are of him, a reflection of things that are what they are only by virtue of their relationship to one another. Here, most emphatically, the trickster marks the mystery of Being, that indeterminate difference—deferral or *différance*—between what is in us and what is before us, what is human and what is not.

As the trickster's world is one of process and constant change, we are not so much confronted in this world with things we manipulate as we are enveloped in a presence as vital and unpredictable as our own, with which or whom we communicate by means of a *techne* that begins, we might say, by way of our reckoning with Eshu. Of course, much depends on the meaning of "*techne*," especially its meaning in rhetoric. We have discussed Plato's opinions on this matter, who claims that rhetoric must be an art to be legitimate, guided by a *techne* precluding anything irrational (*Gorgias* 645a). His "rhetoric," as his philosophy, is therefore shaped by his purpose to control language, to curb its powers, for he sees them, at least as they were extolled by the Sophists, not only as irrational but dangerous, destructive of souls and the *polis*. And we have discussed Gorgias's response, who says the art of rhetoric is concerned with language, though he seems at a loss, or not to care, to offer anything more precise. In his guileless way he claims simply that rhetoric is "the art which secures its effect through words" (450b—c). So the point remains, if rhetoric is this and nothing else besides, its power does in fact reside in language, just as Gorgias has claimed all along, and is not completely the result of a *techne* we devise and apply to language to control it in some deliberate, rational way. So, in this perspective it would seem that language is not essentially an instrument of human currency but does indeed emerge from a power that at crucial times brings to presence the mystery of the "nonhuman" or "uncanny," as it then ironically contains within itself the very experience of being human. At any rate, this is the nature of the great *dynastes* as Gorgias presents it, where language possesses a power we can at times evoke but never master, the belief that seems to lie at the source of the Sophistic *techne* of rhetoric.

But given all that it entails, it is little wonder that Plato claimed the Sophistic art was mean and ignoble, dealing in appearances rather than reality, purely a matter of cosmetics that "cheats" by shapes and colors (465b). More simply, the lasting complaint leveled against the Sophists is that in their rhetoric the speaker uses language to trick us, to deceive. Yet here, even for Plato, with his attacks on the Greek *paideia* and *doxa* in general, the real issue seems to have little to do with the speaker of language and much to do with language itself, for it deceives, "shapes and colors," on its own account. In fact, insofar as we are mindful of the Gorgian world view, deception is what language does intrinsically, as there is no stable reality within nature or otherwise, no eternal forms to which language can faithfully correspond. Here, indeed, *logos* is free, and we are left with language and its deceptions or, as I say, with the antics of the trickster. Always our relationship to Being is crucial, of its fluency as an ongoing event, as language itself. It reveals and conceals, loves to hide, as that is the necessary effect of this conception, of this verbal environment in which we dwell. In these circumstances deception is inherently an effect of the alliance of language and Being, in the end a most essential expression of the trickster who in light of his *raison d'être* appears more and more *as* that alliance.

In any event, such is the source of Sophistic rhetoric's "*techne* of *apate*," as it thereby excludes "the pre-givenness of the actual" and "an accurate representation of the facts" as necessary conditions of rhetoric. In their place, as Poulakos says, Sophistic rhetoric affirms, "the fact of language and its impact on people" ("The Possible" 219). This is in keeping with Gorgias's contention that the first priority of an art of rhetoric is appreciation for the great *dynastes*, and that, in turn, rests upon an openness to the power inherent in language. Indeed, a more complete Heideggerian interpretation of the matter, seeming to diminish the role of the speaker even further, would place the emphasis in rhetoric, as in language, squarely upon listening and in that sense upon thinking as an expression of Being, thus giving shape to a *techne* that emerges from the powers of language itself. Again, the idea of a *techne* is clearly problematic in this context, as this conjunction of language and Being, given the way we interpret the latter, places *techne* under the influence of the uncanny, of the trickster, and thereby removed from the safe and comforting assurances of both the practical and more analytic methods of philosophy. But that could well be the case in any event, given rhetoric's long, uneasy, and often adversarial relationship with philosophy. The differences break along lines already discussed, of essence and appearance, truth and opinion, and ultimately what the philosophical method indicates to be the differences of rational and irrational. The latter lie within the trickster's sphere, of a *logos* of language and therefore, in various ways, of the irrational. Or what in certain contexts is precisely the "uncanny" to Heidegger. Under the trickster's tutelage, in any event, we can come to a new and fuller appreciation of Gorgias's contention that through the power of language the deceiver is more honest than the non-deceiver, and the deceived more wise than the non-deceived. Once again, deception in this case is not the result of malice on the speaker's part, nor of malice at all, but of the very nature of Being as the "quiet power of the possible." In this process of deception, reality emerges, shifts and changes, appears and disappears, phenomena determined by the deceiving powers of language.

In the Heideggerian perspective this is, in essence, the power tapped by the rhetor, that can be expressed through him if he but first listens to language and in his openness allows the "letting be" of Being. Most decidedly, then, this sense of language as the "house of Being" is clearly distinct from the "prison house of language" to which we are presumably condemned by Gorgian rhetoric. Here the promise of possibility intrinsic to language redeems the Gorgian world view of its harsher aspect. Indeed, the Gorgian Affirmation is given clearer definition and cogency in Heidegger as he claims the horizon of our linguistic world is not a prison that encloses us but a gateway to the "non-established" of Being, to the "abyss" of possibility, where "we do not go tumbling into emptiness" but "fall upward, to a height" (*Poetry* 191). Not only is this belief far removed from the "prison house of language" but in the Gorgian view we achieve a wider, more embracing perspective than anything we could possibly gain from Plato's, condemned to be chained in his reality, as most of us are, in the dark and dampness of the cave. In this context Gorgias himself is the ultimate trickster, or he sure enough becomes one in light of the long, entrenched conventions of philosophy and the iron rule of Plato.

If the trickster goads us, he empowers us as a result, even as we seek the comfort of a well ordered and regular world, constantly striving to make it so by reducing things to our measure and control. Surely this purpose of keeping things orderly is the function of philosophy and reason; and, in the specific case of philosophical rhetoric—in our efforts to reduce the use of language to a discipline—this function is served by a *techne*. Even as Plato says he is unable to give the title of *techne* to anything irrational, here it means far more than merely giving a rational account. It is the express opposite of an openness to the uncanny, ultimately joined to Plato's idea of Being as stable and enduring reality (*Gorgias* 465a). It is the "pre-giveness of the actual" in its exemplary Platonic aspect, where our openness is preempted, and Being, except as a lifeless abstraction to round out his philosophic scheme, ceases "to be." Or is forgotten. With his commitment to the dialectic, his means to know true reality, Plato's depictions of *techne* in the *Phaedrus* are no different than those in the *Gorgias*, even in that most salient passage in the *Phaedrus* concerning rhetoric, *techne*, and *kairos* (271d–272b). Essentially, as J. E. Tiles says, Plato's *techne* is a model of critical judgment involving both goal and method in the intelligent application of thought to practice. "The man with *techne*," Tiles says, "grasps the principles which govern his activity and relates his procedure to those principles" (55).

Janet Atwell offers a more comprehensive view of *techne*, while at the same developing its place in rhetorical theory. It is therefore instructive to note that in her explanation *techne* is formed by the effort to mitigate or circumvent the forces of necessity, chance, and nature (719). The proviso here, and a crucial one, is that in rhetorical theory *techne* is defined in contrast to *physis*, even as the latter is ultimately the means or knowledge in the Heideggerian perspective through which we allow its power, ironically by virtue of a *techne*, to be expressed through us. Thus, whatever else *techne* might entail as both a model and means of knowledge, in traditional rhetorical theory it is precisely the agency to preclude the unexpected and unseemly in language, to mitigate or overcome completely the vicissitudes of nature, fate, or chance through the studied application of human will and reason (719). On these grounds, of course, Plato takes Gorgias and Sophistic rhetoric to task, and we might then conclude that these notions of *techne* are both the source and legacy of philosophical rhetoric, and that their effect was definitely to banish the likes of the trickster and all his pomps from rhetorical theory.

Again, insofar as the meaning of *techne* is conceived in ways to reduce things to the "calculable and calculated," it is condemned by Heidegger. Insofar as *techne* is joined to philosophy at all, to metaphysical thinking in particular, he condemns it as taking on implications of management, control, and use, much as some would see it now in a long descent as commensurate with technology. "Technicity" (*Technik*) is the word Heidegger uses to designate *techne* in this pejorative sense, as it indicates to him the effort to "dominate the earth."[17] In this way we distance ourselves from Being and thus, as Heidegger claims, from the primordial sense of *techne* integral to language itself. In his effort to retrieve the original sense of the word, Heidegger's tells us that *techne* is evident in our openness to Being and, by virtue of that openness, manifest in our handicrafts, various skills, and most especially the arts of the mind. And it is more. To him, *techne* is essentially a matter of ontology rather than epistemology, a matter of coming to terms with the power of language rather than one primarily of human will and reason. It is indeed a knowledge but a "knowledge" in ways recounted earlier, as that sense of *thanc* he traces back to a capacity for appreciation and wonder, back to the basic premise of *dasein* itself, that sudden and profound sense of "being-there" forced on us by Being (*Introduction* 164). Here, most crucially, a work of *techne*—a work of art—is not essentially an art because it is wrought, made, or produced by our hand but because it "brings about" Being in a thing, brings Being "to stand," stabilizing it "in something present (the work)." Thus, in its essential sense, *techne* "brings about the phenomenon in which the emerging power, *physis*, comes to shine," to be beautiful, to appear and be resplendent in the work (*Introduction* 159). So, by way of this *techne* "self-emerging" and "appearing" are one.

This peculiar definition of *techne*, especially in its likeness to *physis*, is one of the more striking features of Heidegger's interpretation, and quite remarkably he claims it lies with Aristotle as much as any of the pre-Socratics.[18] On Aristotle's authority he claims that both *techne* and *physis* are an "opening up," a revealing, a "bringing forth." As *physis* brings forth on its own account, as a rose would grow and blossom of itself, *techne* brings forth through the work of another, the craftsman or artist, yet it does so by virtue of *physis*. In this "primordial" sense of *techne*, the artist does not make, manipulate, or manufacture an object but participates in the power of *physis* to bring that power to stand in the object, to bring it into appearance in "aspect" and "matter."[19] What is very significant, at least in terms of rhetoric and the Gorgian *kairos*, is that here *techne* is linguistically based, utterly grounded in the power of language. In one sense *techne* is thus the way of *legein*, precisely the congruence of "to let lie" and "to tell," though, as we shall see, it has aspects that seem hardly in keeping with the "quiet" of "the quiet power of the possible." And here the difference between *techne* and *physis* is apparent, at least initially, in the former being an openness on our part, a means of knowing whereby the power of *physis* is then expressed through us, most particularly in the exigency of the right moment. It is a process that brings things about, brings them to appearance expressly by language, that Heidegger describes as knowledge "experienced in the Greek manner," as a "bringing forth of beings [things]. . . *out of* concealedness and specifically *into* the unconcealment of their appearance" (*Poetry* 59). Thus, in its relationship to *physis*, *techne* is an awareness, a projection of *aletheia* as such, or, as we see it here, an attentiveness inherent in the possibilities of *doxa*, bringing forth appearances culled by language. Again, pivotal to this *techne* is that curious connection of "self-emerging" and "appearing."

In various contexts the Greek term for this power is *poiesis,* and to Heidegger *poiesis* is not simply a body of verse or means to criticize it but the "poetized thinking" intrinsic to a *logos* of language. It is thinking joined with listening rather than speaking, as it was discussed previously as wedded to the Saying of language. Here, *techne* remains a knowing, but one retaining an aboriginal bond with *poiesis*, in fact an identification with it.[20] Such a bond would seem inevitably to be the result once *techne* is conceived as being in the sphere of language's preeminence, as a method displacing our command and control of language. Obviously, any conventional sense of a rhetorical *techne* provides a systematic understanding of language, and from there deals in strategies concerning how it can most effectively be applied to serve our needs, persuasive or otherwise. Just as Atwell informs us. However, perhaps not so obviously, the place of *poiesis* in a rhetorical context is just as worthy of our attention. In the connection of *physis* and *doxa* that Heidegger suggests, *poiesis* epitomizes the powers of language engulfing us, in which we ourselves have our being, leading imperatively to the realization that what language "brings forth" on its own account is as significant and worthy as any of our aims and purposes in assuming that we use language as our instrument. What is decisive in *techne*, he says, does not lie in making and manipulating but in revealing and making aware (*Question* 13). Again, we are presented with an ontological consideration of language that, in rhetoric especially, we are transformed in our relationship to it. In a sense we lose ourselves in this relationship; we are overwhelmed, overpowered by it. In this view all things are animate, vibrant, and alive in the expression of language, and we therefore wield this power only to the degree that we are first overpowered by it, deceived by it, participating then in a *techne* of *apate* whereby we overpower our listeners only to the degree that language first overpowers us.

This idea of *techne* embraces a process every bit as complex and difficult to accomplish as any conventional *techne*, even as it involves a "letting be" of language. Here we serve as its means of realization, expressing through our relationship with Being what achieves its power and integrity only in our relationship to it. This sense of *techne* is thus grounded in our awareness of this power, not in usurping it for our use nor in bearing witness only, but in bringing about the realization of this power by virtue of our "Being there," of human being as the "place" of its disclosure (*Introduction* 163). Thus, the "poetized thinking" constituting *poiesis* is similar if not identical to what Heidegger calls "authentic thinking," once again to that sense of *thanc* occurring by virtue of our relationship to Being. In this perspective the better part of wisdom, and of rhetoric in terms of this peculiar *techne*, lies in our openness to this power and our acceptance of language as ascendant, as the expression of *physis* itself. And here language is indeed the great *dynastes* that Gorgias claimed it to be in his various exaltations and homages, tantamount in all crucial respects to Being, to this idea of earth and thus to language in Heidegger's thinking.

Not only is this view of *techne* different from conventional views but seems in ways directly opposed to them, as it opens us not just to the power of Being but to its capriciousness. In conventional views we seek to "humanize" language, to domesticate it in order to bring it to bear in service of rhetorical purposes; while the other marks the way to the "nonhuman." Here it is not for us to know and manipulate language by means of a *techne* but rather to appreciate by this Heideggerian *techne* that always behind the facade of human will and reason lurks the uncanny as Being itself, the nonhuman, the terrible secret of the irrational presenting yet other facades, fathomless on the whole. The abyss. These are the traits abiding in the person of the trickster, and if they, as the trickster himself, confront us as chaos and clutter, by the same token they confront us as creative possibilities insofar as we are open to them as such. This dynamism is crucial, as it puts Heidegger's sense of *techne* into sharper focus still. Here, *techne* is interpreted by Heidegger as confronting *dike*, the latter meaning to Heidegger the "overwhelming" or "overpowering" prerogatives of the power of *physis*.[21]

Indeed, Heidegger speaks of the encounter as violent, as an "assault" or "attack" of *techne* upon *dike*. Yet they are antagonists of a very peculiar sort, being complementary aspects of essentially the same phenomenon, completing one another, neither having meaning nor integrity apart from the other. Therefore, the violence in this clash, he says, need not be considered in terms of some rancor or another, as the result of a viciousness or brutality; but is an ardor or passion that is not precisely ours alone, for it arises in the "reciprocal confrontation" of *techne* and *dike* (*Introduction* 161). In the end this confrontation is itself a manifestation of Being and our "Being there," and in ways the entire motif fits the Sophistic dynamic, specifically as it is expressed in Gorgias's rhetoric. If *dike*, this power of the "overwhelming"—of *physis* or the "divine" to the ancient Greeks—was expressed as violence in its encounters with mortals, in the Gorgian perspective, according to Mario Untersteiner, it was a violence "sublimated" in the transition of *logos* from the transcendental to the immanent, in that manner encompassing utterly the "magic violence of speech" (105–07). This is surely the case as this idea of *logos* is played out in Gorgias's "Helen," where persuasion by speech is equivalent to abduction by force. In Heidegger's very similar delineation, the confrontation plays out as yet another variation on the ontological difference, leading indeed to the ontological dimension of *techne* itself. It thereby leads as well to a most critical result of the process, of finding our way into the overpowering, to that sense of the nonhuman, by means of this confrontation and therein finding ourselves (*Introduction* 156–57).

Thus, to Heidegger, *techne* is the "violence of knowledge" (165), and as it is fundamentally a linguistic enterprise in any event, it becomes the means of our involvement and participation in the overwhelming power of *physis*. Through this sense of *techne* we not only participate in the "overwhelming" but express its power by means of language, gathering and bringing to "manifestness" (150) the "essents"—the things—of our world. That is the nature of "poetized thinking," this peculiar sense of listening to the voice of Being. The idea has its viability in the alliance of *techne* and *poiesis,* as the motif of violence is retained, in ways sublimated as *logos*. By "the violence of poetic speech" we are thus the pathway of this power, "of thinking projection, of building configuration, of the action that creates states." It comes not by virtue of faculties we possess of ourselves but as a "taming and ordering" of those powers through which "the essent opens up as such when man moves into it" (157). Of course, we can never master the overpowering but the "sapient man," Heidegger says, "tears it open and violently carries [B]eing into the essent [thing]," though the "sapient" man himself is undoubtedly Being's most remarkable expression. Or, to state it otherwise, by virtue of "poetized thinking" we "wrest" Being from the mundane, even as poetized thinking is not ours alone.

Again, we are in the thick of Heidegger's jargon, and his perplexing discourse, but I think here of "The Captives," those figures sculpted by Michelangelo. They are but half formed and otherwise sealed in stone, struggling to be released from cold marble blocks, to emerge from concealedness into appearance. They seek to realize that possibility, indeed they embody it, and the passion present to this day in their struggle is surely that experienced by their creator, these "Captives" seeming to carry the violence of that struggle against *dike* as much as he. And so, yet again, the creature and creator are one. But the measure of success is necessarily meager, so even here in the august sense presented by Heidegger, or Michelangelo, we realize the presence of the trickster. Maybe especially here. In these matters of the boundless, chaos is an ever-present possibility. Therefore, "tear" as he might, "the sapient man is tossed back and forth between structure and the structureless, and order and mischief, between evil and the noble." But in the end these features of violence, chaos, power, and creative possibilities come together in a statement as concise as Heidegger seems to offer on the subject of *techne*. "The violent one," he says, is the "creative man, who sets forth into the un-said, who breaks into the un-thought, compels the unhappened to happen and makes the unseen appear. This violent one stands at all times in venture" (*Introduction* 161).

The violence of this encounter of *techne* and *dike* might be seen as either the counterpart or precisely the violence brought about in "undergoing an experience with language" that Heidegger says we "suffer" and "endure" as it "overwhelms and transforms us" (*Way* 57). Here the emphases may differ considerably, but the nature of the encounter remains the same. In this "reciprocal confrontation" language is, most significantly, the very nature of our being. In the "de-cision" that results, it is the "violent one" as much as we ourselves, as in this confrontation we *are* language. At such a moment we do not create but experience creation, just as we might say the artist did not create the sculptures but experienced their struggle to be released, to be "carried into the essent [thing]." The very struggle that is "The Captives" and that remains with them—*as them*—to this day.

Though none of this is as concise as we would like, this interpretation of the "violent one" is essentially as I have described the Gorgian *kairos,* and it is stated here in terms of a *techne* not only congenial to its nature but, as the case may be, a *techne* virtually indistinguishable from it. The emphasis is on the creative possibilities inherent in the "overpowering," present as the un-said, un-thought, unhappened, unseen that we wrest from Being to make manifest through language. As an ontological matter, as the "violence" of knowledge, *techne* is for these reasons a dramatic process of knowing. Rather than involving issues of invention essentially as ways and means we devise and apply, in its ontological aspect *techne* comprises a "venturing forth" where, as Heidegger says, hazards of dispersion and instability, disorder and mischief not only await along with creative possibilities but very often await *as* creative possibilities. Herein, language in its essential sense, or at least as we encounter it in the experience of *kairos,* does not reside in documents or speeches, in words on a page or even those spoken to others, subject in all cases to our culling and application at the opportune time. On the contrary, it is manifest in our "Being there," as words embodied in the breach or borderland of our openness to Being, in that place of Being's disclosure where, indeed, in their "de-cision" words and "Being there" converge and the "quiet power of the possible" is realized.

As *techne* in this interpretation is ontological, "ad-ventive" rather than inventive, it follows that the emphasis is on the immanence of Being and therefore specifically on *physis* as an emerging power, shaping the clash of *techne* and *dike* in what can properly be understood as a physical encounter. As Heidegger says, *dasein*—"Being-there"— means to be at the breach "into which the preponderant power of [B]eing bursts in its appearing, in order that this breach itself should shatter against [B]eing." If the explanation seems rife with ambiguity, even contradiction, that is as it should be, for what Heidegger is saying, or so it seems, is that we exert the power of the overwhelming only as we are overwhelmed by it. The idea gives further definition to *techne* as a dramatic process of knowing and creating; and, as I say, an ontological issue that is essentially an effect of the ontological difference itself. It is "ad-ventive," Heidegger says, where "the violent one is the uncanny." He is the "strangest (man)" who "cultivates and guards the familiar, only in order to break out of it and to let what overpowers it break in." Here, "Being itself hurls man into this breaking-away," and he is thus driven "beyond himself to venture forth toward [B]eing, to accomplish [B]eing, to stabilize it in his work" (*Introduction* 163). This sense of *techne* as venturing forth, however cluttered with Heidegger's jargon, is nevertheless the paradigm of the Homeric heroes discussed previously, who in their openness to *physis* allowed its power to be expressed through themselves, if only then to "shatter" against it, sealing their fate as tragic; as it is the paradigm, I believe, of those others, heroes and otherwise, discussed in the same context. As it is surely the paradigm of *kairos* as a *techne* that we can draw from the example of the Athenian warriors of Gorgias's "Funeral Oration."

In any event, the antecedents of this or a quite similar sense of *techne* are apparent in Gorgias's treatises and fragments, even as they are tacit, assumed by Gorgias rather than given critical development. It does seem that his art of rhetoric, insofar as he is able to express it at all coherently, is limited to mimicry of the great *dynastes* that we understand—or misunderstand—as word play, mere tricks of style. Nevertheless, given his conception of language, it should come as no surprise that he can offer nothing other than his contention that the art of rhetoric has to do with words and is unable to offer anything more precise, even when further pressed by Socrates to do so. Yet, predictably, words alone are quite enough in his world view. Here the "magic" of *logos* is in play, with words bewitching and overwhelming the speaker himself, prior to any effect they might have on his listeners. That language is then more significant in what it does to us rather than otherwise is, as I say, specifically the paradigm indicated by the Gorgian *kairos*. Surely as Gorgias presents it in his "Funeral Oration," *kairos* obtains its ontological emphasis in this manner, in the experience of men *in extremis* that through the channels of "de-cision" are delivered over to their fate. In the same sense it also obtains its equivalence to a *techne* as a "venturing forth," apparent as a power manifest in language that comes from Being, even as *dasein*—the "breach"—"shatters against it." The transition from an understanding of *techne* as a "rational application of thought to practice" is evident. More precisely, in Gorgias's treatment *techne* provides a most likely instance of the "breach" of our "Being there," the implication of the ontological difference as the aperture through which the great *dynastes* obtains expression in the "Saying" of language: in a *kairos* joining language to our being in as close to a perfect union as we can possibly experience in this life. By means of this *techne,* and thus by *kairos,* we indeed ride the back of the tiger that is language, using language effectively only to the degree it uses us, achieving in that manner the "perfection of argument."

Untersteiner tells us that *kairos* is the basis of Gorgias's ethics, aesthetics, and rhetoric (161), and in his analysis *kairos* seems to have such a strategic place only by virtue of language conceived as an "indigenous field," consummately the equivalent of Being in this interpretation. Again, Heidegger makes the connection explicit. In fact, once his retrieval of *techne* is complete, he returns to what is most essential in his thinking and, it would seem, in Gorgias's as well, claiming that "language is the primordial poetry in which people speak [B]eing" (*Introduction* 171). *Doxa* is then the measure of Being in each case, being the very event of its immanence; and, in each case, if *doxa* persists as the prescribed and habitual, it is nevertheless more than a collection of beliefs, of mere opinion we accept or reject at our discretion. As it harbors the "primordial poetry" of a people, the "Saying" of language, there is surely survival and endurance in *doxa,* but for this very reason there is also creative flux and even turmoil, a "bringing forth" that is revelation and insight, a "para-*doxa*" that is *kairos.* In this dynamic a peculiar rhetoric of the earth is evident, mustering a *techne* joined with *physis* that "guards the familiar only in order to break out of it," thereby to let the "overpowering" or "overwhelming" break in.

These "peculiarities" come about not only in contrast to Plato's ideal realm of Being but, because of that, in contrast to conventional interpretations of the rhetorical *techne*. Particularly telling in this context is Plato's view that words at best provide only a marginal correspondence to Being, and at worst provide none at all but are purely "deceptive" in some grievously pejorative sense. In fact, Plato sees deception precisely as a question of morality, and though his influence is pervasive, the issue has a deeper history and significance than even he provides. In Greek lore, as we have seen, Stesichorus typifies the attack on language and myth, condemning them as both cause and consequence of *apate*. Each is deceptive pure and simple to Stesichorus, whether issues of literary expression or not. So he rejects Homer and essentially all Greek poetry and myth as a compost of artifice, various canards, and lies, and then censures even the gods that he claims implant and nurture such deceptions.[22] This pattern of morality, with its firm standards of right and wrong, and firmer still of true and false, is fostered by the imposition of a *logos* meant to preserve us from the ravages of language, and by means of which mystery is destroyed or displaced. By its authority we cease believing in the power of language, in its magic, and all instances of its workings then become, by and by, artifice and shame, tricks that deceive us. Thus, magic is overcome by a *techne* more analytical by far than the Gorgian *kairos,* such as those intended to mitigate the effects of chance and nature, such as Plato's dialectic.

Yet the idea of *apate* was hardly rooted out entirely, nor was it even consistently condemned, especially among the Greeks. The kinship of poetry, deception, and imaginative power remains far from broken, and, as the example of Stesichorus himself indicates, myths have a way of co-opting attacks upon them, embracing them as yet other myths. Though marginalized as demonic and irrational, in that very capacity *apate* was seen as creative, bringing strife to what is otherwise sound and sensible. In fact, the conflict is evident in the dispute between Plato and the Sophists, is *central* to it, turning on issues of whether language is rational or irrational, mundane or magical, passive or dynamic, even violent. But these oppositions, and definitely the idea of *apate* itself, are far more ancient than those between Plato and the Sophists. They are "primordial," stemming from the "basic unity" that is ambivalence itself, manifest as the mystery or uncanniness of Being, and in that respect reflecting what is most essential in the paradigm of the trickster. In his person *apate* has a vitality still, surely as we see it manifest in Heidegger's interpretation, where Being is portrayed in the belief that deception is not the result of our deficiencies but of Being's mysteries and ambiguities, of a boundlessness that necessarily runs counter to reason, ultimately encompassing reason itself. Accordingly, Heidegger's sense of *techne*—essentially his *Augenblick* recapitulated and generalized—affirms this primordial state of affairs, as Being reveals itself, prior to all our efforts to shape it to our measure. And so it seems with Gorgias, in whose view we do not rule the world with our *logos,* but, as Untersteiner says of Gorgias, "the *logos* of the contradictory world" rules us "in its contradictory way." In this we do not create the world with our mind, nor are we "capable of endowing it with unity and harmony" (122).

So again, to conceive Being in this way, as changeable and unsettled, as a panoramic display of appearances brought to the fore by language, is to conceive it also as naturally and necessarily deceptive, often felicitously so. *Es gibt*—"it gives"—Heidegger says of Being, and for this reason harbors within itself, as language, the "contra" to each and every "dictum"; while at the same time, as Being, it preempts by virtue of its boundlessness any intervention coming from the "outside" as a corrective, Plato notwithstanding. *Apate* is therefore a crucial feature of Heidegger's sense of *techne*, as it is of Gorgias's, owing to the fact that deception is a direct consequence of language whether conceived as Being or the great *dynastes*, each by nature defying categorical moral codes we impose through our sense of correctness, of right and wrong, true and false, proper and improper.

Thus, however else Gorgias failed to measure up to Plato as a philosopher and rhetorician, he possessed a wisdom that Plato did not. He had the capacity, as Untersteiner tells us, for "the kind of thought which unties the hidden knot of reality to discover its irremediable contradiction" (114). And as I have attempted to develop it in this section, the Gorgian *kairos*, unlike Plato's, is disposed utterly to that wisdom. More essentially, Gorgias understood that the power of Being, insofar as we can ever experience it, resides in language, and he attributed to poetic phenomenon the "motive of *apate*" as comprising "its basic creative agent." In so doing he recognized that poetry is "a confession of the non-rationality of the universe," that the "contradictory multiplicity of the real world . . . raised to validity the theme demanded by *kairos*" (113). If, then, Gorgias's mimicry of the great *dynastes* is so hackneyed and awkward to be ridiculed even in his own day, it is important to realize that his *techne* was not in essence a matter of style at all but purely of *kairos*.

Untersteiner traces this aspect of *kairos* to the Pythagorean doctrine of opposites, where they are nonetheless "bound together by harmony" giving "life to the universe." Here a "form" is imposed on a *logos* or its opposite, not by our will or desire but by *kairos*, and forthwith, by *kairos*, that form is imposed on us as well, the result being "deceptive" on all counts (110–11). And here the nature of *techne*—in both its Gorgian and Heideggerian emphasis—rests on a peculiar dynamism of being used by language and using it, of Saying and the human response resulting from our participation in the event of Being. Again the presence of the trickster is apparent in his hostility to boundaries we would draw between ourselves and Being, between ourselves and language, to achieve an objective stance. And if this *techne* displaces or drastically alters rhetoric's relation to reason, at the same time it invigorates rhetoric's relation to language and thereby our relation to rhetoric. Acceptance of this attitude is to acknowledge that we are in the grip of powers often beyond our control, that they seemingly defy any conventional *techne*, yet it is a *techne* that is as much an expression of wonder, strength, and resilience as resignation. Most critically, it provides the way to the power of possibility, a way of invoking these powers insofar as we shared in them, thereby realizing them in ourselves. And that, too, reflects the nature of the trickster. He is the master of crossroads, the penetrator of thresholds, the "living copula," as Henry Louis Gates says. He is a figure of "doubled duality, of unreconciled opposites living in harmony," treading the razor's edge (30), as does Eshu, as does the Greek mythological figure Kairos himself. There is more rhyme than reason to his universe, but an undeniable symmetry, both balance and harmony. A music to the spheres, mystic chords though they may be.

TIME AND THE UNCANNY: THE EMERGENCE OF THE "NONHUMAN"

Plato tells us the soul is for all eternity, that it partakes of the timelessness of his sense of Being. Yet Heidegger's purpose was to place Being within the horizon of time, his effort straightaway resulting in the thesis that Being *is* time. Equating the two, says Gadamer, "burst asunder the whole subjectivism of modern philosophy," thus annulling also the metaphysical imperative of equating Being with the permanence and stability of what is present (*Truth* 228). In ways, this absence or "decentering" of the subject is compelling enough in its own right to inspire the awareness of mystery, but there can be no surer expression of the mysterious nature of Being than time itself. Heidegger says authentic Being is necessarily "Being-towards-death" (*Being* 279, *passim*), that in the absence of pain, death, and love time is destitute (*Poetry* 97). Thus, in this transition from a belief in the soul as a timeless subject to the idea of *dasein*, time is apparent essentially and authentically only in terms of momentous existential concerns. And, very critical to our purposes here, to affirm Being as time is to Heidegger to affirm it as immanent by vesting it within the corporeal reality of language. Not only are we then dispossessed of immortal souls, but also in Heidegger's analysis all that philosophy condemns in temporality and physicality comes to pass in the union of the two in language.

As this analysis is pursued in his later works we come to understand language in terms of its unqualified ascendancy, a process taking place in and around persons, at times "bewitching" or "appropriating" them, utterly determining their nature. Our "authenticity," then, is the result of these gainful encounters with language, never of our soul, personhood, sincerity, or whatever the wherewithal we assume to possess apart from language. In this manner Heidegger's work in its entirety prepares a way to understand the Gorgian *kairos* as utterly an event of language, albeit contingent on the subject's nature as the relationship to Being, for in this context *dasein* is expressly that place or focus of the "de-cision" of the Saying of Being and human response. Indeed, one of the more critical passages dealing with *dasein* in this way is found, improbably, in *Being and Time*. Again, as Heidegger's concern in this work is principally the authenticity of *dasein*, that concern, even here, is joined with the authenticity of language. In crucial respects that seems to be the opinion of Derrida as well, especially in his explication of a relatively significant passage in *Being and Time*. In a different context I discussed the passage before, but now I extend those remarks, again through Derrida's commentary, as here I am interested in Heidegger's idea of listening and language in terms of the "uncanny," the "nonhuman." In the passage Heidegger says hearing is not only "constitutive for" discourse but that it "constitutes the primary and authentic way in which Dasein is open for its ownmost potentiality for Being—as in hearing the voice of a friend whom every Dasein carries with it" (*Being* 206).

The latter half is particularly significant. As we have seen, this "voice of the friend" is incidental to a distinctly pre-Socratic idea of *logos*, and if Heidegger never seems to come to this conclusion in *Being and Time*, Derrida does it for him, claiming the passage is based on Heidegger's interpretation, or more precisely his "destruction," of the Aristotelian interpretation of *logos* ("Heidegger's" 172–3). Heidegger's purpose here is to tap the primordial sources and experiences obscured by the philosophical tradition, ultimately to focus on the experience of wonder (*thaumazein*), on truth as *aletheia* that comes prior to things being settled and conceptualized. The emphasis is again on origins rather than termini, in that process so aptly described by Gadamer as a "teleology in reverse" (*Truth* 277). In this context, then, Derrida claims that Heidegger solicits what is prior to Plato and Aristotle to declare a more "originary *logos*." He links the early Heidegger of *Being and Time* with the later Heidegger where the nature of *dasein* is presented essentially in terms of language. In fact, Derrida claims that "destruction" quite simply means listening, but listening means to be changed, affected, influenced, to be "moved around" by language ("Heidegger's" 180). Thus, Derrida returns to another theme previously explored, to that of Heidegger's equation of hearing and belonging, with Derrida's interpretation extending the connection to embrace an "originary belonging" so intense that listening is tantamount to obedience, even submission or subjection (174).

So it follows, as Derrida says, that the "ear's opening" is *dasein*'s opening to Being (178), with the result that listening is essentially the way in which *dasein* is open for "its own most possibility for Being." Thus, it is "through the ear" that *dasein* "carries the voice of the friend" (178), where the voice is "situated neither inside nor outside of Dasein, neither near nor far, a voice that participates as it were in the opening of the Da- of Da-sein" (175). The idea of self-presence is displaced in this analysis; there is no preexisting priority of the self to be assumed whether by virtue of the soul or other first principle. "The call comes *from me* and *yet from beyond me and over me*," Heidegger says (*Being* 320, his italics). In response to this call, this voice, Derrida's explains we are simply *bei sich*, carrying the voice "beside ourselves" (178). In this analysis language is inevitably linked to the immanence of Being, to what Heidegger comes to call "earth," where the voice of Being, as it is heard and heeded, is the imperative ingredient of the "situation" in that the voice wholly comprises *dasein* as that opening, the very "thereness" of Being.

Things here, of course, become more and more peculiar, but the perspective Heidegger develops simply evolves from (and to) the primacy of language—this "originary," "aboriginal," or "pre-Socratic" *logos*—the basis of his determination of this immanence, his belief that we dwell in language. In one sense the ontology that results is surely a matter of "spinning language out of our being and ensnaring ourselves in it," and is here specifically apparent in the presence of *dasein* as a result of language. Given this of idea of *logos* we indeed carry the voice of Being "besides ourselves"; and, as *dasein*, we *are* only insofar as we are the "outsider within," in our authenticity being made of words in a process where, at the right moment, it is impossible to distinguish the dweller from the dwelling, the inside from the out. And it remains, to be sure, a world far removed from the well ordered state of affairs coveted by those believing in the lawfulness of the flux, those given to that sense of *logos* insuring a perfect fit of the subject and outside reality. In a sense this world is simply a more or less sophisticated reiteration of the trickster's, where the "originary *logos*" is our realization of his rule or misrule. If never the term, definitely the phenomenon of the trickster is central to Heidegger's writings.

He calls it "the uncanny" (*das Unheimliche*), "the strange." As we have seen, it is initially experienced as "the overpowering" (*Uberkomnis*), as the comfort and security of "average everydayness" is shattered and *dasein* feels "homeless" or "unhoused." Heidegger says, "We are taking the strange, the uncanny, as that which casts us out of the "'homely' life" (*Introduction* 150), where "home is dominated by the appearance of the customary and commonplace, the familiar and secure" (169). Everydayness is situationless. Time as abstract and routine. Destitute. Everydayness is "precisely that Being which is 'between' life and death," he says (*Being* 276), a timelessness where our lives are mere commodities of the customary and commonplace, as we would then indeed measure them out in coffee spoons. Thus, the uncanny is not only the "happening of the unfamiliar" but the breach or outright dissolution of everydayness, what Heidegger significantly calls the *"apolis"* (*Introduction* 167).

We might say that Heidegger's analysis at this point bears similarities to "absurd existence" heralded by existentialists such as Camus. But his sense of the uncanny is more complicated and perplexing, entangled as it is in what he sees as the essential failing of Western philosophy, the "forgetting of the question of the meaning of Being." Again, knowing is a matter of questioning to Heidegger, and he therefore advocates "a more original form of inquiry," an enduring search for an answer that, significantly, "does not crave an answer" (*Nietzsche* 2: 192–93), and so once again the refrain, "what we know authentically, we know only questioningly." Surely Plato's doctrine of the soul marks an essential step in the forgetting, in particular given the premise in his philosophy that whatever is moved from within has a soul and what from without is soulless. Such explanations undermine the nature of *dasein* as the relationship to Being, therefore the relationship to language; and such doctrines lie, as Heidegger says, at the beginning of "that peculiar turn and intensification that consciousness has taken" in Western philosophy (*Poetry* 108).

More generally this turn is apparent in the stance we assume in opposition to the world, to the earth in this context, where our "humanness" is enhanced, even exalted by such a stance, as in the last analysis it is the source of consciousness itself. But absurd existence or not, there is nothing of the uncanny in this attitude, at least not in Heidegger's sense of the term. In fact, he claims that we thwart the uncanny by isolating ourselves through this intensification of consciousness, thereby obscuring the mystery of Being by relegating all that is "external" to our consciousness as separate and objective. In this way we presume to know the ways of the world by delivering it, and Being too, over to ourselves. We humanize it, placing it completely within the horizon of human experience, making of it an exclusive human concern and prerogative, so that in this perspective we are indeed the measure of all things. We thereby not only deprive ourselves of our relationship with Being but of Being altogether. Again, such is the substance of Heidegger's assertion that "the higher the consciousness, the more conscious being is excluded from the world" (*Poetry* 108). In effect, there is no "Being-there" insofar as we are determined by a consciousness based on the premises of traditional subjectivity, or as we are "moved from within" by a soul.[23] Being is excluded or usurped in either event. It is forgotten.

Thus, the uncanny in Heidegger's formulation is hardly the result of our forgetting of Being but leads, most explicitly, to our awareness of Being, to our openness to it. Here the uncanny indicates more than what is arcane and ancient, though it is surely both. It is nothing less than the effect of what Heidegger designates as the "nonhuman." We are not the measure here, as our nature as *dasein*, as "Being-there," necessarily extends beyond the horizon of exclusive human experience and awareness. Quite ironically, then, Heidegger tells us "the question concerning man's nature is not a question about man" (*Discourse* 58), claiming what concerns our nature is not a matter of who we are in body, mind, or spirit but who we are in terms of the vitality of our relationship to Being, to that which is decidedly "nonhuman." It is what allows Heidegger to affirm, just as curiously—and very remarkably—that *dasein* is "essentially an eccentric being, since it is nothing outside of the *relation* to the other than itself" (Raffoul xxiv; Heidegger, *Essence* 99). As such, the nonhuman is paradoxically an integral part of who and what we are and thus, to Heidegger, our essence as human beings is never a factor of who we are isolated in any way from the nonhuman. On the contrary, as our "essence" is resolved only in terms of our relation to Being, we are inevitably creatures of the nonhuman, the uncanny.

As the relationship is foremost, the issue is never one of "what" but always of "how," and always of process. The relationship is not only reciprocal but is fulfilled by language, for language is "the primal dimension within which man's essence is first able to correspond at all to Being and its claim, and, in corresponding, to belong to Being" (*Question* 41). The "whatness" of things, of them existing independently as objects in their own right, is therefore derivative and subordinate to the relationship. Just as I attempted to illustrate in discussing the phenomenon of *xa*, there is no separate reality, no such object or thing as the Vietnamese village, but its very existence consists in the revealing and concealing of Being, in the "how" of crucial relationships between language and us, Being and us. So it is even with ourselves. By virtue of our relationship to Being we are made of words to the degree we participate in the nonhuman expression of Being, brought to immanence in language precisely as the peal of "not anything human" (*Poetry* 207). In this context, then, it is not "man" but human aspiration that is the measure, our relationship to Being bestowing the "nonhuman" to authentically constitute the human, providing the full measure of our being. Here it is not proper to claim we serve some purpose higher than ourselves but that, through the nonhuman presence of Being, we are the purpose higher than ourselves. The paradox seems fully in keeping with the way of the trickster, and is precisely "the quiet power of the possible."

Again, traditional notions of the self, soul, and consciousness are undone in this process, as they are indeed creations of the staid mentality of everydayness. Perhaps any departure from everydayness to events more calamitous would lead to a similar result, where the "other than ourselves" exerts a telling effect, often with sufficient force, surely, to make us "realized" as something other than what we might assume ourselves to be in more placid times. The critical point is that the idea of the self in no small part comes about as a result of our attachment to everydayness which allows us to foster illusions of the importance, permanence, and power of self. In just such a context, then, it is important to understand that one of the more distinctive aspects of the

Gorgian *kairos* is this sense of the moment as the experience *in extremis*, that "de-cision" resulting in the loss or decentering of the self, an effect seemingly obliged in the alliance of Being and time.

As Being reveals and conceals, it is not only identified with time but, as such, is to Heidegger "that which dispenses to all beings [things] their time of appearance and disappearance." Therefore, he claims that, to the ancient Greeks at any rate, it was the right or wrong time intrinsically, "the appropriate or inappropriate time" in every case, something that "in its way bears beings, releasing them, and taking them back" (*Parmenides* 141). In this perspective *dasein* does not exist in time but, in its relation to Being, *as* time (*Concept* 6E), and when we think of the birth and maturing of children, the passing of family members and friends, or simply the comings and goings of people and things in our daily lives, conceiving of time in terms of these continual leave-takings and reunions seems to make complete sense. Indeed, in many ways what fails utterly to make sense is conceiving time as otherwise, as homogeneous and abstract, as a sequence of indistinguishable and indifferent now points, that "empty frame of progression of occurrences one after another" claimed by Heidegger to be inauthentic (*Parmenides* 142).

From the beginning Heidegger sought to establish connections between Being and time, and throughout he determined no greater evidence of their accord than death itself. *Dasein*, then, is not just the entity open to time and therefore death, but it is the event brought about by each, and through each bears the mark of the nonhuman by virtue of being the relationship to Being. Whatever our preoccupations, digressions, or good intentions, most obviously in matters of time and death, Being cannot be reconciled to our needs, to our all-too-human image and likeness; and if our illusions lead us to believe otherwise, Being as the uncanny is there to disclose the error of our ways, to bring havoc and turmoil in the guise of the trickster, often as the harbinger of some tragic, sorely gained wisdom. He is the necessary wisdom of all cultures, the nonhuman in human form, being us and yet other than us, disrupting the self-centeredness of humankind and bringing us to account in disrupting time's orderly sequence, the routines of everydayness. So, whether he is Being or merely its messenger, principal or agent, in his presence there is the promise of redemption, an enlightenment or awareness even as it comes invariably with the realization that something other than man is the measure of all things. Especially as it comes coupled with the measure of the nonhuman provided so starkly by death that is, at the same time, so paradoxically the most certain measure of the human.

In these matters the trickster is an abiding presence, and in his name we necessarily pay homage to the mystery, that great universal obscurity he personifies. He manifests a peculiar yet pervasive sense of autochthony, being the giver and taker in many things that matter, most especially life itself. He therefore finds his ultimate expression in the power of nature, of *physis*, of earth opening for us to emerge and appear, then to return there, to be concealed once again. His presence is then profoundly apparent at times when the nonhuman is most keenly felt, at the point of the coming together of alpha and omega, of beginnings and ends and all points in between when Being breaks through everydayness and makes us aware. The trickster can then be considered the mythological source of *kairos*, at least as Gorgias presents the phenomenon of *kairos* in his "Funeral Oration," insofar as we can consider that presentation in all its doxic pitch as anything other than mythological itself. Indeed, as the Gorgian *kairos* is a linguistic process in origin and nature, the "irrational *logos*" itself is the trickster's stock and trade; but more importantly, in Gorgias, *kairos* is so felicitously allied with death that it is the "perfection of argument" by virtue of the bridge between the human and nonhuman, "of men possessed of a courage that was divine and mortality that was human."

In shattering the spell of everydayness, the trickster's purpose, upon which his many and various traits converge, is to force the confrontation with himself, with Being as the other, thus literally with ourselves as the other in our relationship to Being. Again, he is the measure of the uncanny. Like the two farmers in the Eshu trickster tale, things for us are seldom as they seem, and hardly ever as we would expect or wish them to be. As the nonhuman personified, the trickster is not only a definite and enduring confirmation that we are not the measure of all things but that, more significantly, we are not even the measure of ourselves. Like the voice of Being we carry, the trickster is with us, though neither wholly within nor without, but *bie sich*. We can then conceive his presence as a *topos* or place, as the situation. Or it is the "district," Heidegger says, where "the uncanny shines explicitly and the essence of Being comes to presence in an eminent sense." But it is just as surely an event, so intrinsic to the ontological difference that the promise for some accord in the Saying of Being and human response is apparent. It is then a "gathering," Heidegger says, "a holding together of what belongs together" in the nexus of the human and nonhuman, as that is explicitly the nature of the uncanny in Heidegger's formulation (*Parmenides* 117). Here, as Being and *dasein*, Saying and human language are manifest only by virtue of the ontological difference; they are neither separate nor opposite but complementary aspects of their relationship, an intimacy or balance of "what belongs together."

This "holding together" is the essential expression of the Gorgian *kairos*, and here it is important to understand its connection to Heidegger's interpretation of Greek thought as it was in its "original beginning" rather than what it became, with Plato and Aristotle, in its "incipient end." In the former, time moves in contrast to "everydayness," yet however it is perceived or measured, its association with language and rhetoric is especially vital. Here I have in mind a most commendable essay by Richard Enos, an analysis dealing with the emergence in fifth and fourth century Greece of artificially imposed time constraints on rhetorical activities. In the broadest sense, of course, time is the exclusive province of *physis*, apparent in the turning of the seasons, phrases of the moon, duration of the day, circumstances in ways mitigated by our technology though never ours to command and master. However, Enos's contribution is to trace the relation of discourse to "immediate time," the units of minutes and hours that to him measured the "momentary" in Greek rhetoric. This era of Greek history therefore witnessed the appearance of sundials and other devices to measure time, in particular water clocks used to allot time for pleadings in the courts and other venues, providing measures exact enough to allow equal time to accuser, defendant, and judges to render verdicts ("Inventional" 82).

But what is most interesting in his analysis is neither the invention nor use of these devices, but what comes to pass by virtue of each; namely, the invention of "immediate time" itself, so that by these devices we might indeed bring time to account, impose our will upon it in a manner to make it amenable to our purposes and goals. We then master time as we do the rhetoric it measures, thereby endowing the self with an integrity apart from time and language as a means of governing each. Here the invention of "immediate time" seems to give credence to rhetorical *technai* that account for interpretations that put the moment at our disposal, such as that given in the Platonic *kairos*. Equally significant in this process are Enos's closing remarks, that constraining rhetoric by the "concept of immediate time—the presence of the moment—was not only a part of the mentality of classical Greece, but was a real factor in the shaping the relationship between orality and literacy" (79).

This difference between orality and literacy, in their "relationship" or otherwise, is in some sense what this book of mine is all about. For his part, Heidegger would dispute neither the need nor expediency of "immediate time," though to him it would be abstract, derivative of what is more authentic and essential, time that is not a matter of our measuring but once again the right or wrong time "primordially," the appropriate or inappropriate time "in every case." In other words, time that is more in keeping with the phenomenon of *kairos*. Again, the presence of the nonhuman accounts for the latter, and we can then return with increased attentiveness to the thesis that Being is time, seeing both in terms of their connection to language. Here, also, time's relationship to rhetoric is apparent, though in ways remarkably distinct from that presented by Enos.

In one sense the relationship is obvious, the fact that time and language are expressed chronologically, being as one in comprising the narrative, the unfolding of a story. But more essential to this relationship and critical to Heidegger's approach, I reiterate, is his joining of language and Being, where he depicts the "Saying" of language, this voice of Being, as the surest confirmation of the nonhuman, the uncanny. It is here that language speaks as "the peal of stillness" that "is not anything human." And in this relationship we must be resolute, keeping our silence to hear the voice of Being and thereby remain open to it, to the possibilities of Saying where the influence of the nonhuman is not only conspicuous but essential. Once again, we are constituted by our relationship to Being and thus by language, "needed to bring soundless Saying to language" (*Way* 129). By virtue of this relationship, we are "linguistic," Heidegger says. We are "given over to speech" (*Poetry* 207). Here, most clearly, we are the "irruption" of language, *dasein* brought into its own in language, being "given over" to it. In contrast to the more mundane aspects of life, we then confront the uncanny in the sheer sense of "appropriation," appropriated by language that at once means to be appropriated by our mortality, as each is apparent in the narrative of our journey from birth to death. To deny this appropriation by language, as was Plato's wont, is not just to deny appropriation by death but in that very effort to deny appropriation by life as well. Here the latter is the "most uncanny," the consummate effect of language as the immanence of Being. Here we can take to heart the imperative of George Steiner's view, given by way of Nietzsche, that language is the "life lie" and, Plato to the contrary, any "leap out of language" is inconceivable, for that is death itself (*Babel* 227, 111).

Here, as well, to be "given over to speech" means that *dasein* should be considered as not just the place of the confluence of events but as the event of their confluence. In the context of the uncanny, *dasein* is an "openness" to Being that leads to the clearing, the self-disclosure of Being in the event of appropriation by language. As early as *Being and Time* the motif is apparent, Heidegger saying that the "caller" of the call of language is *dasein* itself "in its uncanniness," that "the call is unfamiliar to the everyday they-self," that "it is something like an alien voice" (321). Again, the distinction between inside and outside is obscured in this interpretation. As we have seen, in its resoluteness *dasein* is itself "per-cepted" in "ad-ventive thought," a phrase William Richardson applies in "the double sense of open-ness to the ad-vent of Being which [itself] determines this openness" (420–21). Thus, in this sense the situation marks the "arrival" (*ankunft*) of Being as that place or event where "Saying" occurs, prompting a "venture into language" that is the event of *dasein* in its authenticity. Such a venture is one manifestation of the Gorgian *kairos*, providing not just a departure from the "everyday they-self" but a new sense of our being that prevails in full realization of the nonhuman as that "alien" voice. The venture also attests to a very ancient sense of our being, comparable as it is to the *psyche*'s coming together with *physis,* as the *psyche* is then "transfigured" by the word in a fitting expression of *kairos* in its own right.

Thus, *kairos* is the event of Being coming to immanence in language, connecting two seemingly distinct movements, the Saying of language and human language that comes in response. It follows that Heidegger's sense of Saying is vital, as it "brings the unspoken word of Being to language" ("Letter" 239). Yet the human response is as vital, for it is the touchstone of authentic speech. As Dastur tells us, there are not two different languages at issue here, "the language of Being, on the one hand, and the language of mortals on the other." All things are simultaneous here, the voice of Being and human speech, the listening and responding. There is only one language, she says, "neither human nor non-human which is the *topos*"—the point or place of the difference—"of the intimate self-differing of responding" ("Language" 365). In this converging the *topos* is "Being-there" in its authenticity, realized in the very moment of the "intimacy" of Saying and the human responding. *Topos* is thus the uncanny as well, being neither Saying nor responding alone. Nor is it both in any essential sense, but is again the relationship preceding each, that in propitious moments prompts the Saying of Being and human responding seemingly as one. The conception is confounding, excruciatingly so in places, but its nature and nuance are wonderfully captured in the image of the trickster, the man of borders. In Eshu, or as aptly in the trickster Kairos himself, these realms of Saying and responding are bridged, as the trickster treads the razor's edge, bringing them together in "de-cision." Here *kairos* is realized in the unfolding of *physis,* a "bringing forth" in language coming about of its own accord, using us as its means of realization.

Thus, in this interpretation of the Gorgian *kairos* it is important to understand that even as Heidegger refers to the "uncanny district" as a place, he claims it is not one located in homogenous space, not a place fixed by coordinates but a situation that in this perspective is an event driven by extraordinary circumstances, encounters with the unfamiliar and unexpected. Here the idea of the "uncanny" is not just indicative of the trickster's nature but altogether expresses his mettle, a process put in motion through this emphasis on Being as time, specifically on "authentic" time, the "qualitative right time" and *kairos* as Heidegger develops them.

The matter is hardly so spirited in every case, seldom entailing obvious matters of life and death, but nevertheless remains most evident in the more dramatic aspects of the experience. So we might say, in *kairos,* we close with Being to reach a flash point, and in an instant so intense as to tarry indefinitely, language is made intrinsic to time, that "qualitative right time," for surely in that instant each approaches Being in its wholeness. In this event time loses its human vector, all connections to everydayness and *chronos,* and in contrast embodies Being's vitality and power expressly as language. Time is animated by language, "authenticated" by it. Time *is* language in this sense. In *dasein*'s openness, Heidegger says, "the Present is not only brought back from distraction with the objects of one's closest concern, but it gets held in the future and in having been." The "Present" is this resoluteness, this awareness not exclusively ours alone, determining time as language. "Rapture" (*Entruckung*) is the word Heidegger uses to designate this aspect of the *Augenblick*, occasioned by a "nearness" of Being and *dasein* where the tension between the two is most keenly present. In Heidegger the *Augenblick* is then understood as "*ek-stasis*," the "resolute rapture with which Dasein is carried away to whatever possibilities and circumstances are encountered in the Situation as object of concern," a rapture "which is *held* in resoluteness" (*Being* 387). It is then, as well, a "nearness" that does not depend on conventional spatial-temporal conceptions but is independent of time and space the way each is construed by the "calculating mind" (*Way* 102–03).

Indeed, time in the distraction of everydayness is so ephemeral as to be non-existent. The past is gone, the future is not yet, and the present lacks temporal extension altogether, at best a marker and itself outside of time as utter *stasis*, the "standing now" (*nunc stans*). As we have seen, such a marker is indicative of individual consciousness, simply the soul itself or "the ego's primal form of Being" as Robert Dostal says (148). In the mandate of transcendence, temporality as an integral feature of *dasein* is but an illusion, necessarily so in this perspective and vanquished altogether as a matter of mere appearances. Heidegger, of course, sees things otherwise. Though the past, present, and future are necessary fixtures of "objective" time, to Heidegger they are derivative, contingent themselves on authentic temporality where time is Being, embracing before all else our mortality in "Being-towards-death." Here, "time times simultaneously," he says (*Way* 106). The vital direction of the movement of language is, as Being and time itself, from the past and future through the present, and in the power of the *Augenblick*—the moment—language exerts its claim to become manifest. The present becomes the portal or focal point, the *topos* or place of the expression of language in *kairos*. In the *Augenblick* past, present, and future coalesce in "*ek-stasis*," in that tension played out in the ontological difference.

As Derrida suggests, we might better understand the event by thinking in terms of an "originary," "pre-Platonic," or "aboriginal" *logos*, in a listening that is a belonging in our being claimed by language. In the gathering, this "holding together of what belongs together," time is as integral to *dasein* as it is to *physis*, as natural and unaffected as the cycle of the seasons. Language, then, is measured by time only in the derivative sense, for in the phenomenon of *kairos*, once again, time is language, or language time. In Being they are one, providing especially in this context one of the clearer delineations of

Heraclitus's dictum that, once we heed the *logos* speaking through us, it is wise to say that all things are one. At once, the same yet different, for these are the dynamics involved in the "intimate self-differing of responding" of which Dastur speaks, of the *Unter-Schied*, that interface of Saying and responding whereby we undergo an experience with language and speak authentically only as "language brings itself to language." Here "we let the soundless voice come to us," Heidegger says, "then demand, reach out and call for the sound that is already kept in store for us" (*Way* 124). We thereby "put into language something that has never been spoken" (59). Thus, the *Augenblick* stands at the threshold of possibility, marking the orientation towards "that which remains to be said." It is indeed the aperture of Being and time through which the world of the speaker opens, where, as a result, the speaker himself is the aperture, being at this moment at one with both language and situation, where the un-thought, the un-said, is uttered and becomes manifest.

Earth, *Doxa*, and Para-*Doxa*: Translating Yesterday's Words into Novel Utterances

Heidegger claims Saying "rises from the earth," bringing "the unspoken word of Being to language." Thus, it is manifest as *physis*, in its most embracing and vibrant sense as earth itself; and, as such, manifest not just as the verbal environment of *doxa* but also as the "para-*doxa*" it contains. In this context paradox is not *doxa*'s opposite, as the conventional is opposed to the unaccustomed and strange, but in the revealing and concealing of Being it is *doxa*'s correlate or counterpart, its complementary aspect. And here, too, the image of the trickster treading the razor's edge is fitting. In the "irrational novelty of the moment" paradox comes to pass in the departure from everydayness, and through paradox the insight and wisdom gained on the boundary, the razor's edge in the "de-cision" of Saying and human responding.

As we have seen, the path of Heidegger's thinking inevitably leads to the retrieval of some very ancient beliefs concerning language, and when he thereby celebrates the worthiness of the Greek language for the disposition of its speakers to "dwell" in language, he retrieves not only that archaic sense of *doxa* comprising the verbal environment in which we live but, in light of this idea of *doxa*, retrieves as well an understanding of "earth" as the source of our nature as "dwellers" in language. In this sense there is nothing in Heidegger that is not apparent in any sincere appreciation for language, in literature or rhetoric, because this relationship of dwelling in language goes beyond a merely objective study or application of either, embracing instead the imperative that language is not or seems not be our own, as even we, in a sense, are not our own but creatures of language. Were it otherwise, were *doxa* as aimless and unavailing as Plato describes it, perhaps his ire would not have been so roused to condemn it as vigorously; nor then—God help us all—to consign the poets and rhetoricians to the lower rungs of Being, down next to the tyrants.

Indeed, earth is the essential expression of the immanence of Being, as here we cannot presume to frame Being to our liking and project it to some transcendental realm where its influence cannot be so forcefully felt. Yet earth is neither nature nor physical reality alone but embraces both womb and grave, the whence and whither of our lives. And it is still more. In Heidegger's interpretation, earth can only be understood in relation to his peculiar sense of "world." As we have seen, world is the web of relations connecting us to things, the network of signs and significations that establish our sense of reality. It is then a creation of language, not only establishing the nature of human experience but constituting our horizon of understanding, "a practical everyday understanding of things available for some use," putting them at our disposal (Haar, *Song* 10). Earth is otherwise, Heidegger says, essentially different from world yet they "are never separated." Curiously, "the world grounds itself on the earth, and the earth juts through it" (*Poetry* 48–49). If world is the place of "everyday understanding," of what is inherited and accepted, earth is the enigma, a place of concealing. In its relation to world it is the mystery of "non-sense," a "universal negativity," John Sallis claims, seeming to comprise a "monstrous contradiction and doing so especially as it intones "the song of the earth" in the "corporeity" of our own voices ("Forward" xi—xiii). If earth is not Being itself, it is the experience of becoming aware of Being, what "immediately claims and moves us." Thus, it is not a ground in the sense of a "securely established foundation," Karsten Harries tells us, but the *Abgrund*, the abyss ("Heidegger" 322). Or it is what Haar calls the "nocturnal ground of the world." Here the uncanny is apparent in the "collapse" of the everydayness of world as earth indeed "juts" through it in a process putting us in relation with the sacred, with the gods (*Song* 42). So, as we enter into the abyss, we do not tumble into emptiness but, as we have seen, "fall upward, to a height" (*Poetry* 191).

Whatever it may seem, this process of earth and world is hardly the result of some convoluted philosophic maneuvering, as it is far more common and accessible than we might suppose. Consider, for instance, the distinction Flannery O'Connor makes between mystery and manners, as she chooses to emphasize in her writings those instances when the "manners" of a culture, its customs of typical social patterns, are punctured by the unexpected and inexplicable, as "mystery" juts through this veneer of manners, whether revealing itself as divine or diabolical depending upon the degree to which there is a difference in her works. In a similar sense we might appreciate Heidegger's distinction between earth and world, though here, of course, the significant point is that the creative power of earth is manifest in its difference from world, in the interface between the two. Thus, earth's power, its "self-assertion," is never "a rigid insistence upon some contingent state" but is the result of the "surrender to the concealed originality of the source of one's own being" as "each opponent carries the other beyond itself" (*Poetry* 49). As we have seen, any genuine work of art is a result of this peculiar strife, being to all intents a reiteration of the difference between *dike* and *techne*. As the source of the uncanny lies in this opposition, so in this context it lies in the rift between earth and world, in the *topos* or borderland, the razor's edge between the two that occasions the event of *kairos*.

The vitality of earth, definitely the awareness it inspires, is a result of this tension. As I have stated, language itself prevails *as* this tension, so that the "voice of Being," in a sense that of nature itself, is culturally manifest. Earth then embodies spiritual or divine attributes by the very fact of its corporeal nature, expressing most facilely these connections of *physis* and *doxa*. In the aspect of world alone, *doxa* is opinion, the result of human preoccupation; but in the tension of earth and world, by dint of Being's immanence as earth, *doxa* is glory, or can be, as we ourselves then embody the nonhuman, the sacred. Thus, even as we dwell in language, in that verbal environment of *doxa,* paradox abides as *doxa* itself—as glory within the mundane, the unfamiliar within the familiar—just as the trickster would have it. By virtue of his presence, things by nature are steeped in ambiguity and conflict, a premise in good accord with Heidegger's emphasis on awareness as the play of the revealing and concealing power of Being, *aletheia* and *lethe* respectively. All things are "conflictual" in this process. "*Un*-concealedness makes manifest a conflictual essence," he says. "It is unconcealing when in it something comes to pass that is in conflict with concealment" (*Parmenides* 18).

This peculiar notion of awareness, contingent as it is on this conflict between revealing and concealing, characterizes awareness as a process in which we are involved rather than a quality we possess. Its "truth" is then conceived in terms of *aletheia* rather than *veritas*, and at this point is most critical in our understanding of the Gorgian *kairos*. In a paradigm gathered from the concluding lines of Plato's *Republic*, Heidegger speaks of the concealed and concealing as the "Field of *Lethe*," a region of "emptiness and abandonment" due to pervasive concealing. We might say it is a state of perpetual everydayness, but here there is neither earth nor world, clearly no "jutting" through of earth, and all things dwindle to absolute non-Being, the Field of *Lethe* preventing every disclosure of things, even of the ordinary. Here it is apparent that awareness is not ours, at least not ours alone. As "outside reality" vanishes in this Field of *Lethe*, so does our relationship with it, with *physis* as such. Consequently, so do we in the utter absence of awareness. *Lethe* can thereby be conceived as the opposite of *physis*, as apparently Plato does himself, describing the Field of *Lethe* as being completely bare of everything the earth lets spring forth.[24] In fact, in ways it seems as if the Field of *Lethe* is as close to the "natural order of things" as we can ever possibly imagine. In that perspective we might also come to appreciate the question that to Heidegger is the source and nature of authentic thinking, of why there is something rather than nothing. Thus, in the truth of *aletheia* we not only gain deliverance from *lethe* through awareness and revelation but realize *aletheia* as the uncanny indeed, an emergence from the natural order of things only as it is inherently part of it, and that we are, as a result, the truth of the uncanny ourselves, as it "envelops the course of mortal man" (126). It is an emergence Heidegger often identifies as divine, though it is an extraordinarily earthly conception of the divine to be sure. In fact, in the workings of the Gorgian *kairos* there is the influence of the divine, but only in what Heidegger claims to be the "Greek" sense. Here, any notion of the divine as transcendent is precluded, surely as we would presume to witness its descent in displays of great special effects suspending the laws of nature—say, a parting of seas or staying the sun in its course—because in the Greek sense, as Heidegger construes it, the divine is a matter of nature, of our awareness of Being in the here and now based simply and immediately on the emergence of "the uncanny in the ordinary" (122).

The trickster lurks here, this being his work, and thus providing the key to understanding the nature of the Gorgian *kairos* specifically in terms of the Yoruba's invocation of Eshu, of he who "translates yesterday's words into novel utterances." My argument, of course, has been that much of the Sophistic perspective can be understood in terms of connections such as these, that they in particular determine the nature Gorgian *kairos*. And if this perspective remains obscure, perhaps it is so for having to do with the transformation of *logos* from language to reason and our subsequent reluctance to any longer believe in the powers of language, definitely not in its powers as the great *dynastes*. We have apparently lost the capacity of "dwelling in" language that to Heidegger once gave the Greek language its power. Yet in the course of these changes language itself is no different, only our perspective toward it, and that in itself is a change of the most profound sort. We no longer believe in its powers but presume to appropriate them, gathering them to ourselves by means of a rhetorical *techne*; and thereafter, by virtue of such a *techne*, the expression of these powers is often seen only as matters of guile and artifice, trickery and deceit on our part. In this transformation, as well, the Sophistic sense of *doxa* is faulted and enfeebled, and any sense of "dwelling" in language—most assuredly in terms of the primal accord of *physis* and *doxa*—is dismissed altogether.

Here, in particular, central to the issue of *doxa* is the phenomenon of revealing and concealing, the source of the Sophistic ideal of *apate*. It is not an ideal that reveres falsehood but proceeds necessarily from the recognition of what is most fundamental in the nature of language, its ambiguities and equivocations that in the power of *apate*, whatever its snares and iniquities otherwise, presents a vast array of creative possibilities. As Steiner says, these powers and possibilities lie at the center of our need to "unsay" or "gainsay" the world, allowing us to imagine and speak it otherwise, as more gratifying and benign (*Babel* 217–18). They are indeed the means of our believing in the "life lie," as it is sheerly the mystery of Being itself. Again, what is at issue is a radically different attitude toward language, inherent in its nature as the great *dynastes*. Here the issue is not, nor should it ever be, stylistic tricks or artifice, but instead concerns both speaker and listener being bewitched by language, ultimately by that power of dwelling in it. The attitude is not only intrinsic to *logos* as language, in lieu of most descriptions of it as reason, but it accounts for the nature of our being as the uncanny itself in the revelations and awareness it brings about. Here we may presume to use language specifically to serve noble purposes, and well we might, but in this perspective this power resides ultimately in language itself, whether we realize it or not. Indeed, it resides in the viability of *apate*, where the issue is not simply the realization of the world we assume to be reality but is the reality that language carries within itself, worlds other than the one currently revealed that language can reveal and elicit in turn, calling them to the fore.

In these ways, along with others I have suggested, the nature of the trickster seems most congenial to Gorgias's world view, and to Heidegger's too, as the phenomenologies of each, particularly in light of the uncanny, meet and often seem to merge in the emphasis given to language. In Gorgias we might best understand the uncanny in the way Untersteiner previously expressed it, claiming that we do not rule the world with *logos* but that the *logos* of the contradictory world rules us in contradictory ways (122). In this process not only does language reveal and conceal, undermining the sense of a stable reality, but in these ways controls and governs, as does the trickster. *Apate* is surely his specialty, an event neither good nor evil though it often yields, as Gorgias says, the most profound wisdom. As language harbors by its very nature the "contra" to each and every "diction," the trickster, as Lewis Hyde says, "leads the way or leads astray" (121). And often he does neither one nor the other but both simultaneously, for in the trickster's world, surely in Eshu's, to lead astray can be precisely to lead the way. In any event the uncanny is intrinsic to the *logos* of his contradictory world, so the danger is not necessarily in being lead astray but choosing to ignore the trickster's authority altogether. And that is to preclude awareness, very often as a matter of method.

Thus, to say with thoughtless self-assurance that contradictions cannot be, is to say that Eshu is arrayed in white alone. Or only in black. It is refusing to reckon with Eshu and that, in essence, is refusing to reckon with the powers inherent in language. Perhaps we do so in the supposition that there is some power outside of language, most likely something above and beyond this earthly realm, such as the Platonic Forms, insuring a perfect correspondence of symbol and symbolized. But to see the issue otherwise is what Gorgias taught and in many ways, millennia since, Heidegger as well. What the trickster has taught all along. He is the god of connections, of crossroads and thresholds. In this light we see in him the realization of immanence, not just in customary ways of embroiling us in the foibles and frailties of the flesh but in manifesting crucial connections between us, language, and the here and now. He is then the consummate emissary of the earthly realm in which we dwell, the reality continually brought to mind at his prodding. He stands at the edge of our awareness, or he is that awareness himself, as human reckoning is determined or displaced by language. He is the nonhuman personified as human, that power of earth expressed through us most essentially as language. We would then do well to heed this power, at least to appreciate the gravity of the Yoruba supplication to Eshu, "Do not falsify the words of my mouth."

Here the trickster embodies the boundlessness, hence the capriciousness, of earth itself, and therefore in our dealings with him we cannot avail ourselves of the usual binaries, least of all those of truth and falsehood, even those of good and evil. Deception is here simply the issue of language and Being, evil only to the degree that it is good, false only as it is true. In all this the trickster is the manifestation of Being or of its "nearness" as flux and instability, obliging the departure from everydayness, provoking the movement from the all too human desire for fixed realities to an openness where language is joined with these shifting appearances, a departure that other than through this openness can only be understood as deception in some purely pejorative sense. This departure marks not just the transition from *doxa* to para-*doxa* but, so to speak, *doxa*'s apotheosis as para-*doxa*, so that the old and the new, the habitual and the unfamiliar, are neither contradictions, opposites, nor mendacities of any sort but purely of the nature of revealing and concealing of things already present in the largesse of Being. Just as Eshu did not cause the rift between the farmers but simply disclosed an animosity already there, the phenomenon is wholly in keeping with the ruses of the trickster who always provokes, as Jacques Monod says, "*revelations* of something already present, not creations of something new" (Hyde 120).

Francoise Dastur says essentially the same of Heidegger's sense of the uncanny, linking it with his emphasis on our dwelling in language. She claims, as we have seen repeatedly, that "proper dwelling does not mean familiarity" but "requires the ability to see the unfamiliar in the familiar" ("Language" 363). In that paradox persists in the "conflictual essence" of *doxa*, our language is creative and insightful insofar as we remain open to its power, to its "Saying." Then we can see paradox manifest in ways similar to the revealing and concealing power of the trickster, to the power of language and Being itself. But whatever his ruses and wiles, it is important to understand that the trickster is as much the tricked as the one who tricks, as much the human as nonhuman, always the outsider within. In this sense we are in fact the most uncanny creature of all, and the trickster thereby comprises who we are as much as anything we might confront that is separate and complete. The tension prevails whether we assume it to reside wholly within or to impinge from without, whether such distinctions are valid or otherwise. It prevails as the ontological difference itself, and as the trickster provides a completeness to the human experience through the investiture of the nonhuman, he is essentially the mythological antecedent of the ontological difference. He personifies the nonhuman as an integral part of who and what we are, paradoxically affirming in his very person the nonhuman and human as complementary aspects of our being. We must therefore pay particular attention to this paradox of the nonhuman, the "uncanny," entailing as it does the unfamiliar within the familiar, these "revelations of things already present rather than creations of something new." This specific sense of the uncanny is a prominent motif of the trickster's canon, and at this point provides yet another and very critical means to understand *kairos* in its peculiar Gorgian and Heideggerian emphasis.

Nevertheless, we desire to avoid the mystery of the nonhuman, surely as it comes coupled with "earth," and seek to be delivered from that vexation by way of philosophy's time-honored purpose to care for our souls; and in the interim the meaning of *doxa* persists essentially as Plato described it, as "mere opinion," a prescribed and habitual way of believing based on the deceits of stories, legends, and myths. However, if "nature" speaks in these stories, if they are the "voice of Being" rising from the earth that in our listening constitutes *doxa*, then *doxa* is a way of believing that claims us as believers, determining who we are, *that* we are, in this mystery of Being rife with the ontological imperatives of the uncanny. In this sense, we might say para-*doxa* is to *doxa* as earth is to world, but must always be aware of the fundamental inseparability of each, that they are mutually implied, as the familiar and unfamiliar. But insofar as we see *doxa* in this connection to earth and language, it is one with the play of appearances, of all that is boundless, shifting and changing. Preeminently it is the "uncanny," the experience of being "unhoused" as a source of rebirth and constant renewal, and for that reason alone is at times sufficient to connect us to earth (Haar, *Song* 62–63). Here the voice of Being does not speak in monotone, nor in staunch and wearisome consistencies, but breaks the spell of everydayness to reveal "the uncustomary within the habitual" (43). In this perspective the Gorgian *kairos* is therefore "de-cision" conceived as an utterly linguistic phenomenon, not just because there can be no awareness apart from language but at certain propitious, paradoxical moments we are awareness itself, the "irruption" of language. The latter is expressly the nature of the outsider within, the human and nonhuman simultaneously that defines the trickster's sphere, surely Eshu's, given the impeccable rendering in that tale of the unfamiliar within the familiar, of the trickster who "translates yesterday's words into novel utterances."

As I have suggested, the "de-cision" of Saying and human response undermines the conventional sense of a subject. But this understanding hardly "rehabilitates" Sophistic rhetoric on that account, much less redeem the various "iniquities" attached to it, those of *apate*, indulgences of "mere" opinion, and the apparent disregard for "truth." But we might surmise, insofar as these remain matters of error, untruth, or evil, they are invariably rooted in an all too human perspective that is necessarily incomplete, inauthentic for the very reason that it is only "human" and that alone. Indeed, such a perspective often seems fragmented and jaundiced, in the way so nicely illustrated, say, in the self-absorbed storytellers in Akira Kurosawa's film, *Rashomon*. They have witnessed or heard testimony of a possible murder and, as a result, experienced the dissolution of everydayness. They are "unhoused," left abandoned in the *apolis*, all of them locked into their own individual perspectives on the event. They gather at the gate of Rashomon, seeking shelter from a drenching rain, and tell their stories and listen to those of others, yet each is hopelessly the hero or heroine of their own peculiar story as, in a sense, they necessarily must be. Thus, they are "unable to tell the truth even to themselves," and so they speak falsely, having failed, we might say, to reckon with Eshu. On this account all of them in the end are revealed to be selfish, for that is inevitably the nature of the human perspective.

That also, in a sense, is as it necessarily must be. That is the tragedy of the tale, the "outrage" that the characters lament. That is the tragedy of knowledge. Here, revelations of "novel utterances" cannot transpire in terms of the isolated ego but invariably involve something larger. But the storytellers *are* at the gate, and their deliverance begins in their acceptance of the outrage, leading to that "de-cision" where wisdom lies. If they do not move beyond the human perspective, at least they realize its limitations, realize that to embrace life and language in their fullness entails the joining of the two in an event enkindled by the distinctly "nonhuman." Such is literally the experience of one of the characters, a woodcutter, whose story of the killing is as likely the lie as any of the others, but who makes of it the "life lie," and at story's end decides to take and care for a child left abandoned to die at the gate of Rashomon. That alone suffices for deliverance in this tale, as at that moment the woodcutter is not simply the one telling his story but is the story he tells, the story he becomes in his awareness of the earth, if you will, but awareness nevertheless of the "nonhumanness" of Being that is the purpose higher than himself. Here, in the face of the nonhuman, the unforgivable sin is the certainty we tell the truth. The tragedy of knowledge, then, is that we do not know, cannot know, yet move beyond the human perspective to the paradox affirming that we are the story not of our telling alone, being in that sense, indeed, the "irruption" of language.

We might understand "yesterday's words" as culturally bound, for they are *doxa* epitomized, but for that very reason can be seen also as an expression of *physis* or Being, of the powers of earth. As such, yesterday's words need not be conceived as frail or restricted but in the vitality of language as vibrant and innovative, as the catalyst of novel utterances rising from a *logos* moving beyond "purely phonetic data" and the work of "physical organs" to embrace, in Heidegger's words, the sound of language, its "earthyness" held in harmony and attuning the "regions of the world's structure, playing them in chorus" (*Way* 101). He speaks of "Saying" as this earthly source of the sound of language, and in this particular context of the care incidental to belonging to a community, to "a people of poets and thinkers" through whom Saying is manifest and in that sense constituting a culture, a tradition, a *volk* (*Parmenides* 112). Again, language is primary. Only "in the disclosiveness of word and legend," he says, do we truly experience Being, with the result that all works of art "exist only in the medium of the essentially telling word," in the realm of the legendary, in *mythos*. The approach is crucial, accounting for the preeminence of the word among all arts and cultural practices (114–17). Again, we belong to such a community insofar as we hear the word and are claimed by it, insofar as we experience hearing and belonging as one, just as each is apparent in the interplay of *hören* and *gehören*.

In this peculiar emphasis on community we are in essence concerned with the Sophistic *doxa*, more specifically with the Greek *paideia* as sum total of the literary tradition, as the very élan that determines our being.[25] In this context yesterday's words can indeed become novel utterances, as the latter come about not essentially in our wielding of language but in the power of *doxa*, in the legendary word as the measure of the boundlessness of Being. In such a context Heidegger claims that Being is the Greek equivalent of the divine, an earthly conception of the divine that in this sense is the mythical. It is the uncanny within the ordinary, he says—"the appearing as what is perceived in appearing"—the coming together of "that which is to be said" with "that which is said in legend" (*Parmenides* 112). This coming together, this "de-cision," within the legendary word is a defining characteristic of the Gorgian *kairos* much in keeping with the paradigm of the trickster, for in the event of *kairos* we are so intimately involved in the event that we are quite literally "beside" ourselves, dispossessed of the objective perspective that would separate us from our experience. We are the "god-sayers" but therewith singularly unable to make of this confrontation with the uncanny an exclusively human concern or prerogative. So, to the extent we are wise, we listen to the voice that comes neither from within nor without, to the "alien" voice of the trickster or clearly one, as Heidegger says, that participates in the opening of the "*da*" of *da-sein*, the "thereness" of Being. We listen and thereby belong as the measure of our authenticity, and in *kairos* embody the "irruption" of language that creates us as surely as it does the language we speak.

Once more, to appreciate this notion of the unfamiliar within the familiar—these revelations of things already present in the boundlessness of Being—we are inevitably concerned with language in its essentially ontological dimension. Thus, we are concerned with language as we embody it, in the *mythos* engendering in this perspective those "irruptions" of language that we are. And this dimension is basic and omnipresent, even in ordinary times and places when we assume language always to be otherwise and are thus otherwise unaware of the impact it bears. But even then we might hear its call and respond, as we do at other times far more profound when we are griped by its awareness, perhaps even to the extreme of the *Blut und Boden* of the sacrifices praised in funeral orations. Here we understand Saying to be essentially the legendary word of *mythos*, and in the "glory" of *doxa* we live it out, at times fulfilling and even enriching further the *mythos* by virtue of our embodiment of it. Through this *doxa* language "speaks" by Saying, and, Heidegger claims, "what it says wells up from the formerly spoken and so far still unspoken Saying which provides the design of language" (*Way* 124). This is a recurring theme in Heidegger, definitely in his later works. It is, moreover, the essential linguistic manifestation of the unfamiliar within the familiar, most explicitly his rendition of translating yesterday's words into novel utterances.

Such is the sense of *doxa* as the "mystic chords of memory" of which Lincoln spoke and, of which he himself, in his answering of Saying, can be considered a most striking example of its embodiment in the legendary word. And, again, much the same could be said of Martin Luther King, who in Lincoln's symbolic shadow he stood in advocation of the dream "deeply embedded in the American dream." In each case the ontological force of language is significant not only in revelations of what is most cherished in yesterday's words but more compelling still, by virtue of the sacrifice of each, the consummate embodiment of novel utterances. In this perspective there is no transcendence in language, only its opposite, a peculiar rhetoric of the earth joined to a *mythos* that Plato quite characteristically condemned as the harbinger of death. As it is the *mythos* Heidegger quite characteristically embraced, yet with a similar logic, telling us the nature of language draws us into its concern, confirming our mortality by virtue of our connection to earth. The soul was to Plato the ultimate hedge against the greatest evil to be inflicted by earth, namely death itself; and for this same reason death was to Plato, no less than Heidegger himself, joined with language. They belong together. And as language "rises from the earth," claiming us as mortal, Heidegger says our awareness of the phenomenon does not necessarily come in the shock of imminent death but emerges over the long and passionate attentiveness to the uncanniness of our mortality. In this sense he claims Saying needs mortals (*Way* 134), expressly as we see language as ontological and life as the unfolding of its narrative. Only mortals can then experience death as death, and in that realization language reaches out, he says, and the "essential relation between death and language flashes before us" (107).

In no other manner does earth enter into world so manifestly, "jutting" through it to open as native soil. If the various elements of Heidegger's thought sometimes seem but nuances needed to complete an elaborate and complex conceptual scheme, then native soil is an exception, being earth in the most tangible sense we can know. Yet it is neither matter nor nature in an exclusively physical sense, not political territory nor least of all a place of racial or biological exclusivity, but it is a rootedness that, *Blut und Boden* notwithstanding, prompts an awareness of life in its authenticity as the borderland, ultimately a poetic dwelling apparent in beginnings as much as endings, an "authentic mythos of autochthony," as Charles Bambach says (65). Thus, we are part of that native soil but inevitably *are* that soil, as we live in the awareness of the enigma that it literally will open to receive us, and in between, in the borderland, is the abyss comprising the nature of our ontological relationship with earth. And here, John Sallis says, earth as native soil is both given and chosen, not a factual place of birth but the homeland one has learned to come into one's own ("Forward" xi—xii). But in every sense it is what one has gained through sacrifice in the sense of a giving of one's self, of finding one's way into the "overpowering" and therein finding one's self as a result (*Introduction* 156–57). Indeed, that is the essential message of the funeral oration, but it pertains not just to warriors, and in ways least of all to them, but to all who stand in awareness of the abyss. Of those who, in the peculiar transformation sacrifice brings about, are the abyss. Who, in this sense, are language itself rising from the earth, altogether as native soil.

And that, too, is the message of the funeral oration, at least in its emphasis on the narrative of who and what we are, insofar as that narrative is expressed in the *mythos* of a people. Heidegger explains the process more precisely, claiming that even as Saying "appropriates" us, at the same time "it releases human nature into its own," but it does so, Heidegger continues, "only in order that man as he who speaks . . . may encounter and answer Saying, in virtue of what is his property," namely his response in the "sounding of the word" (*Way* 129). In this way the latter is dependent on our mortality even as, in *kairos*, it may be said we speak the language of gods. The notion that we transcend death only by affirming it embraces an irony as old and venerable as Greek tragedy, and older still, as primeval as the trickster. There is transcendence of death if only in language, yet by virtue of *kairos* it is the language of the gods we speak, and with every breath we take such language is life itself. That is the glory of Sarpedon and those other Homeric heroes, even that of the "storied crest" of Little Round Top where "proud young valor" rose above the mortal only as it was mortal after all, for life is most intense when lived attentively in the shadow of death, prompting awareness in some well-nigh perfect expression of *doxa*. Again, to live at all is to live in awareness of the abyss, and in that awareness of the nonhuman genuinely to be the abyss. If any leap out of language is death, as Steiner says, then also by this awareness the essential relation between the two "flashes" before us.

Here, due to their sheer corporeality, blood and earth are spiritual forces by virtue of sacrifice, affirming our being utterly in terms of language and mortality. At issue is the most curious aspect of this process: the sacrifice of selfhood for the sake of self-possession, for what is communal in the values and beliefs of a people of poets and thinkers, ultimately for what is most vital to its nature as earth, namely the expression of Being as Saying. So, again, the full measure of sacrifice is scarcely the warrior's alone but consists in the giving of *oneself* that, in accordance with Heidegger's explanation of the process, means to be appropriated by Saying and then to respond to the sounding of the word, to express by means of our corporeality those "novel utterances" inherent in Saying. In this perspective our mortal coil is our essence, but for that reason it is also the sign or sacrament of our being one with the earth. We can then affirm in its full significance Campbell's description of sacrifice as a "fresh enactment" of the sacrifices experienced in the beginning by the gods themselves through which they "became incarnate in the world process" (181). Here the idea of a dying god puts the meaning of sacrifice straight away on our mortality, as we quite literally sacrifice our immortal souls for the sake of keeping body and soul together, joining the divine to earth, making it "incarnate in the world process." Thus, acts of sacrifice do indeed make sacred the earth, most expressly in this process of immanence, investing Being as earth.

If the human and the divine inevitably remain separate, they are always contingent on one another, complementary aspects of what is the same, as in this perspective we are the embodiment of the divine, the way of its being. So there is redemption in this, an awareness allowing the one who by means of self-sacrifice—whether in the appearance of the warrior, hero, saint, even of the Christ or some likely other—"comes into his own" through this transfiguration of Being to native soil and thereby, also though his sacrifice, to achieve the same for his people. Native soil is then homeland, and earth has a cultural dimension in the mythology that constitutes *doxa* itself. Here by virtue of *kairos* earth and world are mutually implied, seeming to merge in the moment to embrace both life and death simultaneously and thus become *doxa* itself, being as much the sacred narrative as reflecting it, being as much of the earth as the words it speaks. Earth as abyss, there being no ground. So it is most profoundly the "life lie." And here, quite simply, timing is everything, just as we have it in Gorgias's definition of *kairos*, of those men inspired by God but possessing the destiny of death who, in their sacrifice, took it as their choice the perfection of argument in believing the universal divine law to say and refrain from saying, to do and refrain from doing, the right thing at the right moment" (Untersteiner 177; Freeman 130).

On the other hand, this talk of life and death, especially as it is laden with the jargon of the "uncanny," is never so abstract in the person of the trickster. In his person the boundary between life and death is eased, seemingly erased, as they meet, circular like, and beginnings and endings become one. In his person death is simply the "compost for life" as Robert Pelton says (252). If the trickster in countless ways is our undoing, he is also, by his very nature as trickster, our salvation. He is the copula in this sense, like Eshu wending his way between the farmers, defining the boundary between them even as he erases it in the very act of definition. In the duplicity of his two-colored cap, he ironically creates and fosters a harmony or balance of opposites, embodying the separateness of each farmer in his wholeness. In this function the mark of the trickster is most apparent.[26] He is "double-voiced," the "god of indeterminacy" (21), or the "master of roads and crossroads" (31), as Henry Louis Gates says of Eshu; or he is "the bond between the divine and the human worlds," as Robert Pelton says of the trickster Legia (108). He is at the gate, the copula in either its utter sexual emphasis or otherwise, but a "penetrator" or "crosser of thresholds" in any event, the means of exchange between discursive universes. As his nature is manifest in wanton sexuality, the phallus itself is the harbinger of the divine, yet being both male and female in bringing balance to the world, as in his person the "outside" meets and merges with the "inside." In his capacity as messenger of the gods he joins the "wild with the well ordered," the "cosmic dialectic with the social process," taming the terrors of the former and giving stability without rigidity to the latter (109). This emphasis on the trickster's powers of mediation between separate entities, indeed his embodiment of the meeting of each by virtue of his immense sexual prowess, allows us to understand his presence as more the relationship than things related—the "copulating copula," as Pelton says (26)—even as he invariably brings with it the gnawing sense of indeterminacy. In fact, in this blending of ambiguity and balance, he is the embodiment of the ontological difference. But by virtue of his presence there is greater accessibility, certainly greater congruity or connection, to the esoteric aspects of the "uncanny district," giving it form and flesh in the unfolding event of the legendary word.

Always our attention is drawn to the ontological difference, that coming together in "de-cision," the sort of intimacy suggested earlier in perspectives on the cemetery and the likes of Momaday's story of Abel and Francisco. The emphasis is on the breach or borderland, the "in-cident" or falling in-between. If still it mystifies, consider how the difference is dramatized for the ages in Michelangelo's centerpiece on the ceiling of the Sistine Chapel, the "Birth of Adam." The figures do not touch, but nearly so, and in that moment of nearness, as their joining seems nigh, we might see them encompassed by a metaphor not quite referring to either but appropriating each and therefore, in a sense, more real than each. In that instant language turns in upon itself where the distinction between image and reality, sign and signified, does not hold, certainly not in conventional ways. So still it mystifies. Again, the abyss. The hermeneutic circle. *Différence.* Some dancer and the dance indeed, mutually determined in their nearness, each being of each yet different. Neither God nor Adam is at its center, each ceasing to be except for the vitality of the event that transpires between them, bringing each into being. If the figures were ever to touch, there is no difference. No borderland and then neither life nor language. Definitely no "life lie." So the mystery of Being persists, "the simple nearness of an unobtrusive governance," Heidegger says, occurring "essentially as language itself" ("Letter" 212). Thus, to perceive either the Godhead or ourselves as being prior to the metaphor joining them is to see Being, to see God Himself, as we see ourselves, as soul possessed, an entity or ego separate from the process of Being.

But if God works in mysterious ways, the trickster *is* mysterious ways, metaphor as pure event or activity, the process of the nonhuman within that comes to pass in the tension of the ontological difference.

This is the domain of the uncanny, all hints and veiled contingencies, the scene of the "de-cision" of the two where the trickster has his place and purpose. He himself can then be understood as the *ek-stasis* of language, as the relation of all relations coming to fruition in cultural practices having tangible reality as *physis*, of Being's immanence as the earth itself. Ultimately the experience precludes deliberate, direct action on our part if we are to act "authentically," and through "the quiet power of the possible" we are carried along more than we can ever realize in this manifestation of para-*doxa* within *doxa*, the unfamiliar within the familiar. Or we might simply say the trickster is *doxa* animated, the crosser of thresholds and thereby the bond, the de-cision, between the divine and the human worlds who in that capacity translates yesterday's words into novel utterances.

Coda

It seems to me very curious that language should have grown up as if it were expressly designed to mislead philosophers.

—G. E. Moore

Among the panels of Orozco's magnificent fresco lining the walls of the Baker Library at Dartmouth College, there is one variously called "Gods of the Modern World," "Knowledge Stillborn," or simply "The Academics." Six figures, cadaverous and horribly grim, lurk in the background against roiling sheets of fire, while another labors up front, seeming to deliver his offspring from a skeleton stacked on piles of dusty tomes. It is "knowledge stillborn," and in the spitting image of its sire, that is then, accordingly, embalmed and stored in a bell jar to be strewn atop the books and bones with others like itself. True to their calling, all seven figures are draped in academic robes, quite smug and cocksure in the business of discovering truth. Indeed, to all appearances they are utterly possessed of the arrogance of positive right, unencumbered by tricksters, contradictions, or terrible secrets of the irrational. There is no tragedy of knowledge here, for the simplest of reasons that there is no life at all. In this sense these gods of the modern world are frightening specters, so taken with the cold, curious tools of analytic inquiry that they are bereft of flesh and blood, skeletal themselves, composed even unto death itself in their vanity and self-satisfaction. Yet, their menacing ways, too, could be simply a ruse to cover or compensate for their hollowness. Whatever the case, we might say a visit from the trickster is in order, these gods needing nothing so much as a pie in the face, and in his scathing portrayal Orozco obliges in earnest.

Heidegger, I suspect, would have taken some satisfaction in it. Surely he would have understood. Mystery, especially that of Being, does not fair well in the philosophical tradition, where its suppression is most often the goal in quests for answers, certainty, and truth. No doubt any work of human knowledge, especially as it achieves the eminence of art, is not a thing detached from human experience. Yet knowledge is depicted in Orozco's mural as so detached as to be displayed in bell jars, and more significantly the people are depicted no differently than what they gather, one as grotesque as the other. Some would say the power of the mural rests in its blatant exaggeration, but exaggeration or not, it points irresistibly to the impression of Heidegger's thinking, perhaps even to its very nature. He considers the "Greek experience" the source of "authentic" thought, but to him the "*trans*-lation" ("to carry across") of that experience miscarries completely. He claims the betrayal begins in the Latin sense of *veritas*, of "*Roman thought*" taking over "*the Greek words without a corresponding, equally authentic experience of what they say*," and leading now to the utter "rootlessness of Western thought," a dispersion of authentic thinking through arrogance, presumptions of human mastery, and pursuits of certainty that cleaves thinking from authentic human experience by way of the universal cast of objective, analytic thought (*Poetry* 23, his italics).

Heidegger's description of a Van Gogh painting provides a fitting contrast. It is of a pair of peasant shoes, though hardly one of pastoral simplicity. It bears instead the undeniable presence of human passion, ironically made all the more compelling in Heidegger's interpretation by the presence of the "nonhuman." Again, self-possession and authenticity are essentially expressions of the nonhumanness of Being, and so "from the dark opening of the worn insides of the shoes," Heidegger says, "the toilsome tread of the peasant stares forth," her "accumulated tenacity in the slow trudge" through the furrows of fields "swept by a raw wind." Appropriately in this context the revealing and concealing power of Being is likened to the turning of the seasons. So throughout this portrayal, he claims there "vibrates the silent call of the earth," its promise of ripening fields in summer and its "self-refusal" in the "fallow desolation" of winter. In this sense, if in this sense only, the pair of peasant shoes is truth, the peasant herself as much as the shoes being of the earth. That is the nature of her "primordial experience" of earth, of her being "privy to its silent call" (34).

Humility is implicit in this silent call of the earth, of the very nature of our response to it. Always the emphasis on earth and language is manifest, and in this "Coda" I bring this project to a conclusion with that emphasis, summarizing those particular elements of Heidegger's thought that seem to bear most significantly on the nature of the Gorgian *kairos*, where earth and language are not only conspicuous but integral to its meaning. Of course, there are obvious reasons why this idea of earth in particular has become such a well-traveled route of outrage for those seeing connections between demonic political movements and Heidegger's thought in general. All too often in his prose the idea is presented so lavishly arrayed with talk of struggle and native soil, of language rising from the earth, that it would seem indeed to conjure images only of violence and bloodlust, images as bizarre and menacing to some as the Spartoi springing fully armed from the sown teeth of dragons. But here, in contrast, Heidegger's focus is not only on the humble disposition of a peasant woman but, more to my purpose, on the very means or methods of investigation implied in this idea of earth.

I begin with a focus on Heidegger's conception of *techne*, apropos of both the arts and sciences in his canon and, in mine, of rhetoric. A conception of *techne* not just "authentic" to Heidegger but benign, even benevolent, because of this idea of earth. In his view scientific explanation or investigation, what he calls "calculative thinking," does not disclose "reality" to any greater extent than any other type of explanation nor even provide information "objectively" or independently more reliable. As early as *Being and Time* he claimed that any interpretation, scientific or otherwise, moves in a circle, "already operating in that which is understood" and nurtured within our "common understanding about man and the world" (194). And, as we have seen, that insight itself is not all that original, coming from Kant's thesis that "reality" is shaped by our methods of analysis, that in a sense reason creates the world it investigates as it discovers only what it first places there through its own methodology (21; Preface B xiii).

Of course, none of this is to malign either the purposes or results of scientific interpretation but to appreciate its nature in terms of this hermeneutic circle. Insofar as objective knowledge exists at all, it does so only in terms of its accord with the method used to discover and designate it as objective, and never in terms of some independent reality existing apart from the method of discovery itself. Heidegger's purpose is to identify these interpretations as interpretations and thus to claim they have their source only within particular contexts, those "common understandings" of which he speaks. So he endeavors to retrieve what to him is authentic in early Greek thinking, from whence he asserts his own thinking draws its inspiration and in no small part its substance, most often endeavoring to secure each by retrieving specifically those "Greek words" and the "authentic experience of what they say." As we have seen, one of these words is *techne*, and from that discussion we learned that it is neither science nor even technology that Heidegger condemns but certain aspects of technological culture; and he condemns them for essentially the same reasons suggested in Orozco's mural, for the arrogance and complete absence of awareness occasioned by the work of these gods of the modern world. Heidegger specifically condemns the likes of these gods for excluding the question concerning the meaning of Being from their deliberations, for their commitment to *veritas* in its place and the resulting conception of truth as a correspondence to some independent reality apart from language, any cultural context, or seemingly any worldly context whatsoever. Thus, he condemns them for confusing their arrogance for a fail-safe means to universal truth, whereas to Heidegger the only universal truth lies in the realization that what we know authentically we know only questioningly.

Herein lies the peculiar truth of *aletheia*. It is the event of Being's revelation, of *logos* as the "aboriginal utterance" of Saying and thereby an event to which we are in many essential ways merely incidental. It is not the truth of philosophers, perhaps of poets only, but truth in contrast to the universal sanctions of *veritas*, emerging from the *logos* of a people in their openness to Saying and faithful response to it. If Saying is comprised of the proverbs, rituals, and oral traditions of a people, at the very juncture of their response to it, Heidegger says, "the battle of the new gods against the old gods" transpires, so that it is here, precisely in this context, that every living word fights the battle and puts up for decision what is holy and what unholy, what is great and what is small, what brave and what cowardly, what lofty and what flighty, what master and what slave" (*Poetry* 43). If the account merely restates what we have seen before, it nevertheless puts this idea of struggle, of war being "the king and father of all," into a far more useful perspective. To begin, there are winners and losers in this struggle, new gods replacing the old, but here we see the clash in terms of the borderland, precisely as the place of de-cision. In this place the oppositions of great and small, brave and cowardly, and all of the others define one another, being complementary to one another, thus bringing about a balance, even a harmony, as new gods replace the old, attaining in the process a purpose very congenial to the trickster's.

But what is also significant in this account is to see the clash as engendered and shaped by language, comprised wholly by its preeminence. "*Language speaks*," Heidegger says. It speaks "within *its* own speaking" and "not within our own" (*Poetry* 190–91). Thus, we see the struggle as something we undergo but never one we wage on our own in any essential sense. Being is the measure here—*logos*. Not man. As such, we as humans participate in the struggle only to the degree we participate in the nonhuman, as participants made of words ourselves, and thereby to see the struggle as something that takes place wholly within Being. The issue is not only what can be known but more importantly the manner of knowing it, how knowing is possible at all. Here the paradox of the hermeneutic circle is foremost, not only entailing the inclusion of the knower within the known but the active participation of the known in any process of knowing. Language as the "relation of all relations" makes it so, making of the other an active, living presence of what is within, insofar as we listen and belong.

Thus, if we participate in the process at all, we do so only in deepest humility, say, in the reverence and respect implied by the pair of peasant shoes. This is the nature of Being, its measure, making of knowing an ontological process, one of our participation and involvement, of "being there" and a resulting awareness of the self utterly in these terms. Here, to reduce everything to the human measure is ultimately to reduce us to the less than human, a fundamentally vicious proceeding that debases not just knowledge but yields knowledge that, in the absence of any purpose higher than ourselves, reduces us to the ogres Orozco depicts. As what we know is less than who we are, so in the question concerning the meaning of Being there is no ultimate, unequivocal meaning. As such, the truth of *aletheia* takes place wholly within the power of language, in the persistent and often intense revealing of Being. In this sense *aletheia* is not something we discover or determine on our own but what, in our openness to it, we bear witness to. Yet we are necessary to *aletheia*, participants in the process but only as we, as *dasein*, participate in Being. We are necessary conditions to the process, as the philosophers might say, but not sufficient.

Of course, these things have been the basis of this interpretation of the Gorgian *kairos* all along. In fact, if this study yields little else besides, it could be that Nietzsche is right in claiming that "truth" is the essential prejudice of philosophers, and surely right in claiming that it is a prejudice. In a way it is inevitably so. After all, the mystery that there is something rather than nothing, the motivation lying at the heart of Heidegger's thought, is never so mysterious in the strictest analytic sense. We look to what is given, inevitably to "what is." And if there is something, there are those to be aware. If not, there are not, for the only thing to perish on the Plain of *Lethe* is awareness itself. And at that point everything perishes, for awareness is that relationship between the knower and the known, neither one nor the other but the relationship. Hence, to be truly "philosophical" regarding the issue, the distinction between "what is" and "what is not," between something and nothing, scarcely matters, for that distinction can have no basis in reason, being prior to it. Yet Heidegger's emphasis is incessantly on the question of Being, and on that basis his judgment of those who would presume to have us live by the rule of reason alone is as dire as Orozco's. And often as caustic. Here the mystery of Being is embedded in the natural order of things or, as earth itself, *is* the natural order of things. Thus, the trick is not routinely to seek resolutions, but to be aware of the mystery, open to its revelations and thus to appearances all around, as Heidegger would have it. The trick is to be aware of the mystery even as it lurks in prosaic and practical circumstances, as the trickster would have it. In this perspective on the Gorgian *kairos* my purpose has been to see *kairos* as vibrant and compelling in the most practical of endeavors, in rhetoric, and here especially to see it as para-*doxa* within *doxa*. We might say mystery is simply the realization that there is a power beyond ourselves, superior to ourselves; and that through the Gorgian *kairos* that power achieves not only its expression but also its dominant place in rhetoric as language.

We then follow through with what has been emphasized from the start, that Gorgias's beliefs concerning the power of language constitute in total his "philosophy" of rhetoric. Of course, the discipline of rhetoric necessarily concerns language, but here the power of language is not just the basis of its peculiar *techne* but its very essence. Equating rhetoric with language alone is another of Nietzsche's beliefs, one that comes about when language is perceived as possessing great powers in and of itself, rather than a tool we wield through the careful and practiced application of a *techne* for the sake of some truth or purpose external to language. Echoing Heidegger, Gadamer puts the difference in the starkest terms, saying the transition is from a "complete unconsciousness" of language that is itself the result of a mentality having its source utterly in language, to that of language's "instrumentalist devaluation" in modern thought, a devaluation of language from a power comprising everything that can ever be an object to an object itself (*Truth* 365). In the first instance Gadamer speaks of language as joined to Being, as he claims it was in early Greek thought, in contrast to traditional theories of rhetoric, surely those of "philosophical rhetoric," where language is invariably an instrument to be used. And language, given its unpredictable, even brutish nature in the philosophic view, is often a power to be feared, always one to be domesticated and controlled by a *techne*, consigned to do our bidding like a well-trained horse. Therefore, as Gadamer says, it is obvious why Plato espouses the dialectic as both means and goal to free ourselves from the power of language (422) and, as we have seen, why Plato himself contends philosophy and the quest for truth rightfully begins when we stop telling stories and embrace the dialectic and discover truth (*Sophist* 241a–243a). So, again we look to what is given and assume this curious stance regarding very tangible things, those so close and immediate as seeming to be matters of the sheerest common sense. It is our disposition, Heidegger says, to accept the "unquestionable character" of these things, to turn towards things "most readily available" and turn "away from the mystery" ("Essence" 135, 138). Thus, the more obvious and explicit they seem, the more we come to know them from the outside, seemingly as we must, because in this matter we assume we must stand outside ourselves for the sake of ourselves, for the sake of our very souls. Otherwise we might be forced to accept the mystery that we cannot distinguish between what is in us and what is before us.

But that is the nature of earth and language, of the creative possibilities that result in imagining the two as one. In this perspective *techne* is not a matter of adjusting language to our purposes but realizing ourselves in the purposes of language; and according to Heidegger, and Gorgias, that process begins in our openness to the power of these possibilities, allowing them to be expressed through us. Here, as we have seen, *techne* is "ad-ventive," essentially concerning not what we do but the ontological imperative of who and what we are as *dasein*. Indeed, if nature in the Greek sense of *physis* is a physical, tangible thing, it is the corporeity of language that makes it manifest, bringing Being to immanence most apparently as earth. And, as we have seen, for soil to be native, for it to be earth, it must literally embody the beliefs that embody a people, the *doxa* that is the mystic chords of memory. It bears the stamp of autochthony, the myth of rootedness in earth's "nourishing soil" that is truly the latter only as it is sustained by "a people of poets and thinkers." And that mystery is the work of the legendary word. Of *doxa*. Of earth. Of *logos*. Of the story. But as Plato himself has taught us all so well, language is vague and indirect, unruly and even at times sinister. Inevitably it discloses, at times very conspicuously, the mysteries of human events, their ambiguities, uncertainties, and paradoxes. Not only does language have a will of its own, but within its boundless élan resides the unfamiliar and unknown, and that, in a sense, is its very essence. As such, it is not primarily informational but withholds, "conceals far more than it confides," as George Steiner says. It is subterfuge and immeasurably dense, "full of mirage and pitfalls" (*Babel* 229). Here the manifold senses of meaning that emerge from the story requires the questioning stance mandated by our "immersion in a wording movement" (Bruzina 197; *Way* 131), and thus what we know authentically is embedded so deeply in our personal and cultural narrative that what is revealed therein can be known only questioningly. Thus, meaning is seldom explicit and never objective in these matters, for that is the nature of Heidegger's "Song of the Earth," of Being expressing itself through us, manifestly as language.

In this context Heidegger condemns "exercises in esthetic judgment" where a *techne* comes to dominate the experience of the legendary word, where literature is taken as an artifact that we objectify and submit to technical scrutiny. Such an approach precludes our "belongingness" to literature and other works of art by the very methods we apply to understand them, as distance and detachment is an essential component of those methods. Just as we witness its effects in Orozco's mural, in the mindless routines and haughtiness of these gods oblivious to earth's native soil. Everything about them is desiccation and decay, of the arrogance that precludes openness to Being, the rootedness in native soil reaching down to realms of life and growth. Heidegger's phrase is "losing the earth," meaning to lose any sense of who we are to this attitude disrupting and then destroying the unconcealing "silent dialogue" the word is meant to evoke. And yet it is only through such a dialogue that the claim of Being exerts itself, that it can be "taken up by man into dictum and legend" (*Parmenides* 114–16). And, in turn, it is only through the "legendary word"—through *that* story—that we can come to know Being at all. But, again, the issue is not to know Being but the way of knowing, the nature of this "ad-ventive" *techne* coming about in our openness and participation in Being. Whatever the case, insofar as we are true to the "Greek experience," we are the beings that emerge in our essence by naming and saying. In that sense Heidegger asserts we are the "god-sayers" (112).

He says we cultivate and guard the familiar only in order to break out of it, a perspective seemingly obliged by a *techne* that reveals the unfamiliar within the familiar, para-*doxa* within *doxa*, the nonhuman within the human. The perspective not only applies to the Sophistic sense of *doxa* but accounts for the Gorgian *techne* as that art of rhetoric having its effect through the power of language alone. Another way of understanding the nature of this "great *dynastes*" is metaphor, pure metaphor, where vehicle and tenor become one and language as the "relation of all relations" subsumes all that otherwise seems so terribly apart from it. In this perspective metaphor is no longer metaphor in any conventional sense, for it turns language perpetually inward, deeper and deeper still, even prior to distinctions between literal and metaphorical. Ultimately the way leads back to an "aboriginal" *logos*, to that "immersion" in a "wording movement" of which Heidegger speaks. Not only are conventional distinctions between literal and metaphorical undermined in this perspective, but so are those between knower and known, sensible and nonsensible, rational and irrational, ultimately those of body and soul, for in the essential nature of language these would-be dichotomies are complementary aspects of what is the same.

So we conclude with our current theme, that of the trickster and reckoning with his presence. If he is the bearer of chaos and mischief, the personification of the struggle of the outsider within, he teaches us at all times of the limitations of the human perspective and conventional notions of *techne*. Most cogent here is the coupling of his presence with language, especially as it sheds light on the nature of the Gorgian *kairos*. If his purpose is preempted as *logos* comes to mean reason rather than language, it would seem that it had not yet come to that in the world view of Gorgias. Here, to equate language with the trickster is to bind it fast to the earth, with the very tangible realities of Being, so that Being itself moves to the cadence of language, *being* language, to prompt the flourishing and vibrant play of appearances in which we dwell. By virtue of his presence we dwell in language, or by his presence become aware of the fact, and our dwelling is thereby flush with mysteries, cabalas, riddles. If the trickster "is all metaphor, all ambiguous oracle," and we understand metaphor as turning language inward, we might see metaphor then more readily in terms of the trickster's modus as "copula," a peculiar spiraling inward—or outward, if you wish—to what is boundless, seeming to connect all references there to deception, irony, and paradox. He is, in fact, pure metaphor in that the only reality to which he refers is himself, a trickster's realm where language, as language alone, creates reality as much as reflecting it. Thus, he is also rhetoric, as rhetoric is conceived in its essence as joined utterly to the power of language. He is, therefore, in this very context the voice of the friend we carry, yet always an *alien* voice, as truly he must be, calling us back from our separation from Being to our "own-most-possibility-to-be." The trickster dramatizes the process, embodies it as language. Being both us and the other, he personifies the uncanny as the voice we carry with us.

And now, one final example. Robert Pelton speaks of the healing ceremonies of the Yoruba, where the trickster's purpose is realized in a "sociotherapy" to make known and visible the "unknown and invisible agents of affliction." Here the paradigm of the uncanny is turned inside out, vitalizing it all the more, in a process of "divination" that in many ways is remarkably similar to Western approaches to therapy stressing *kairos*. In this process, Pelton says, "an inestimable moment comes to pass when language is enlarged to name and to humanize an otherwise unintelligible and therefore unassimilable event," a moment of "enlargement" that is "also a transformation of non-sense to sense, impasse to passage" in which the Yoruba perceives the hand of Eshu. Significantly, this movement is always part of a "semantic structure" that reaches into the "realm of the sacred" and then "in some mysterious way mirrors the human world" (143). What is most important here is the sense of language taking us into the swirl of the divine, being in this context the very nature of the "wording movement." The motif not only provides a particularly significant instance of self-possession as the essential expression of the nonhuman but most directly casts us in the role of "god-sayers." Again, the specter of Being is raised as the "relation of all relations," the "simple nearness of an unobtrusive governance" where the divine mirrors the human world, where Saying mirrors the human response. If we are never gods ourselves, in our role as "god-sayers" we are the gods' only access to the human world. And, in this interpretation, that is *kairos*.

The process is key in revealing the uncanny as *dasein*, the nonhuman as human being itself, affirming it in our relation to Being, in the "simple nearness" of that "unobtrusive governance." Or, that is to say, we are not God but without his nearness, the proximity of his touch, neither are we ourselves as our "own-most-possibility-to-be" is never realized. The process indicates, as well, a peculiar equation of the trickster and *kairos*, as the trickster in this perspective is the human face of the nonhuman. Though the message he bears is humanized by the simple fact of coming from his person, and thereby from us, still he remains the emissary of the nonhuman, a paradox in keeping with the purpose to mirror the human world "in some mysterious way." Here, the "semantic structure" is crucial, in essence being "earth," and we can then extend the irony, to realize not only that language is the nonhuman but, in being so, is more human than we are ourselves apart from language. Among many other things, that is Heidegger's admonition. Ultimately it is to live by the paradigm of the trickster, thus to see *kairos* as a way to harmony and balance achieved in the relationship with Being, the "quiet power of the possible" as it is expressed through us in our response to Saying.

In the end, this peculiar world view—the separate mentality of the "not-yet" and "no-longer" metaphysical that Heidegger offers—shapes the conception of the Gorgian *kairos* as something other than a procedure through which the speaker adapts his speech to the occasion and audience, to what is fitting in time, place, and circumstance. This latter conception, the Platonic *kairos,* is based upon the most transparent presumption of our mastery of time and language. In contrast, Heidegger's emphasis on the uncanny, of something coming from beyond the realm of our reckoning and control in the form of "Saying"—of earth "jutting" through world—constitutes a conception of language and *kairos* not only at odds with Plato but more in keeping, I believe, with the pre-Socratic mystery that in this interpretation wholly envelops the Gorgian *kairos*. The mystery is precisely that language is antecedent and dominant, and at certain crucial times—in the impetus of the right moment—makes a claim upon the speaker, adapting him to *its* will. *Kairos* can then be seen as a purely linguistic phenomenon, manifestly a function of the "semantic structure" that is our very being. In that sense it is the progeny of the trickster, surely of the phenomena he represents, as is so clearly apparent in the mythological figure of Kairos himself. If language is hardly the measure of all things, in this view it is surely the measure of ourselves. Here, to be human is not just to be "released" but to partake of the nonhuman in being "appropriated" by language; in effect, to become our words as language moves from everydayness to authenticity in marking the moment as *kairos*. It was, of course, Plato's purpose to deliver us from iniquities of the trickster's sort, his mischief and deceit, those iniquities epitomized by language. In this sense the Platonic *kairos* is not only at odds with this perspective on the Gorgian *kairos*, but as a *techne* to control and use language for the sake of decorum and propriety, it is to all intents the antithesis of the Gorgian *kairos*, of that *techne*.

Though Gorgias and Heidegger are well matched in any number of significant ways, essentially it is the *Augenblick* that provides one of the clearest, most accessible means of understanding the Gorgian *kairos*. If Gorgias, especially by way of his vision of the great *dynastes*, plays the trickster to Plato's philosophy, then Heidegger, as he stands at the gateway to the poststructuralist movement, does the same for the philosophic tradition in place since Plato. In both Heidegger and Gorgias there are games and riddles, the word play intrinsic to the expression of their thought, but the salient point in this interpretation of the Gorgian *kairos* is that both offer views of language embodying imperatives of Being's immanence—those of earth, in Heidegger's interpretation—that define our being in very essential ways in terms of language. No less than Heidegger, Gorgias offers a rhetoric of possibility on the basis of this view of language. The rhetoric of each thereby endeavors to move present understanding beyond habitual, commonsense views of truth preserved in language: Heidegger in terms of overcoming the inauthenticity of "idle talk" and Gorgias in terms of the para-*doxa* that overcomes the "arrogance of positive right" and the "rigor of the law." The rhetoric of each displays a similar view of the rhetorical situation, as a dramatic yet fundamentally mysterious "event" that prompts disclosure by virtue of language of something more essential and basic than what is available to conventional understanding.

Thus, the *Augenblick* and the Gorgian *kairos* in their rhetorical implications are justified temporally, where in the focus of the moment Being is experienced in the call of language, in the form of the unsaid exerting its claim to become manifest. Once this emphasis on language is seen as a process of revealing on the temporal level of the here and now, where both Heidegger and Gorgias decidedly place it, then on this matter of *kairos* Heidegger and Gorgias are in alliance, and the one's call of Being and the other's irrational *logos* can be seen as one. *Kairos*, then, designates the instant which to Gorgias triggered the force of an irrational *logos* free from the privileged representations of "things that are"; and, to Heidegger, *kairos* structures "authentic" language as a response to an "aboriginal" or "originary" *logos*—to "Saying" that is veritably the voice of Being—whose call comes in our openness to it, in the *kairos* of a primordial hearing, prompting in the right situation the creation of language itself. Thus, we heed the novel utterances of the new gods through whom we hear perpetually the voices of the old. This ontological conception of language—this spontaneity in its own creation—seems to tap, and best explain, the more mysterious facets of the Gorgian *kairos*.

Notes

Chapter One

1. Heidegger, *Introduction to Metaphysics* 189. I have relied throughout on the translations of others, as that above by Ralph Manheim. Only in those rare instances when translations were not readily available, or when some "fuller appreciation" for a particular nuance or another seemed to require it, have I ventured translations of my own or, indeed, have combined mine with those of others. These instances are indicated in the text or endnotes, and include translations of German only.

2. See Johnson xv. Interestingly, Scholes quotes the same passage to take to task deconstruction and seemingly the poststructuralist movement in general. He claims that deconstruction critics "erase" the world by means of the "mirage of language," thereby providing no pathway from text to world, nothing upon which we can, for instance, "ground an argument for evolutionary biology as opposed to fundamentalist creationism"—nor, presumably, nothing to ground any number of arguments in disputes between science and pseudo-science (92–99). His point is well made and engaging, but to see the issue in the context of Heidegger's application of Kant's explanation of the Copernican Revolution is to understand that any interpretation, or perspective, presupposes some prior understanding on the part of the interpreter. Such a formulation leads in turn to Heidegger's conception of the hermeneutic circle, through which, Mark Okrent says, the "practical context of activity," apart from any vision of reality as it "truly" is, grounds the interpretation, supplying it with "certain evidential conditions" that must be met if an interpretation can be judged as successful (167–68). In other words, contrary to the way Scholes presents it, the question is not whether creationism is true or false but whether it is successful in meeting, within the scientific perspective, stipulated "evidential" conditions that in these cases are set by the scientific perspective itself. And creationism—or "intelligent design," as it has come to be called—clearly does not. Not only does it fail to meet the evidential conditions necessary to qualify as science, but by first endorsing a supernatural view of creation and then abruptly shifting to a

"scientific" perspective to "prove" it, creationism stands the scientific method on its head. In short, it posits a conclusion and then seeks evidence to support it, in this way failing to constitute a scientific perspective at all. But the basic fault of fundamentalists, very evident in the arguments of creationists, is not only that they fail to understand science but more grievously fail to understand religion, to appreciate the mystery of origins and ends to which authentic religious experience can so eloquently speak. They abandon what is most valuable, and essential, in the religious perspective in an effort to accommodate that perspective to the methods of science. The same effort, of course, could be made concerning the Ptolemaic view of the universe, insofar as anyone would wish to support it through an application of the scientific method. Thus, what is at issue concerning the Copernican hypothesis is whether different perspectives can be equally valid under certain given conditions, never arguments fashioned to meet evidential conditions within any single perspective.

3. Derrida, along with a number of less significant others, would disagree. In his essay, "Structure, Sign, and Play," Derrida argues that the purpose and prominence given the nature of Being in Heidegger's philosophy makes him (Heidegger) not a poststructuralist but the last of the metaphysicians. He accuses Heidegger, as Heidegger accuses Plato, Descartes, and Kant, of being driven by a desire for a center, of using the idea of Being as a fixed origin that would orient, balance, and ground a philosophic system. Elsewhere, Derrida says that Heidegger "leaves intact, sheltered in obscurity, the axioms of the profoundest metaphysical humanism" (*Of Spirit* 12). Whatever the case, my purpose in this book is to establish a series of connections between Heidegger and the Sophists in order to shed some light on the rhetoric of the latter; it therefore may not be pertinent to argue that Derrida's interpretation is based on a peculiarly skewed presentation of the Heidegger's idea of Being. Yet Derrida's criticism is telling, indicative of something essential in both Heidegger and the Sophists. Heidegger's idea of Being begins with what the Greeks called *thaumazein*, a sense of admiring wonder prompted by the very mystery that we are, the wonder that inspires in turn the most basic and persistent of questions, of why there is something rather than nothing. Being in this Heideggerian perspective is thus neither a state nor condition of anything fixed and unchanging, transcendental or otherwise, but embodies a process of immanence, of Being itself keyed by our awareness and appreciation of its mystery, ultimately to what Heidegger called "earth." The point that I will attempt to make in later chapters is that Heidegger's sense of Being is aligned with physical nature, with the pre-Socratic "*physis*" as he himself would have it, and is therefore hardly metaphysical in the classical sense of transcending the physical realities of the here and now. It is obvious, however, that Heidegger, especially in his later works, invokes in countless ways the work of the poet in his quest for Being,

often seeming to abandon altogether in the process the methods of the philosopher. In light of this purpose he claims the thinker's responsibility is to "poetize on the riddle of Being" (*Early* 58).

Derrida, on the other hand, is methodical and, relative to Heidegger, lucid to a fault. He is in fact an exemplar of the rational man in every way, and thus of a similar frame of mind to the system builders he so avidly deconstructs. Indeed, in matters concerning the idea of the trace, the metaphysics of presence, and the methodology of deconstruction in general, Derrida's thought seems as a series of more or less consequential footnotes to Heidegger. But to put Heidegger's sense of Being under erasure, to "deconstruct" it to that degree, is to ignore the question concerning the meaning of Being for which Heidegger indicts all philosophers, from Plato to Nietzsche. Indeed, to deconstruct Being in the manner of Derrida is to place one's discourse in some utterly practical, nuts and bolts never-never land that is still wholly groundless for these very reasons, resulting in the *reductio ad absurdum* of the thinker's own project in the remarkable achievement of deconstruction deconstructing itself. So, in the end, spurning the turn of the poet, the ultimate victim of Derrida's deconstruction must inevitably be Derrida himself, and I doubt that Derrida would see it otherwise. Thus, what is most curious in Derrida's approach is that in extending the methodology of the Western philosophic tradition to its breaking point, he ends up with a discourse that is every bit as "mystical" as any of those he deconstructs. Here, to say nothing of Derrida's advocates, Derrida *would* no doubt see it otherwise. For a more extended discussion of this issue by one who may or may not see it otherwise, depending upon the reading given him, see Magliola, *Derrida on the Mend*, especially Part 1, "Between the Tao: Derridean Differentialism."

4. See in particular Hannah Arendt's presentation of Plato's conception of *aneu logou* in context of her discussion of "Metaphor and the Ineffable" in *The Life of the Mind*, Vol I, *Thinking*, 110–25. Whether Heidegger's ontology can also be included under this rubric, as Arendt herself seems to suggest (122), will figure significantly in the final chapters of this book.

5. Whether we read them as ink blots or not, fragments are texts embedded in other texts, "epitomes" of the lost originals, given in contexts most likely serving the purposes of those who reconstruct the fragments rather than their "original" authors. As Catherine Osborne points out, problems also arise when a particular text (fragment) is given in two or more ways in two or more subsequent texts, a process, she says, aptly demonstrating that the division between "text" and "interpretation" "cannot be maintained" (3). In many cases, then, any sense of the "original" is of dubious currency, and the critic is faced with deciding whether she is dealing with different versions of one fragment or with as many fragments as there are versions of it. The process illustrates handsomely the issue at stake in the hermeneutic enter-

prise, as "reality" necessarily succumbs to the various version of it given in particular interpretations or perspectives.

6. The term of course is George Kennedy's (*Classical* 41–60) and is used to designate the rhetoric of Socrates, Plato, and Aristotle. To all three, Kennedy maintains, the essence of any genuine art of rhetoric was based upon truth(s) or probabilities derived through the methodical procedures of the dialectic. Kathleen Welsh extends Kennedy's definition to defend Plato against what she considers to be the excesses prompted by the burgeoning interest in Sophistic rhetoric: "Plato attacked Sophist rhetoric not only because it denied his conception of reality but, crucially, because it is inherently passive and therefore allows the soul to atrophy. Plato praised philosophical rhetoric because it depends on the active use of the dialectic. Passivity precludes dialectic. The activity, the interdependent exchange of ideas and emotions, the push and pull of spiraling intellectual and psychological inquiry, constitute Plato's conception of philosophical rhetoric in *Phaedrus*" (4–5). Welsh's statement not only offers a good delineation of the battle lines drawn between Sophistic and philosophical rhetoric, but her suggestion that there exists a symbiotic relationship between the dialectic and Plato's conception of the soul will be dealt with at length in this study, for it seems to be not only the foundation of Plato's view of rhetoric but of philosophical rhetoric as such. Moreover, the belief that Sophistic rhetoric allows the soul to "atrophy" is an especially interesting observation and will also be discussed, at length.

7. The idea of Plato's *techne* or art of rhetoric will be developed in the next chapter, but the source of this *techne,* to whatever extent it can be developed in rhetoric by Plato's lights, lies in philosophy, as the *Phaedrus* indicates (especially such passages as 270d—e). On this point see also Hunt, "Plato on Rhetoric" 36.

8. Among those others who agree are James Kinneavy ("Relation" 14), Richard Engell (178), and George Kennedy (*Art* 66). It is important to note that all extended treatments of Gorgias and *kairos* owe a debt to Untersteiner's enormously influential book, *The Sophists.* See also Bizzell's and Herzberg's book of readings, *The Rhetorical Tradition,* where the doctrine of *kairos* is said to be central to Sophistic rhetoric (23).

9. Two of these books warrant special attention. Consigny's is not only the most recent but surely most cogent and comprehensive in dealing specifically with Gorgias and his rhetoric. He classifies approaches to Gorgian rhetoric into subjectivist, empiricist, and "antifoundationalist," disavowing the first two and endorsing the latter as the ideal model in view of Gorgias's advocacy of an array of maneuvers or tropes for use in "the sanctioned agons of the culture" (60). Vitanza's *Negation, Subjectivity, and the History of Rhetoric* is a wonderful ride of a book, most compelling in featuring "Negation" as the "real substance" of rhetoric, an idea that, in the person of the trickster, will be addressed in the final chapter of this book.

10. John Smith makes the same point concerning *kairos* and scholarship as they pertain to philosophy. It is worth noting, however, that Smith establishes the "natural habitat of the concept of *kairos*" not only in the traditions of classical rhetoric but in the same passage from the *Phaedrus* (271d–272c) where Kinneavy finds its specific exemplification in rhetoric (52).

11. Poulakos's account is reiterated by numerous critics of Heidegger, such as Robert Bernasconi, who writes of the kinship in Heidegger's thought of "the not-yet metaphysical and the no-longer metaphysical" (24). In various places Heidegger himself introduces the issue specifically in these terms, claiming there is a kinship, but that the "no-longer metaphysical" is more problematic, as metaphysics is not so easily and quickly dispelled given its three thousand year legacy (*Heraclitus* 75–76). That such a kinship exists—albeit in the rather nuanced sense Heidegger presents and Bernasconi explicates—is necessarily the basis of my own approach to Heidegger.

12. Heidegger's sense of "destruction" is the prototype of "deconstruction." Or, as was once suggested to me by Calvin Schrag, the relationship between the two is that of genus to species. The most general distinction I make is that to Heidegger destruction is a means to an end, the end being the unveiling of Being in a process leading back to that original "wholeness" of Being; whereas deconstruction, at least in the hands of many critics, encompasses both means and ends, involves pure process that would deconstruct Being, Heidegger's or anyone else's, in "deferring" its meaning or placing it under "erasure" altogether as metaphysical. Yet, given his peculiar sense of Being, Heidegger himself was the first to "defer" its meaning, to place it under erasure—quite literally. See *The Question of Being*, especially 81–89. Heidegger's purpose here, it seems, was to indicate that attempting all too specifically to designate or define Being was to make of it an object, to so restrict it as to make it something other (much less) than Being. Thus, it could be argued that in either case, destruction or deconstruction, Being is reduced to nothingness, to the abyss. Nevertheless, as we shall see, in Heidegger's view Being remains insofar as "the question concerning the meaning of Being" remains.

13. In this context *logos* is conceived as language rather than reason or anything other than language. According to Theodore Kisiel, *logos* in this sense is "the indigenous field in which man lives, moves and has his Being" (91). That phrase, of course, has it source in scripture (Acts 18.28), where, given this context, it quite fittingly designates none other the Godhead.

14. Johnstone (68) is not alone in linking Heidegger specifically with the "rebirth" of rhetoric. Schrag's article, "Rhetoric Resituated," is pivotal, his argument on this point similar to Johnstone's, claiming that rationality is most aptly understood as the result of "communicative praxis rather than as a preexistent logos that antedates and governs it" (172). See also Heim, Worsham, Michael Hyde, Cascardi, and Gadamer, "On the Scope and Func-

tion of Hermeneutical Reflection" in *Philosophical Hermeneutics*. Heim's argument—"that the existential ontology of Martin Heidegger is the basis upon which the distinction between philosophy and an ultimate rhetoric becomes superfluous" (181)—is also one of the more pertinent approaches linking Heidegger's thought with rhetoric and will be considered further in a subsequent chapter.

15. To Heidegger, "hearing is constitutive of discourse" (*Being* 206). The idea seems explicit enough, but in contrast to the "sight metaphor" that Heidegger says is constitutive of metaphysics (whereby *logos* is linked both metaphorically and literally to vision rather than hearing), it is integral to the interpretation of the Gorgian *kairos* I am presenting here.

16. What many perceive to be the shift from Heidegger's early works to his later is called the *Kehre*, the "turn" or "reversal" in his thought that marks generally the transition from what these critics would see as the emphasis on *dasein* and the "residual subjectivism" of *Being and Time* (Zimmerman, *Eclipse* xxv), to the more direct encounter with Being and language in his later writings that leaves *dasein* a "transparent space" (Magliola, *Phenomenology* 64). In his "Letter on Humanism" Heidegger himself was the first to signal this turn in his thinking, but he vigorously denied the significance and many of the implications that a number of his critics attached to it. In his "Preface" to Father Richardson's *Heidegger: Though Phenomenology to Thought*, Heidegger explains that "a good number of years are needed before the thinking through of so decisive a matter [as Being] can find its way into the clear" (xvi), and he therefore condemns the "groundless, endless prattle" on the part of some of his critics about the *Kehre*, claiming that he never abandoned the fundamental issue of *Being and Time* and that the entire book was "outside the sphere of subjectivism" (xviii). Whatever the case, I hold with Bernasconi's endorsement of Heidegger's statement on the issue, that "only by way of what Heidegger I has thought does one gain access to what is to-be-thought by Heidegger II" (94). Indeed, Richardson's summation seems particularly apropos, claiming that "the whole of Heidegger II to be a *re-trieve* of Heidegger I" (625). None of this is to suggest that Heidegger's thinking has been uniform throughout, or that there is no difference between Heidegger I and Heidegger II, but I see no reason to dispute his own position on that matter, and that is to say his basic philosophical convictions were set down in *Being and Time* and these were never abandoned. And, with Paul Ricoeur, I agree that one of the most basic and enduring of these philosophical convictions presented in *Being and Time* was "aimed at destroying the claim of the knowing subject to be the measure of objectivity. What we must take over from this claim about the subject is the conditions of dwelling in the world in terms of which there is a situation, understanding, and interpretation" ("Task" 153). However, insofar as one may persist in seeing the essential message of *Being and Time* in terms of a reverent regard for *dasein* as

subjectivity, I do not feel that the three themes from Heidegger that I use as a mode of analysis are affected in any critical respects by the supposition of a *Kehre*, even one of cataclysmic proportions, in Heidegger's thought.

17. The idea is in accordance, I think, with Edward Schiappa's argument that Sophistic rhetoric is a mirage, "something that we see because we want and need to see it." But I do not agree that, like a mirage, Sophistic rhetoric thereby "vaporizes once it is carefully scrutinized" ("Sophistic" 5). Schiappa says that any account of Sophistic rhetoric necessarily begs the question of its own subject matter because it presupposes on doctrinal grounds who should be called a Sophist (8). I agree, and make no secret as to the "doctrinal grounds" I take in presenting this study. The Sophists, of course, did not define themselves. Plato did. Indeed, Schiappa says that Sophistic rhetoric is a fiction invented by Plato for his own ends (16). And that is surely the case. But for that reason Sophistic rhetoric is hardly a construct, as Schiappa says, that we can do without, any more than Plato's philosophy is a construct we can do without. Here, we might say, we beg the question in any event, as that is the nature of the hermeneutic circle. My quarrel, then, is not with the basis of Schiappa's argument, which I endorse, but with his failure to complete it, with his selection of Sophistic rhetoric as the sole or even the most likely target of that argument, as it can be extended with equal validity to Plato's own philosophy. Once this is done, the argument's most critical implication is that Sophistic rhetoric is no more a fiction than Plato's philosophy itself, and therefore, by virtue of Plato's philosophy, every bit as extensive and complete as Plato's condemnation of it. In other words, they are clearly complementary ideas, contingent on one another, there being as sure a sense of Sophistic rhetoric as there is of Platonic philosophy. Nevertheless, Schiappa's arguments remain engaging, and for obvious reasons have created a stir in neo-Sophistic circles. See Vitanza 27–55 and Consigny 11–14, two others who find it necessary call into question these arguments before proceeding with their own.

18. These stories of German soldiers packing copies of Heidegger have achieved legendary status and come from various sources, such as the one above from Peter Gay, who quotes Paul Hühnerfeld, once a student of Heidegger (81). Many of them, as this one, pairs Heidegger with Hölderlin, and they have their source, I suspect, in efforts to equate Heidegger with the likes of Professor Kantorek, the stern and pompous schoolmaster in Remarque's *All Quiet on the Western Front*, corrupting his charges with talk of blood, sacrifice, and the glory of the Fatherland. And it could be Heidegger himself shares in the creation of the legend, as he claims "during the First World War Hölderlin's hymns were packed in the soldiers' knapsacks together with cleaning gear" (*Poetry* 19). Perhaps carrying Hölderlin's hymns—or, as the case may be, copies of Homer or Hesiod (Hughes 316)—is a real possibility, even for soldiers on the Western Front whose martial illusions had in due

time surely been dashed; but as anyone with some experience of humping a rucksack in combat might tell you, far better use of a soldier's energy and endurance could be made than carrying a tome of the weight and nature of *Being and Time,* especially as "reading" material of a lighter, far more stimulating nature was always so readily available to the soldier in Vietnam and most other recent (and not so recent) wars.

Chapter Two

1. The statement is S. H. Butcher's, quoted in Kinneavy, "Relation" 13.

2. See Kittel, who echoes Butcher's point that there is no certainty as to the original meaning of *kairos,* but who argues that the "basic sense is that of the 'decisive or crucial point,' whether spatially, materially or temporally" (455).

3. Though the purpose of the *koyemshi* is far more complex than smashing taboos, in various, significant ways the paradigm of the trickster can be vital to our understanding of rhetoric, *kairos,* and the relation of each to Heidegger's view of Being. These connections will be explored in detail in the sixth chapter. However, Hill's discussion of the *koyemshi* is fitting as the trickster enjoys an especially active presence in practically all American Indian cultures, in fact in all "primitive" cultures as of yet not bewitched by the flagrant promises and trickery of reason and technology. The *heyokas,* the "sacred fools" of the Lakota, play a role in many ways similar to the *koyemshi,* though the *ethos* of the *heyokas,* also in accordance with the purpose of the trickster, seem to place a greater emphasis on harmony and balance than the *koyemshi.* See Black Elk concerning the *heyokas* and Radin's *The Trickster* for a more general view of the trickster and American Indian cultures.

4. The coming of the Christ is, of course, prophesied in the Old Testament, but it is the New Testament that more fully establishes the distinction between the coming of God's time (*kairos*) and the mere *chronos* of passing time. See Glover for what could be construed as a contrary view, especially his chapter, "*Kairos* and the Old Testament." Other than the passage from Mark and Thessalonians, see also Matthew 16.2 –3, Acts 1.7, and Galatians 4.4. Indeed, Kinneavy says the word *koine,* the Hellenistic variation of *kairos,* appears 86 times in the New Testament, to include the passage from Mark ("Relation" 15). For a complete survey and analysis of *kairos* and scripture, see Sipiora, "*Kairos:* The Rhetoric of Time and Timing in the New Testament."

5. Galatians 4.4–5 is especially pertinent to Tillich's distinction between *logos* and *kairos.* Saint Paul uses the coming of Christ in the "fullness of time" as the means to dissuade the Galatians from observing Jewish law. The discipline of the law is seen as the period of tutelage or preparation leading to freedom from the law through Christ. *Logos* is thus identified with the

static moral code of Jewish law and *kairos* with the dynamic, creative force of Christ.

6. In terms of its specific philosophic roots, existentialist psychotherapy is indebted to Kierkegaard, Nietschze, Sartre, most particularly to Heidegger, and other representatives of the Continental movement identified with existentialism and phenomenology; but insofar as the issue is joined to the phenomenon of time, and specifically the importance of *kairos*, the influence of Tillich is pervasive. See, for instance, Sonnemann's *Existence and Therapy* and May's "The Origins and Significance of the Existential Movement in Psychology."

7. One of the best and certainly most concise treatments of the Gnostic attitude concerning time is Henri-Charles Puech's "Gnosis and Time." In his analysis time was "taint" to the Gnostics. In their belief we are "plunged into [time] and participate in it through our body, which, like all material things, is the abject work of the lower Demiurge or of the principal of evil; in time and by time our true 'self,' spiritual or luminous essence, is chained to a stranger substance, to the flesh and its passions, or to the darkness of matter." That such a conception of time could only be promulgated in defiance of the doctrine of the incarnation is obvious, but the idea that time's essence lies in its "rhythm of death" (65–66) seems, as well, to be the particular attribute defining the malaise of the hero of Proust's novel. Interestingly, this motif of the "darkness of matter" has resurfaced of late in the Gospel of Judas, the essential argument being that the human body is a prison from which the spirit must be released. So Judas is redeemed in this gospel, as Jesus instructs him to sacrifice "the man that clothes me." To many, this symbiotic relationship between good and evil is as vital to authentic religious experience as it is the world of the trickster. See, for instance, Kazantzakis, *The Last Temptation of Christ*, where the complementary relationship between Christ and Judas was explored decades before the discovery of the Gospel of Judas.

8. According to Rollo May, the Anglo/Continental division in philosophy is present in psychology as well. As existential psychotherapy is Continental in origin and application, the American and English contribution has been in "the behavioristic, clinical, and applied areas" in the field of psychology and in "drug therapy and other technical applications" in psychiatry ("Origins" 9). May says that these methods of treatment constitute the most crucial source of resistance to existential psychotherapy. Indeed, as he recounts in a conversation with Tillich, even the translation into English of the basic papers and premises of existential psychotherapy may be impossible (May, Angel, and Ellenberger vii). Kockelmans points to an area of related concern, and raises the issue as to whether there exists any sound theoretical basis for the tenets of existential psychotherapy at all. He argues that any attempt to establish an existential therapy must begin with the premises of a "knowing subjectivity" or a locus of consciousness. To Kockelmans such a

premise, at least on the basis of Husserl's notion of intentionality as the basic structure of consciousness, is inherently invalid because "the purely psychical" cannot be sufficiently separated from the physical, and is therefore not "an empirical object." Existential psychotherapy is thus necessarily deprived of a subject matter, namely consciousness as "an empirical object" ("Theoretical" 235).

The worth of Kermode's book lies in its interesting, even inspiring speculation, but, as mentioned before, its arguments are underdeveloped and the evidence used to support them is very selectively applied. The basic challenge to the work is related to the charge Kockelmans directs toward the theoretical structure of psychotherapy: that is, human being is "necessarily related" to the world through intentionality. Thus, the chaos surrounding us all—the "inhuman reality" of which Kermode speaks—is no such thing, for it is, in fact, just as much a fictional creation as the literary response to it: i.e., the fiction itself. On this point and others concerning Kermode's *The Sense of an Ending,* see Bersani and Byatt.

9. Among these others, and one to be cited again on matters equally important to this study, is Pierre Bourdieu. In his *Outline of a Theory of Practice,* he makes a distinction between theoretical and practical knowledge that is comparable on many levels to Tillich's between *kairos* and *logos.* One feature of Bourdieu's argument is that to achieve an adequate "theory of practice," we must "reintroduce time into the theoretical representation of . . . practice which, being temporality structured, is intrinsically defined by its *tempo*" (8). In this context he mentions *kairos,* but only in the restrictive sense as an element integral to the pedagogy of the Sophists (20). However, in the overall context of his argument it is apparent that *kairos* is one basis of his critique, constituting a specific example of confounding what he terms the "rule-dominated objectivist perspective."

10. In a previous, unpublished version of "Neglected Concept" (14b), Kinneavy says that the phrase, "in any given case," is the translation of *peri hekaston* that constitutes the "*kairos* element" embodying the idea of the "unique situational context" in Aristotle's definition. The argument is significant and will be revisited.

11. See Arendt, where she says whatever differences exist otherwise between Plato and Aristotle regarding language and truth, the notion of *aneu logou* is central to the philosophy of each, indicating for each a "truth that is beyond discourse" (1.114).

12. Pirsig's "former self" or alter ego—his "Phaedrus" in *Zen and the Art of Motorcycle Maintenance*—rebukes Aristotle for claiming that "rhetoric is an art because it can be reduced to a rational system of order" (353). I am not sure that that statement perfectly reflects Aristotle's very nuanced, highly deliberated position on the matter, but Pirsig curtly dismisses it as an "asshole" statement in any event (353). Most rhetoricians, ancient and modern,

would surely find Pirsig's rejoinder arguable, to say nothing of how Aristotle himself might find it. I do not presume to know how Gorgias, whose rhetoric Pirsig is defending, might react. In this, as in all other matters concerning Gorgias, I can only surmise.

13. There are at least three basic perspectives scholars adopt concerning Plato's attitude toward rhetoric: 1) Plato had a consistently negative view of rhetoric; 2) He had a consistently positive view of rhetoric, scorning only certain excesses on the part of the Sophists; 3) He had divergent views, attacking rhetoric in the *Gorgias* and vindicating it in the *Phaedrus*. The first perspective attributes to Plato the absolute priority of philosophy over rhetoric. Ricoeur and Ijsselling are representatives of this view. Ricouer says that, to Plato, "'true' rhetoric is dialectic itself, i. e. philosophy" and that "the *Protagoras, Gorgias,* and *Phaedrus* lay out Plato's uncompromising condemnation of rhetoric" (*Rule* 323–24). Ijsselling adds that Plato's attitude toward rhetoric was "clearly negative" if not "avowedly hostile," though this attitude was "perhaps a little more nuanced in the *Phaedrus*" (7). Black, Hunt, and Kennedy adopt the second perspective, with the latter explicitly endorsing Hunt's contention that Plato's views on rhetoric given in the *Phaedrus* is essentially the same as Aristotle's given in his *Rhetoric* (*Classical Rhetoric* 52–53). Black argues that Plato had a unified theory of rhetoric and that the *Gorgias* was aimed at refuting only Gorgian rhetoric. Otherwise, "*rhetoric in general has not been attacked*" (178, his italics). Murphy, Quimby, and Morrow endorse the last view, maintaining that there was a growth in Plato's conception of rhetoric, from the outright condemnation of rhetoric in the *Gorgias* to its acceptance and development in the *Phaedrus*. Common to all views is that "good rhetoric," to the degree it exists at all in Plato, was "philosophical rhetoric." Indeed, in a sense, all three positions outlined here result from different perspectives on what constitutes "philosophical rhetoric."

14. One who seems to see it other than settled and certain is Julius Elias. In his *Plato's Defence of Poetry,* Elias claims that Plato laid the foundations for a defense of poetry by endorsing (or, at least, not excluding) the possibility of inspiration by divine madness, the "gift of the poets" (32). Elias does maintain, however, that the poets "ran afoul" of Plato's epistemological and moral criteria (13) and that their poetry, once articulated in the here and now, should be subject to the same critical examination of any other discourse, regardless of its ultimate source in some celestial realm (32). As such, Elias' position is consistent with that taken by Havelock and many others regarding the moral strictures Plato would place on poetry. Moreover, Elias says his title is intended to be "provocative" (1), thus leaving the reader to surmise that, however well presented it might be, Elias's is a minority view on the matter of Plato and the poets. For what is clearly the majority view, and more capable presentation, see Kaufmann's *Tragedy and Philosophy,* especially his discussion of Plato's *Republic*.

15. There are a number of editions of Diels and Kranz and the translation from a later edition is not so simple and direct: "*Haben sie nicht mich, sondern den Sinn vernommen, so ist weise, dem Sinne gemass zu sagen, alles sie eins*" (8.161). Here the Freeman translation would seem to offer a more accurate version of B50: "When you have listened, not to me but to the Law (*Logos*), it is wise to agree that all things are one" (28).

16. "*Es gilt unberhaupt fur Heraklit, dass er die Sprache durchaus als ein Empfindender handhabt, sie aber nicht so sehr nach logischen Klarheit hin entwickelt*" (357). In many of the Heraclitean fragments, for instance, the copula is omitted altogether, and sometimes when included, the predicate is placed before the subject for emphasis (see Cleve 46–47). Accordingly, the language would seem to signal a "pre-predicative" involvement in the world. In the Diels translation of the fragments, the attempt is made to simulate the peculiar stylistic effects of the original Greek—an important effort, indeed, in that the transformation of Greek culture from the dominion of poetry to the conceptualizing thought of philosophy was as much a syntactical revolution as it was one of ideas, as Havelock claims (*Preface* 259–63). Unfortunately, the available English translations of the pre-Socratics, either through the choice of the translators or as likely due to the nature of the English language itself, ignore these stylistic features, giving the reader only the "essential meaning" of the fragments, insofar as meaning can possibly exist apart from style in these contexts.

17. The statement is from Havelock, *The Literate Revolution*, 179. See also Plato's *Republic*, V: 476c, and Havelock, *Preface to Plato*, 190 and 238ff.

18. As Ong says, the Platonic Forms (*eidos*) are voiceless, conceived through analogy with visible forms. Indeed, "the term *idea*, form, is visually based, coming from the same root as the Latin *video*, to see, with such English derivatives as vision, visible, videotape" (80).

19. Most of these references begin in contrasting attributes of body and soul, mortality and immortality, invariably giving preference to the latter and resulting in statements such as this from the *Apology*: "Are you not ashamed that you give your attention to acquiring as much money as possible, and similarly with reputation and honor, and give no attention or thought to truth and understanding and the perfection of your soul?" (29e).

20. See also the *Sophist*, particularly 242ff, where Plato affirms the principles and purposes of dialectical philosophizing in contrast to those of myth.

21. The demands Plato places upon the speaker to know the truth on every single subject and to possess knowledge of the psychological disposition of various types of souls are so great, and seemingly beyond the reach of any speaker, that some critics have said that in the above passage Plato meant to be ironic in his description of rhetoric. Wolz, for instance, says the passage represents the philosopher's "devastating attack upon the false pretenses of

the rhetoricians by describing the impossible condition which would have to be satisfied before rhetoric could call itself a genuine science"; that its "implied requirements" illustrate "simply a more subtle version of Socrates' way of reducing his opponent's view to absurdity" (276–77).

22. Other than the *Rhetoric*, one place where the word *"kairos"* does appear in Aristotle's works is in the sixth book of his *Nicomachean Ethics*. As we shall see, this is the reference Heidegger takes to task, censuring Aristotle for his failure to properly distinguish between *kairos* and the "now" of mere succession (*Problems* 288).

23. I have gathered these comments from Kinneavy's CCCC presentation, *"Kairos* in Aristotle's *Rhetoric."* See also his "Neglected Concept" in Moss.

24. Quintilian distinguishes *chronos* and *kairos,* claiming the Greeks meant by the latter a period of time, such as summer or winter. Thus, in terms of the cycle of the seasons, Quintilian makes reference to attendant "problems," such as that "about the man who held high reve in a time of pestilence" (III.vi.26). H. E. Butler, Quintilian's translator, notes that Qunitilian's reference seems to be the *Laws,* where Plato speaks of times of pestilence, the violence of wars, and the hard necessities of poverty coming in a successive periods of "insalubrious weather" (709a—b).

25. See Carter, *"Stasis and Kairos,"* and Dieter for a more detailed discussion of the rhetorical *stasis*. The more familiar classical presentations of *stasis* (the Latin *status*) are found in *Rhetorica ad Herennium* I.xi, 18–19; Cicero *De Inventione,* II.iv—xi; Qunitilian III.vi; and, "incipiently," in Aristotle's *Rhetoric,* 1417b 21–25.

CHAPTER THREE

1. The citations appearing in parentheses refer to the identifiers given in Diels' *Die Fragmente.* Unless otherwise indicated, all quotations from Gorgias and the pre-Socratics are taken from either the Kathleen Freeman or George Kennedy (in Sprague) translations of Diels. Again, for reasons of clarity or context I have in some instances "supplemented" or substituted these translations with those of others or my own, and those instances are indicated as such.

2. For these and other singular details of Gorgias's life, see Diels, "Gorgias" 82.A, "On Life and Teachings." See also Kennedy's translation of "Gorgias" in Sprague 30–67, from which the above accounts are taken.

3. In particular I have in mind aspects of the account of Gorgias and his rhetoric given in Eric White's *Kaironomia: On the Will-to-Invent.* I have no quarrel with White's contention that Gorgias promoted *kairos* as the determining principle of rhetorical invention (17). Indeed, I believe White's contribution is extremely worthwhile, and I fully endorse his designation of

kairos as "the irrational novelty of the moment" (18). However, I am wary of interpretations, such as White's, that cast Gorgias in the role of revolutionary or romantic, and where very specifically Gorgias is portrayed as suspicious of memory, conventional wisdom, and the Greek *doxa* (20). Consequently, not only does Gorgias seem removed from any specific historical, cultural, and intellectual context, but the portrayal that results seems wholly contingent on White's emphasis on human will in ascribing significance to Gorgias and his rhetoric. The prevalence of "will" in the phrases given above (White 16, 21), and its very conspicuous place in the book's title, seem to embrace a perspective not only inappropriate to early Greek thought but, more critically yet, one that results in a misreading of the Gorgian *kairos*. I would maintain in contrast that consideration be given the idea that the Gorgian *kairos* is securely anchored in a very antiquated, thoroughly pre-Socratic way of life and thought, and that Gorgias himself, as far as rhetoric his concerned, is one of the more prominent representatives of this way of life and thought.

4. Gomperz refers to the author—or "authors"—of the two declamations as the "pseudo-Gorgias" (479). However, he offers no rationale or support for his contention, and, as far as I can tell, he is the only authority of any weight to dispute Gorgias's authorship. Segal presents a contrary view, claiming that most modern scholars accept both "Helen" and "Palamedes" as genuine to Gorgias (100). Consigny no doubt provides the most workable resolution to this issue of authenticity and authorship, stipulating the Diels' collection as the touchstone of Gorgias's works (*Gorgias* 22–23).

5. Another summary of "On-Non-Being" is available in (pseudo-) Aristotle, *Melissus, Xenophane, and Gorgias*, 5, 6, 979a11–980b21. However, Barnes claims the text of *Melissus, Xenophane, and Gorgias* is "wretchedly" corrupt and the Sextus presentation and argument are "regularly superior." For these reasons, he says, Diels chose the Sextus account over the *Melissus, Xenophane, and Gorgias* for inclusion in *Die Fragmente* (*Presocratic* 173, 612).

Gorgias's treatise is variously named in any number of sources. The title I use here, "On Non-Being" is the variant used by Segal, Versenyi, and most often by Untersteiner. The full title is "On Non-Being or on Nature," and, according to Untersteiner, is itself an expression of the "tragedy of knowledge" that he claims to underscore the message of the treatise in its entirety. The antithesis given in the title is shaped principally by the second element, "On Nature," which Untersteiner encourages us to understand in terms of the ancient, "dramatic meaning" of *physis*. Thus, "not-being" serves essentially as an advance declaration of this meaning of *physis*, "which involves the theoretical dissolution of every ontological and therefore epistemological presupposition" (143–44). The imperative word is "presupposition," for *physis* is hardly without meaning whatsoever but is, nevertheless, altogether contingent on "Non-Being." Again, see Vitanza for a detailed development of this view, where "Negation" is seen as necessary to the Sophistic view (*et al.*, I would add), not only then but most assuredly now.

6. Diels 82.B26. The translation from the German is mine, following closely Freeman's translation in Untersteiner (121).

7. See "Parmenides" in Diels and Freeman, especially fragments 28.B2 and 28.B6.

8. On this point see Coulter. He argues that the *Apology* was Plato's very astute adaptation of Gorgias's "Palamedes." As Socrates was portrayed in the *Apology* as a man who "transcends all need to employ a rhetoric which aims at imparting the semblance rather than the substance of truth" (68), Gorgias depicted Palamades as a man "compelled to recognize the unique demands of his situation" (50). The contrast offered by Coulter provides some of the more cogent distinctions to be made between philosophy and rhetoric, providing as well the basis upon which some of the distinctions between the Platonic and Gorgian *kairos* can themselves be established.

9. See Diels 82.B4, where the reference is to Plato's *Meno* 76b—e. If Gorgias's position seems bizarre, it could, nonetheless, be quite conventional. According to Kerford, Protagoras and Euthydemus held a similar doctrine of perception with similar implications, and both, like Gorgias, adopted the doctrine of Empedocles (72).

10. For a refutation of this position, see Gomperz. He alleges that Gorgias employs in his treatise the "abuse of identic propositions, the abuse of the copula, and illicit logical conversions." Gomperz does add, however, that these "errors" are typical of the time and are by no means peculiar to Gorgias, occurring very frequently in Plato himself (483).

11. See Diels 28.B7—B8 and Freeman 43. According to Kerford's discussion of fragments from Parmenides, Gorgias attempted to "pull apart" what Parmenides had joined together, "namely [B]eing, thinking, and saying" (99). If true, the "pulling apart" seems ultimately to favor the one, a *logos* of language rather than reason, that Gorgias calls the great "lord" or "*dynastes*." The idea is critical to our understanding of Gorgias and will be treated later in this chapter and more extensively in the following two chapters.

12. Both Heidegger and Gadamer claim that the meaning of *logos* as reason or conceptual thought is not one alternative of many different meanings of *logos* but is derivative of the primary meaning of *logos* as language. See Gadamer's "Man and Language" and by all means Heidegger's *Introduction to Metaphysics* (especially 154ff), a book that, as Werner Marx says, "tries to rethink the sense of *logos* in the first beginning insofar as it meant language." This "first beginning," by the way, has to do with the "intertwining" of Being and language transpiring as a "nonhuman presence" that thus precedes, or transcends, our own (Marx 158; Heidegger, *Introduction* 51–53).

13. This is, of course, with the possible exception of Gorgias's "Palamedes." In terms of their respective representations of *kairos,* the fundamental

distinction between the two declamations is that the "Funeral Oration" deals with the immediacy and tension of pressing circumstances, whereas "Palamedes" deals not with the nature of the moment but the relative calm of a situation allowing time for prolonged and practical deliberation.

14. The translation follows Freeman's in the *Ancilla*, but for the sake of consistency I have substituted her translation (in Untersteiner) for the lines I have previously referred to as the "*kairos* statement." In the *Ancilla* her translation of those lines is as follows: "Often, indeed, they preferred mild reasonableness to harsh justice, often the correctness of speech to exactitude of law, holding the most divine and most generally applicable law was to say or keep silent, do or not do, the necessary thing at the necessary moment" (130).

15. Rodhe applies the notion most broadly, and profoundly, observing that "once reflection on such problems is aroused, life itself, standing as it does on the threshold of all sensation and experience, soon begins to appear no less mysterious than death" (3).

16. I have consulted various translations of the *Iliad* and have regularly used those which best seem to maintain what I judge to be the archaic idiom, the "formulaic character," as Lattimore says, of the original Greek. For these purposes, Lattimore's and Fitzgerald's translations have been useful. For these same purposes the modernized, more "readable" versions of the *Iliad* are particularly inadequate. For instance, translations that endeavor to present "in plain English the plain story of Homer" and thereby strike embellishments of style because they "were meant only to please the ear," or omit the stock epithets and recurring phrases because their meaning "is of no account" (Rouse's translation 1), clearly fulfill certain purposes, but not mine. None of this, of course, is to say that any translation can duplicate or even closely approach what the *Iliad* actually means in both style and content in the original Greek, much less what it meant to the early Greeks. There are, in any event, various texts of the *Iliad*, and I can only image the damage necessarily done to these "originals" by the need to translate at all. But the damage persists, whatever the case. Translated or not, I doubt there are any left to hear and belong to the words in the manner of the early Greeks.

17. See Furley (3), who also references any number of instances from the *Iliad* to support his contention of this curious union of discourse with physical reactions provoked by the stressful situations of the battlefield. His purpose overall is to explicate the idea of the *psyche*, and, as many others engaged in similar tasks, his examination deals inevitably with the *Iliad*. The question that arises, especially in the context of this emphasis on the *Iliad*, is whether the *psyche*, in all its aspects, owes as much to war and the warrior society as to any ancient perspective on human being in general; that is, indeed, to the extent that the two can be separated at all. As I say, the gore and bloodletting depicted in the *Iliad* are incessant and as graphic as any in literature.

Curiously, of a far more recent war, Philip Caputo writes that the sight of war's mutilation burst the religious myths of his childhood, among them the myth that the souls of the dead passed on to another life and abode and, more evidently, the myth that we have souls in the first place. In circumstances of war the body emphatically is not the temple of the holy spirit but merely "a fragile case full of disgusting matter" (128), and the battle dead, rather than bearing the image and likeness of God, "were more the image and likeness of crushed dogs lying at the sides of highways" (179). An obligingly serene and gradual transition from life to death, eased along by fine sentiments intoned very reverently over a corpse, is necessarily denied, exchanged for an experience comprised wholly of the immediate and visceral impact bearing the stark contrast between the quick and the dead. The suddenness and dumbfounding simplicity of the experience are the marks of the *psyche*'s domain. Rodhe says of the *psyche* that it necessarily goes unnoticed during the lifetime of the body, and is apparent only in its absence, when it is separated from the body in death (4). Thus, it could be that war works to force that understanding on us in very profound ways, for to comprehend war's mutilation of bodies is also to comprehend that what is most essential in life is life itself, precisely the thing mutilated in war. So there is a sense in which the idea of the *psyche* is prevailing and ever-present in war, an integral part of war's brutalization, while the belief in a pristine other self that is the soul is just too incredible in such an environment to ever be sustained.

18. See, for instance, Engell, Enos, and Gronbeck. It should also be noted that Segal's article is seminal, and its influence, specifically concerning the contention that *logos* is free, is surely reflected in the work of the most able scholars who come after him. However, prior to Segal, both Untersteiner and Rosenmeyer, whom Segal incorporates into his study, make essentially the same point about the Gorgian *logos*.

19. The phrase is Momaday's, from the title and content of his most significant essay, "The Man Made of Words," and the phrase I invoke repeatedly in the course of this work as vital to my approach.

20. Thus, Consigny's argument is that Gorgias's different declamations are essentially adaptations of different rhetorical forms or conventions, all of them located within a peculiar cultural and historical setting. I am not sure I have a quarrel with his thesis, insofar as we keep in mind that these stylistic proclivities of Gorgias were, as indicated in Diels, not only well known but often censured in antiquity as peculiar to Gorgias. They were well known and especially well censured by Plato and Aristotle. So if Consigny's argument necessarily supports the corollary argument that Plato and Aristotle criticized pre-Socratic thought in general, and Gorgias's rhetoric in particular, then I endorse it without reservation.

21. See Van Hook and Dillon (especially 76–84, 93–95) for two of the better, more extended translations of Gorgias attempting to capture in

English the "tintinnabulation" of his style. Concerning Gorgias's use of homoioteleuton, see Thompson (80). Also of importance is that in some respects Thompson's interpretation of Gorgias's style is less speculative than Kennedy's. It is, for instance, grounded in a series of parallels that he draws between the style of Heraclitus and Gorgias, as he links these similarities to similar conceptions of *logos* maintained by each. Diels is credited for doing justice, in translation, to the form and nuance of the fragments of the pre-Socratics, and anyone who has labored to translate the German of Diels into English would be aware of these stylistic peculiarities.

22. It is curious that both Nietzsche and Heidegger attribute much of the same mystique to the German language. See in particular Heidegger's *Introduction to Metaphysics,* where he claims that the German language, like the Greek, is "the most powerful and spiritual of all languages" (57). George Steiner's analysis is most pertinent here, as he claims Heidegger's contention that the German language bears a "unique affinity with the dawn of man's being and speech in archaic Greece" is, in fact, "a constant theme in German thought and self-awareness" and (in addition to Nietzsche) rests in the belief of such luminaries as Bach, Kant, and Wittgenstein that "it is in the German sphere that the genius of man would seem to touch the summits and to plumb the last depths" (*Martin* 125). Though Heidegger clearly overstates the argument, and it is impossible for most to believe any such thing, it may contain elements worthy of consideration. Given the syntax and inflections available in the German language, it could be, as I have suggested, that the translations in Diels achieve certain stylistic effects that more closely mimic the original Greek than the English translations of those tracts and fragments, Van Hook's and Dillon's notwithstanding.

23. Kennedy says "practically everything said by Socrates in the introduction to the *Menexenus* has some serious basis in Plato's thought. The force of inspiration, which Socrates seems to ridicule (235c), is made quite clear in the *Ion* (535ff), nor can we doubt that the effect of such a speech as that of Pericles might linger for days." Moreover, concerning Socrates' own funeral oration, Kennedy says "the very excellence of the speech argues against a satiric intent" (*Art* 161). On the matter of Socrates left spellbound by funeral orations, de Romilly adopts the same position as Kennedy (31). See also the *Symposium,* where Socrates makes a similar concession to the power of Gorgias's rhetoric. Though here he is more likely satiric than not, claiming that Agathon's imitation of Gorgias's eloquence struck him "dumb as stone" (198c).

24. In the *Gorgias* Plato says he refuses "to give the name of art to anything irrational" (465a). For a fuller examination of this passage and of Plato's sense of *techne* in general, see J. E. Tiles' article, "*Techne* and Moral Expertise," a time-worn but still very useful source in any discussion of *techne*.

25. The statement is attributed to Gorgias by Plutarch. See Diels 82.B23 and Kennedy (in Sprague 65).

26. See also de Romilly's discussion of the *Republic* and the *Sophist*, where she concludes that Plato regards as synonymous the magic of the sorcerer and the Sophist (31–32).

27. There were at least two significant reasons for cremation in Homeric times, each having to do with a belief in the *psyche*. If the *psyche* were to be dispatched quickly to Hades, cremation had to take place, otherwise it might linger indefinitely near the body or hover aimlessly between earth and the nether regions so long as the body were intact. For instance, the *psyche* of the killed Patroclus visited Achilles in the midst of his dream of pursuing Hector around the walls of Troy, pleading to be "allotted his fire" in all haste so that he might pass through the gates of Hades (23.71–92). It follows, then, that the second reason for cremation is for the sake of the living as well as the dead, for it would free the living all the sooner from the hauntings of the dead. On this issue see Rodhe, particularly his discussion of the lamentable state of the *psyche* consigned to the underworld, and the helpless, utterly destitute and miserable thing it is apart from the body (4–28). The discussion provides an appropriate Homeric rejoinder to Plato's many discourses praising the soul's splendor and loathing the body's corruption.

28. The case of cowardice identified with the creation of the soul is made most aggressively by Nietzsche, especially in *Twilight of the Idols* and *The Birth of Tragedy*. As one would expect, Walter Kaufmann's account, *Tragedy and Philosophy*, treats the issue in a far more measured way.

29. Julian Jaynes also wrestled with the issue of the soul and *psyche*, very peculiarly seeing it in terms of consciousness and the "bicameral" mind. In the course of his examination he is eventually confronted with incredible indications that the *Iliad* was produced by "a race of men . . . who were not conscious at all" (47). Maybe. And maybe not. The root of the problem, of course, is that we tend to think of consciousness as a thing-in-itself, linked either to certain metaphysical components of our being or more likely, in this day and age, to certain significant biological or biochemical processes that are part of the evolution of the brain itself. Yet Jaynes' approach is similar to Santayana's as he argues that "the brain is more capable of being organized by the environment than we have hitherto supposed, and therefore could have undergone such a change as from bicameral to conscious man mostly on the basis of learning and culture" (106). Here, a change in perspective is, in effect, a change in the thing perceived, which in a sense is *necessarily* the case, as a thing is what it is depending upon how it is revealed or reveals itself. So here, as well, the issue becomes marvelously complicated, for in these matters involving our very consciousness, the question becomes, at long last, *whose* perspective we are talking about. That is an issue to which I will return, but surely one that will never be answered satisfactorily. Thus, Jaynes's work will

be addressed once more in a later chapter, specifically in the context of Heidegger's conception of the "call of Being."

30. In Plato's works the most celebrated example of the recovery of knowledge innate to the soul takes place in the *Meno,* where one of Meno's slaves, a youth apparently ignorant of even the simplest principles of geometry, is able to demonstrate through prompting from Socrates a relatively sophisticated knowledge of squares and triangles. At issue is whether the youth learned these principles from Socrates or whether he was simply "reminded" by Socrates of knowledge he already possessed. Socrates concludes, with Meno's concurrence, that the youth neither acquired his knowledge of geometry from Socrates nor from anyone else in his lifetime but already had it in his possession prior to this life, that it had been embedded always in his soul and required only the proper stimulation for it to be recovered and expressed.

31. See Wedberg's discussion concerning the aspects of classification and division in dialectical inquiry. His discussion is especially pertinent to the larger issue of whether, and to what extent, the dialectic is a purely deductive science. Wedberg asserts, first of all, that the dialectic is the science that investigates the realm of ideas, that it proceeds by means of reason, as Plato says in the *Sophist,* "to show correctly which of the Ideas harmonize with which, and which reject one another" (253b—e; Wedberg 51). Thus, in Wedberg's interpretation the dialectic is hardly confined to classification and division as some readings of the *Phaedrus* might indicate, but performed an additional task. In later dialogues, especially the *Timaeus* and *Republic,* the scope of the dialectic, at least in its "final stage," embraced a "deductive science which is the logical basis for the entire field of rational knowledge and which derives all its conclusions from a first self-evident principle, expressing the supreme insight into the Idea of the Good" (52). Here, it would seem, the workings of the dialectic are internal in all crucial respects, its goal indeed to uncover the knowledge innate to the soul.

32. See, for instance, *Timaeus* 44d–46e, where Plato traces the stages or increments involved in the soul's evolution, from its encasement in the mortal body to its active presence as reason and intelligence. In the end, an absolute distinction is made between the visible body and the invisible soul, the latter designated by Plato as "the only being which can properly have mind" (46e).

33. The analysis parallels Michael Leff's, ably presented in his article, "The Forms of Reality in Plato's *Phaedrus,*" and credit is hereby given.

34. A peculiar interpretation of the Socratic dialogue results from de Romilly's analysis. Rather than a method that seeks fundamental truths through mutual inquiry, trust, and assent, its power is likened by de Romilly to the rhetoric of Gorgias, as in both cases this power is "bewildering, amazing, magical" (37). If Gorgias represented the deceiving power of style and language, the peculiar magic of Socrates lay in "the way he discusses, clinging

to reason and truth, and trying to do away with all fake appearances, all unsound arguments of definition that they [his listeners] are merely bewildered by the power of thorough analysis. They do not understand what happens to them, but we do: they are just confronted with unyielding logic" (36). That logic should be associated with a compulsion so unyielding as to be equated with magic provides a warrant for the pre-Socratic mindset seldom seen in contemporary analyses, even those of the neo-Sophists. Indeed, it is a most welcome turnabout, if only in this instance alone, of seeing the work of Plato from a "Sophistic" point of view.

35. In Plato the clearest instance of *kairos* depicted exclusively as insight is found in the *Seventh Letter,* where it is contingent upon an "inborn affinity with the subject." Plato begins by describing the levels of knowledge involved in the ascent toward true reality, a process that embraces the naming of an object, its description, its image, knowledge of the object, and, last of all, the actual object of the knowledge which is true reality itself. The process is facilitated "not in sounds nor in shapes of bodies, but in minds" and culminates, if the ascent goes well, in "a flash of understanding" blazing up; then "the mind, as it exerts all its powers to the limit of human capacity, is flooded with light" (341a–344d). Plato does include "benevolent disputation by the use of questions and answers" as part of the process of inquiry, but, as in the *Sophist* (263e), the use of questions and answers is not directed outward toward another but is internal, occurring within the soul and addressed to the object under consideration. Moreover, the flash of understanding is clearly *aneu logou,* "beyond discourse," as Plato, in the course of his discussion, specifically points to the inadequacy of language in capturing or even sustaining the integrity of the process (343a).

36. See Derrida, *Speech and Phenomena* 69. "Sign and the Blink of an Eye," Chapter Five of the same volume, contains his discussion of Husserl's notion of *kairos* that Derrida says is part of the tradition carried over from the metaphysics of Plato and Aristotle (63).

Chapter Four

1. For instance, see Mugerauer, *Heidegger's Language and Thinking,* who claims it is impossible to separate the substance of Heidegger's thought from the style of its presentation, and that, owing to this mix, his language and thinking are the same in that they proceed by way of a "figural unity" rather than conceptually or by otherwise logically established categories traditional to Western thought (x). It is also significant that he deals with Heidegger completely in translation, as must be the case, he says, unless we do all our thinking, speaking, and reading of Heidegger in German. Given this approach the issue then becomes not whether to use translations but "how to use them thoughtfully, as a way of questioning." And to use them thought-

fully necessitates "learning to hear a thought originally," just as Heidegger advises when he speaks in German about Greek or Japanese thinking. Here, "all good trans-lation is an interpretation and possibility of going-over, the same as all genuine reading, learning, and thinking in any language are" (xii). This method and Mugerauer's basic argument throughout seem wholly reasonable.

2. Nonetheless, a likely candidate for a good comprehensive treatment of Heidegger's thought, both in scope and elaborate detail, is Richardson's *Heidegger: Through Phenomenology to Thought*. He undertakes to provide, insofar as such a treatment is possible, a complete and impartial explication of Heidegger's thought in an effort, he says, "to understand Heidegger's notion of thought and nothing else" (xxvii). Richardson achieves this goal as well as can be expected, generally avoiding the temptation to build on Heidegger in order to make an argument of his own. There is great value in doing this, the present interpretation notwithstanding.

3. The phrase is commonplace and equally austere in either language: "*Warum ist uberhaupt Seindes und nicht vielmehr Nichts? Das ist die Frage*" (Heidegger, *Einfuhrung* 1).

4. I have generally followed Seidel's translation of "thing" for *das Seiende*, though I have at times resorted to the more common use of "being" (with a lower case *b*) when the context would otherwise be unclear. Other possible translations of *das Seiende* are "entity" (Macquarrie-Robinson [*Being and Time*]) and "essent" (Mannheim [*Introduction to Metaphysics*] and Mehta). I have bracketed each with "thing" in the passages taken from these sources. Seidel's rationale for the translation is that "thing" brings out more sharply and simply than "being" ("entity" or "essent") the distinction Heidegger draws in the ontological difference between *das Sein* and *das Seiende*. Seidel also points out that Heidegger approved of the translation (5).

5. See Mehta's discussion, where he claims that to Heidegger the ontological difference "meant to suggest that essents [things] and Being are somehow held apart, separated and yet kept together in their relationship with each other, in and by themselves and not merely as a distinction of the intellect" (192). Mehta then makes the point that Being as such lacks a clear determination in metaphysics, as in metaphysics Being is projected not in its difference from things but always in terms of things. He therefore claims that "metaphysics is the movement of thought away from and beyond the essent [thing] as a whole towards its Being, but it is at the same time a movement that comes back to the essent [thing] as its ultimate destination" (194). Again, in either case, Being is "forgotten" by metaphysics.

6. Most often Heidegger seems to make no clear distinction between Being and *physis*, and when he does, his concerns center on the various connotations of the terms rather than any basic difference between the two. Perhaps it can be no other way, though he says the essence of Being is *phy-*

sis, Being as it is understood in its "primordial" Greek sense (*Introduction* 100–02). Indeed, as he deals more and more with early Greek thought, his use of Being is replaced by *physis,* and then very often by the arcane yet far more compelling notion, "earth." For Heidegger the meaning of *physis* was less compromised than that of "Being," though he says *physis* was also compromised, degenerating into prototypes, copied and imitated, at long last designating "nature" in a purely abstract sense (61). And that, in turn, is one reason among others for the subsequent transition to the term "earth."

7. Biblical references to glory designated by *doxa* are abundant. In addition to those above, see Matt. 4.8, 2 Cor. 4.17, and 1 Pet. 1.11.

8. What is shared is *thaumazein,* the sense of wonder that in Plato leads to discovery in philosophic inquiry and is otherwise the wonder that is discovery itself. To Plato, it is philosophy's singular origin, ultimately the province of the dialectician and his method (*Theaetetus* 155d); while otherwise it conveys no clear distinction between what we do and what is done to us, between discovery and wonder, thereby occurring essentially as revelation, as it does repeatedly in Homeric literature. Here, "wonder-struck beholding" in the full sense of *thaumazein,* as Arendt says, was reserved "for men to whom a god appears," and for "godlike" men in particular (1.142). If this sense of *thaumazein* approximates the experience of Being's revelation in Heidegger, there are times indeed when Plato seems to depict the experience of the dialectician, along his rapturous journey to truth's discovery, no differently.

9. The translation is Manheim's of Heidegger's translation of Heraclitus (*Introduction* 103) and essentially the same as Freeman's (29).

10. A full text of the letter is found in Ward, *The Civil War* (82–83). I offer it here with some of the minor editorial modifications used in the television presentation and otherwise incorporating some of those portions of the letter relating to Ballou's commitment to "honorable manhood" excluded from the television presentation. If the letter seems too good to be historically accurate, I know of no challenges to its authenticity; and it is, historically accurate or otherwise, a very fitting example of the idea of *doxa* I am attempting to express.

11. The translation is Freeman's of 82.B6 (130). See also Kennedy's in Sprague 48–49 and Dillon's 93–95 where the translations more faithfully capture, perhaps, the grandiloquence of Gorgias's style.

12. Lincoln is speaking of something other than a political entity, certainly modern ones. In fact, the sense of *doxa* as the perfect cosmos—as the union, place, and purpose of all those constituting a society—is often more apparent in "primitive" cultures, those that are small and homogeneous, such as the Lakota of the northern plains. In his famous vision concerning the fate of his people, Black Elk provides a compelling image of the tribe breaking camp and setting out on the "red road" of peace and plenty. Here the sense of the sacred "hoop" of the nation is conveyed, of the harmony, orderliness, and

timelessness embracing all within that circle. In its "going forth" the tribe was led by the warriors, then came the youths and children, then the chiefs and advisors, and they were followed finally by the old men, old women, and those hobbling on canes. But they were not last. In the distance, Black Elk says, trailing like a fog as far as could be seen, "were the grandfathers of grandfathers and grandmothers of grandmothers." The ancestors. They were also of the circle, of the harmony of this cosmos, mystic chords of memory in their own right. And so over all this, Black Elk says, the great voice of the South proclaimed, "behold a good nation walking in a sacred manner in a good land" (35–36).

13. The difference follows the basic distinction applied by Chatman between story and discourse, a most viable criterion in narrative analysis. The story is the content or chain of occurrences in a narrative, the "what" of the expressed; and the discourse is the "how," the means by which the content is expressed (19). It is indeed an artificial if often necessary distinction, for the particularity of the "what" and "how" of a narrative are never equal to the "event" of the narrative itself. In this matter especially, the whole is always more than the sum of its parts.

14. This spell of "average everydayness" of which Heidegger speaks is quite typically broken in the experience of war, where both reason and habits of mind are dashed. Not just in Gorgias's time, but always this seems the case with war, so long as we do not grow too accustomed to it. Indeed, in very remarkable ways the pure "that it is," broached above, is affirmed in the phrase, "there it is," bandied about by soldiers depicted in Tim O'Brien's novel of the Vietnam war, *The Things They Carried*. It was the diction of the infantryman, his peculiar acknowledgement of war's stark and brutal realities, of war itself claiming *him* as its supreme reality. At the same time it was his bearing witness, "to something essential, something brand new and profound," as O'Brien says, to "a piece of the world so startling there was not yet a name for it" (86).

15. Edward Schiappa, who probably offers as close to the definitive word on Protagoras and rhetoric as we are likely to see for a good long while, says "the reduction of Protagoras's concept of 'measure' to perception alone has been widely recognized as a simplification designed to facilitate Plato's attack on those thinkers allegedly equating perception and knowledge." He adds, however, that Plato's own theory of sense perception is very close to Protagoras's, and that Plato's description of Protagoras's theory was not purely a fiction to be discredited. Thus, to Schiappa, evidence suggests that Protagoras was "a transitional figure" between Parmenides' denial of contrary qualities and Plato's elevation of qualities to the Ideal Forms, while "Protagoras's fragments suggest that qualities are directly perceived or experienced by humans and hence are relative to people" (*Protagoras* 127). However, in his many sided analysis of the doctrine, one of his more satisfying interpretations

is that the doctrine moves beyond (or comes well before) all "subjectivist/objectivist" distinctions. Here we might understand Schiappa as saying the doctrine even bears the influence of those such as Heraclitus, even to the degree that it is manifest purely in terms of "phenomenalism," as Untersteiner has claimed. What he *does* say is the following: "If the subject/object pair is reduced to the more concrete concepts of people and things, then Protagoras's answer is clear: the two are 'bound,' 'things' can be 'measured' by people in contrasting ways (*logoi*), and a dominant experience (*logos*) of a thing is potentially alterable as an interchange or swapping of opposites" (130). Schiappa's presentation is nothing if not detailed and nuanced, but in the contention that "the two are bound," his interpretation of the doctrine might parallel Heidegger's in some significant ways, especially as Heidegger's is developed above in terms of *phainesthai* and on the way things reveal themselves. Then the issue turns on conflicting interpretations of the nature of *logos* rather than on conflicting interpretations of things by *logos* (or *logoi*). Whatever the case, it is difficult, still, to imagine that Plato's presentation of the doctrine in the *Theaetetus* as anything other than unfair and heavyhanded. Schiappa did *not* say that and, I suspect, would not agree.

16. See Gaonkar's analysis concerning Plato's indictment of the "subjective relativism" that Gaonkar, in fundamental agreement with Crowley, interprets as inherent in the *homo mensura* doctrine. Perhaps more critical to the issue of Sophistic rhetoric is Gaonkar's argument that the implications of the doctrine extend well beyond matters of individual perception to embrace moral and political judgments as well. Of course, this premise of subjective relativism, that many see as essential in understanding both the *homo mensura* doctrine and Sophistic rhetoric in general, is extremely problematic in light of Heidegger's interpretation of the doctrine.

17. The translation is Freeman's of 80.B1 (125). The translation in Sprague (Michael J. O'Brien's) is the same except for comma placement (18).

18. Heidegger's use of the word "sophism" is quite typically idiosyncratic, and hardly consistent. Here, and throughout *The Question Concerning Technology* (143, 146, 147), the Sophists are pre-Socratics to Heidegger, but in other contexts (*Introduction* 106, "Letter" 195, "Essence" 320) they are treated as if they were in league with Plato, opposed in crucial respects to the pre-Socratics. However, in his most cogent treatment of the *homo mensura* doctrine, "The Statement of Protagoras," Heidegger associates Sophistic thought with truth as *aletheia*, where "measure" is the "measuredness of unconcealment" rather than a prescription to invest "the self-posited ego as subject" to determine or decide the "objectivity of objects" (*Nietzsche* 4: 95).

19. Heidegger translates *aletheia* as "unhiddenness" or "unconcealment" and claims it to mean "originative" truth that comes in response to the *Seinsfrage*, the primal question spurred by the mystery of Being itself. However profound or critical we judge the "turn" in Heidegger's thought to

be, the theme of *aletheia,* if not always the word itself, has been at the center of the development of his thought from the beginning. Gerald Bruns says that to Heidegger *aletheia* is the way that the space between language and poetry was mediated or breached, involving "a complex pun that preserves the darkness or otherness of truth, its strangeness or reserve, its self-refusal, its untruth" (xv). By way of explication Heidegger himself often presents the notion of *aletheia* in contrast to truth as *veritas.* In place of the unhiddenness of Being, *veritas* designates for Heidegger the intellectual distancing apparent in propositional truth, where the agreement or correspondence of mental concepts with their objects is the sole and only proper criterion of truth—a criterion that has been, according to Heidegger, the bedrock of Western thought since the inception of metaphysics. In particular see his "Plato's Doctrine of Truth" for an extended treatment concerning *aletheia* and *veritas.*

20. In the sense it is acknowledged in Heidegger, consciousness is scarcely "wrong" or "wanton." What he condemns is the self assertive gesture, consciousness in its predatory aspect, a circumstance where "man places before himself the world as the whole of everything objective," where he "sets up the world toward himself, and delivers Nature over to himself" (*Poetry* 110). Heidegger is describing here the anthropocentric attitude as he sees it presented, say, in "Existentialism is a Humanism," Sartre's statement of the human condition and the homage he pays to the integrity of individual consciousness. No less than Descartes', and as Camus', this attitude to Heidegger represents a "subjectivity" that violates the idea of *dasein* as the "ecstatic" relation to Being. See his "Letter on Humanism" and *Poetry* (127ff). The former is of particular importance, being Heidegger's response to those, especially Sartre, who project him as an existentialist and *Being and Time* as in the vanguard of the modern existential movement.

21. Heidegger's idea of "world" might seem even more arcane than the interpretation of *doxa* that I am developing here, involving as it does the mutual implication of Being, human being, and nature. Nevertheless, his presentation of it is compelling: "The world is not a mere collection of the countable or uncountable, familiar and unfamiliar things that are just there." It is "more truly in [B]eing than the tangible and perceptible realm in which we believe ourselves to be at home. World is never an object that stands before us and can be seen. World is the ever-nonobjective to which we are subject as long as the paths of birth and death, blessing and curse keep us transported into Being" (*Poetry* 44). This idea of world will be explored in due course, especially as Heidegger will come to compare and contrast it with earth.

22. An anatomy of the process is given by Garry Wills on a quite likely subject. In his brief but very good book, *Lincoln at Gettysburg*, Wills makes the case that Lincoln perpetrated a "giant (if benign) swindle" in claiming at Gettysburg that the Union soldiers died to vindicate the proposition that

"All men are created equal," when every announced purpose for waging the war and the US. Constitution at the time brooked no such interpretation. However, the Gettysburg Address has become the "authoritative expression of the American spirit," for through the "stunning verbal coup" of the Address, "the Civil War *is,* to most Americans, what Lincoln wanted it to *mean.*" The words of the Address "remade" America by giving the American people "a new past to live with that would change their future forever" (38). Wills' argument provides a practical instance of what I have been attempting to present in this chapter, that *doxa* as both product and process—as the creative emergence of beliefs as much as the beliefs themselves—makes its way by holding in abeyance conventional (Platonic) notions of reality, for in this context events are "real" only insofar as beliefs (and the words expressing beliefs) make them so. Moreover, as we shall see, the Address provides, at least in Will's analysis of a "new past changing the future forever," a fitting instance of Being and *doxa* as the trickster's domain, of the transfiguration of "yesterday's words" into "novel utterances" to be discussed in the last chapter.

23. In one of his first novels, *Going After Cacciato,* Tim O'Brien speaks of *xa.* He composes a fantastic scenario that among other things delineates reasons for the American defeat in Vietnam, the most crucial of these being our lack of understanding and appreciation for the idea of *xa.* Yet, for being so fantastic, the scenario is all the more believable. A very ancient, sage-like enemy in the novel, most significantly a tunnel dweller, goes into some detail in telling a wandering GI that *xa* is the earth and sky, and in certain ways all things in between, though it is in all instances an all-abiding sacredness. The soldier takes from the discussion its most telling lesson, that earth in its wholeness with the Vietnamese people is his real adversary (107), a truth made readily apparent throughout the novel.

24. A barrier to understanding *doxa* in terms of this union of natural phenomena and cultural beliefs is, of course, our own cultural beliefs, in particular our commitment to an analytic strain of thought that excludes such "mystifications" of experience. Such, it seems, is in part the basis of Arnold Krupat's criticism of Momaday's idea of racial memory. Insisting upon a perspective that honors the complete separation of nature and culture, an analytic perspective preeminently, Krupat condemns the idea of racial memory as a "fashion in mystification" that we can quite logically refuse to countenance. He is then in a position to ridicule Momaday's claim that American Indians possess a "memory in the blood" setting them apart as American Indians. Quite logically Krupat responds that there is no such thing as memory in the blood, that Indians have no special "genetic constitution," and that racial memory is, as a result, among other "absurdly racist things" that Momaday says (13–14). That the problem might reside, at least in part, in the body of beliefs that moves Krupat, his *doxa,* and not Momaday's, quite logically escapes Krupat's notice. In any event, Krupat's remarks are akin to

those delivered by H. Tint to the British Aristotelian Society, where he claims that Heidegger's thought, as it deals with the unknown and unknowable elements of Being, not only lacks precision but is snarled in mystery, and, worse still, degrades logic because it maintains that logic is only one path, one avenue or way of thinking among others. In essence, Tint desires unambiguous answers, and he is therefore much dissatisfied with Heidegger's, which are, if fact, more in the line of questions than anything else. More generally he claims that Heidegger's thought amounts to a "betrayal of philosophy," ostensibly because Heidegger is not a British empiricist. It is ironic that Krupat would adopt a similar attitude regarding Momaday, for Krupat seeks an explication of American Indian thought and in his intransigence does not have Tint's excuse of membership in the British Aristotelian Society.

25. See Momaday's essay, "The Man Made of Words," for his extended discussions of "racial memory." See also "The Arrowmaker" in *The Man Made of Words: Essays, Stories, Passages*, 9–12.

26. I revisit this issue of the *Kehre* in Heidegger's thinking not only because it bears so significantly on the differences and similarities of *dasein* and the *psyche* but because it is likely that if Heidegger himself had not called attention to this "turn" in his thinking, it would have gone essentially unnoticed by the critics. It certainly would not have become so entrenched in the canon of Heideggerian interpretation. Thus, it is important to realize that his principal purpose in his "Letter on Humanism," where he announced the *Kehre*, was as much to distanced himself from Continental existentialism, from those who took his work to be "humanist" or "subjectivist," as it was to indicate a turning away from his positions in *Being and Time* to a new emphasis on Being and language in his later works. Again, see Heidegger's comment on the *Kehre* in his "Preface" to Richardson's *Heidegger* (xviii). The "Preface" as a whole provides Heidegger's last and most comprehensive statement on the significance (and lack thereof) of the "turn" in his thinking.

27. This is, of course, the theme I endorse throughout, meaning that Heidegger's thought as a whole can be seen as an effort of retrieval, an effort to recapture primordial insights into Being, language, and human being eclipsed by the work of Plato and metaphysics. The source that most successfully develops this connection in terms most relevant to my effort here is Seidel's *Martin Heidegger and the Pre-Socratics*. In his "Introduction," Seidel puts that matter as so: Heidegger's thought "attempts to think down to the ground, down to the very roots of phenomenology, of history, of language, and most particularly of [B]eing in our [W]estern tradition of philosophy. But as one gets deeper and deeper into Heidegger's attempts to think down to these original grounds of the truly original, all these grounds in one way or another seem to lead back to the pre-Socratics" (2).

28. In fact, the challenges to Gorgias's fitness as a serious philosopher are based in large part on convictions that he was a nihilist who believed

there was no such thing as Being or knowledge based on Being, either of which are convictions that many have thought worthy of consideration. For instance, John Robinson is particularly vehement in his expression of the first, concluding that Gorgias did indeed believe in nothing, and that once we move beyond Gorgias's style, which he dubs as "repellent as it was artificial" (52), we see the emptiness, the lack of substantive content to even the claims of nothingness. However, as I have emphasized throughout, a more productive perspective—and one very aptly addressing both challenges—is that Gorgias was a phenomenalist, as Guthrie, Untersteiner, and Versenyi, among many others, contend. Enos in particular makes it clear that to take Gorgias's proposition that "nothing exists" as a contention concerning existence relative to the physical world "would be both an absurdity and a contradiction of the fundamental beliefs of his teachers and his own empirical observations on sense-perceptions" ("Epistemology" 46).

29. *Ousia* means Being, and Heidegger claims that to the early Greeks and pre-Socratics it meant Being in its more mysterious aspects, in many ways inseparable from Nothingness in being a process of revealing and concealing, being the "creative surge" that prompts things to emerge into presence, to abide, and then subside into absence, into Nothingness. However, in Plato and Aristotle, Heidegger says *ousia* is no longer Being in this primal sense, but is derivative, indurate and static, conceived by Plato as idea and Aristotle as substance. As I have stated, this sort of devolution or "forgetting" of Being is a familiar theme in Heidegger, accounting in large part for his emphasis on *physis* or earth in place of Being in his later writings, especially as in the latter he sees Being's most explicit manifestations more and more in terms of language.

30. Untersteiner's mixing of tragedy and knowledge is, of course, of the very nature of tragedy as such, making of knowledge an ontological rather than strictly epistemological concern. Indeed, as will be explored presently, thinking is here not a matter of "calculation" but, as Heidegger says, of "piety"—of reverence, awe, and respect—where *dasein*'s representation as a rational animal is purely derivative. By its very nature, then, "Dasein understands itself in its Being," as Heidegger says, and "*understanding Being is itself a definite characteristic of Dasein's Being*" (*Being* 32, his italics).

CHAPTER FIVE

1. The distinction between "possibility" and "actuality" is a recurring theme in Heidegger's work, central to his "Letter on Humanism." Here, the "quiet power" is Being itself, and in conjunction with "the possible" means essentially to allow Being to do its work, to "let it be." It thereby "unfolds" in its province or element as it "presides over our thinking" without hindrance or interference from presumptions on our part concerning its nature, as we

would otherwise attempt to bend it to our will, to reduce its validity to a *techne* (196–97). The idea is important in this chapter and, especially as it concerns *techne*, even more important in the next. The most complete rhetorical applications of Heidegger's motifs of "possibility" and "actuality" are given by Poulakos in series of articles, especially "Rhetoric, the Sophists, and the Possible." See also Worsham on this and related themes.

2. As I have suggested, the statement is arguable, though on this particular point Plato says in the *Phaedrus* (261a *passim*) essentially what he says in the *Gorgias*, claiming in the former that rhetoric, taken as a whole, is the art of influencing the mind by means of words, and then going on to indicate that rhetoric, of the sort criticized in the *Gorgias*, is no art all. In particular, see de Romilly's examination of that passage from the *Phaedrus*, where her discussion hinges upon the nature of words and magic involved in "influencing the mind" or "acting through words" (15–16).

3. De Man raises the issue of "rhetorical mystifications" specifically in contrast to tendencies to reduce language to mere grammar and mechanics, those that postulate "the possibility of unproblematic, dyadic meaning"; as opposed to rhetoric, where the emphasis is not only on the complexities of the human *psyche* but on the relationships between signs "giving birth to one another" in lieu of outside reality. As such, "rhetorical mystifications" transpire as "rhetoric radically suspends logic and opens up vertiginous possibilities of referential aberration" (128–29).

4. For many, Aristotle's definition ("On Interpretation" 16a3–8) is the archetype of the logocentric view of language, thus serving as a springboard to alternative views of language. In essays appearing in *On the Way to Language* (97, 114), Heidegger uses Aristotle's definition precisely in this manner, to heighten the contrast to his own view of language. See also Derrida's *Of Grammatology* 11, Bruzina 188, and Kockelmans' "Language" 4, from which the above translation is taken. However, it must be pointed out that "On Interpretation" is a work dealing with propositional truth, a product of *theoria* and therefore has its place in the domain of exact science. For this reason, Aristotle's definition is not, perhaps, an apt reflection of his "theory of language" as it would be presented in tracts concerned with probability or "practical wisdom" such as his *Rhetoric*. Nevertheless, it remains a most likely means to contrast "traditional" theories of language with those proposed by Heidegger and various other "poststructuralists."

5. "Idle talk" and "*doxa*," understood in the conventional sense as "mere opinion," are surely comparable in any number of respects, and critics of Heidegger are hardly unanimous in accepting *doxa* as anything other than in its conventional sense. Consider, for instance, the way in which *doxa* is presented by Michael Haar, who asks, quite rhetorically, "Is not the Heideggerian depreciation of collective existence, 'public' in the sense of simply Being-with, the continuation of a theme obscurely derived from the Platonic

distrust of *doxa*, that is, of the opinion of the majority of the masses?" ("The Enigma" 21). My concern is that *doxa* is not, or is not exclusively, "the opinion of the majority of the masses," and to see it as such, and therefore essentially in terms of its contrast to *episteme*, puts *doxa* in a most pejorative perspective indeed. At the least, such a designation lacks the complications, splendid nuance, and very valuable insights Heidegger gains from the idea in his *Introduction to Metaphysics*. As I will attempt to develop in this chapter and the next, it is particularly crucial in this context to see *doxa* not only in contrast to *episteme* but as complementary to "para-*doxa*," providing yet another viable access to the idea of the Gorgian *kairos*.

6. See Bruns 147 *passim* on this amiable coupling of *das Wort* and *es gibt*.

7. Heidegger says the impasse is marked by "language speaking itself as language," when it "has distantly and fleetingly touched us with its essential being." Curiously, we often know these moments when we *are* dumbstruck, when "we cannot find the right word for something that concerns us, carries us away, oppresses or encourages us" (*Way* 59).

8. The passage (170) brings to focus the unsettling matter of Confederate dead and the American *doxa*. It is difficult to imagine Lincoln's Address being about them, especially at the time of its delivery. In fact, there are no Confederate dead at Gettysburg, at least none in any marked and established locations, as all that could be found were disinterred after the war and removed to southern cemeteries, most to Hollywood Cemetery in Richmond. Gramm addresses this matter as well, alluding to insights of Whitman who, he says, acknowledged the evils of slavery but affirmed "the essential unity of the American character, North and South, the shared beliefs, hopes, and values of each side, the identity of virtues and the agreement on liberty's essential place (35). As it is difficult to be sanguine about "shared beliefs and values" in light of the issue of slavery, the matter remains muddled within the ambiguities and contradictions of *doxa*. Clearly, for these men at that time and place, political motivations, even those concerning slavery, were unimportant. Thus, Gramm makes a compelling point in saying, "the average southern soldier didn't fight to preserve slavery any more that the average Yankee fought to abolish it." These soldiers, North and South, "went to war for reasons neither they nor we can fully understand" (140). The statement seems altogether believable, for this standard—one that we can never fully understand—is truly the measure of *doxa*.

9. On the issue of war and witchcraft see Thomas Merton's "War and the Crisis of Language." The essay is decades old but very timely and addresses specifically the "witchcraft" surrounding the destruction of Ben Tre and generally other matters of language and the Vietnam War that continue to have great relevance to this day.

10. On this subject I again defer to Tim O'Brien, this country's most articulate and astute observer—*listener*—of our experience in Vietnam. In particular see his chapter, "How to Tell a True War Story," in *The Things They Carried*. One of his characters tells of his experience on LP (Listening Post), a soldier who listens and what he hears: "Not human voices, though. Because it's the mountains. Follow me? The rock—it's *talking*. And the fog, too, and the grass and the goddamn mongooses. Everything talks. The trees talk politics, the monkeys talk religion. The whole country. Vietnam. The place talks. It talks. Understand? Nam—it truly *talks*" (81–82). And it truly does, I swear, in this strange, aboriginal *logos* that the experience of war returns us to, cutting through the many veneers of everydayness if we but listen. This, I think, is the sort of listening of which Heidegger speaks, of which various characters in American Indian literature accomplish. The trouble is, the soldier on LP concludes, "Nobody listens. Nobody hears nothin'. The politicians, all the civilian types. Your girlfriend. My girlfriend . . . What they need is to go out on LP. The vapors, man. Trees and rocks—you got to *listen* to your enemy" (83). Most emphatically, this is not GI bilge or lingo, or not that alone, and I think we must take it for what it says, because it could be that failing to listen was why we lost the war, or why there was a war at all. And why, perhaps, we fail to listen still.

11. Connections between Heidegger's theory of language and Asian thought were drawn in large part by Heidegger himself, especially by his essay, "A Dialogue on Language Between a Japanese and an Inquirer" (*Way* 1–54). Exploring these connections has been most common among Heidegger scholars. In particular, see Graham Parkes, ed., *Heidegger and Asian Thought*.

12. I have changed the name, though to one perhaps too familiar on the reservation. But the account is true enough, embellished only to the extent memory might have it.

13. Momaday has said as much, asserting that he used Tosamah to express a number of his own thoughts on language (Personal Interview).

14. Tosamah's duplicity in accusing others of the verbosity he so obviously practices himself is truly of the trickster's nature. What is also pertinent, especially in the thick of his verbosity, are his references to fat. With the understanding that Momaday is Kiowa (the same tribal affiliation to which he assigns Tosamah), there is nevertheless a Lakota word for white man that has great currency to this day: *Wasichu*. Its etymology may be uncertain and could possibly be traced back to mean "buffalo" or "spirit," though in more recent interpretations it tends to mean "fat talkers" or "big talkers," either of which, in the Lakota perspective, could quite aptly apply to the white man. So it is with the American Indian perspective in general, especially as we see it given above, as Tosamah speaks of the white man's words as the fat layered

between Saint John and his God. For these (and other) particular meanings of *Wasichu*, see DeMallie.

15. Again, the religious emphasis is apparent. The statement comes neither from Heidegger nor Momaday but Martin Buber (112), addressing matters very similar to those raised above.

16. See "The Morality of Indian Hating" in *The Man Made of Words* (76). Momaday is describing an annual ceremonial race held in Jemez in February and always run at dawn, a race that is "neither won nor lost," he says, but run as "an expression of the soul in the ancient terms of sheer physical exertion" (75–76). This race is no doubt the inspiration for the one depicted in *House Made of Dawn*.

17. Heidegger says, "Temporality is not . . . an entity [being or thing] which emerges from *itself*; its essence is a process of temporalizing in the unity of the ecstases" (*Being* 377). Richardson explanation of Heidegger's complex analysis of this matter is especially helpful, telling us, among other things, that the ecstases are not separate parts of something but are best discerned as "directions" in which the process of temporality comes to pass (88).

18. On the issue of the "unifying scission" see in particular Heidegger's essay, "Language" in *Poetry, Language, and Thought* (189–210). It is on this particular point of the *Unter-Schied* that Steiner says "Heidegger's terminology reaches new extremes of impenetrability" (*Heidegger* 103). As that might be the case, see also Richardson's relatively accessible interpretation of the *Unter-Schied* (577–82).

19. There may be some truth to Will Durant's judgment, given many years ago, that epistemology with an exclusive emphasis on analytical description has "kidnapped" modern philosophy, and "well nigh ruined it" (xxiii). His *Story of Philosophy* was written as a corrective, and the popularity of the book, he claims, was the result of readers' astonishment "to find that philosophy was interesting because it was, literally, a matter of life and death" (x). He himself, of course, has been criticized for being a popularizer, for treating extremely complex issues in simplified ways, but after all these years his point concerning the epistemological emphasis of modern philosophy is no less valid. Heidegger notwithstanding, the question of Being is left unasked because it is not susceptible to analytic treatment, even to the extent that we no longer seem to know how to think about these life and death matters provoked by the question of Being.

Chapter Six

1. The verses are from the Yoruba *Oriki Esu*, a cycle of praise songs to Eshu (variously "Esu" or "Edju"), that is itself part of a complex of myths, rituals, and histories of the Yoruba. Eshu is an Orisha, a trickster god of the

Yoruba pantheon. Various versions of the *Orika Esu* are found in various places. For instance, see Gates 1, Pelton 127–163, and Lewis Hyde 238–40.

2. The entire account could well be apocryphal, spread by enemies of the Pythagoreans or, for whatever peculiar reasons of their own, by the Pythagoreans themselves. In any event, the story is properly neither true nor false but one gathered here, with very modest embellishment, from Marx W. Wartofsky, *The Conceptual Foundations of Scientific Thought* (81).

3. See *Timaeus* (91b) and Kerenyi's essay in Radin, where Kerenyi highlights this distinction between Plato's concern and the trickster's identity by virtue of the industry of his phallus alone, the latter being, to Kerenyi, "the trickster's double and alter ego" (180 *passim*).

4. The line is from *Song of the South*, Disney's version of the trickster tale, but it provides a most poignant contrast in any event, though Br'er Bear's line precisely as presented here is no where to be found in Joel Chandler Harris as far as I can tell. The attending contrast of Hermes and Hercules I have pilfered from Kerenyi (176).

5. The Eshu story of different colored hats is found in a number of places. See Lewis Hyde 238–40, Gates 32–35, versions of all stripe on the internet, and Pelton, who offers not only an extended analysis of the story but applies it most thoroughly to an overall context of the trickster's purpose (127–63).

6. In most cases and cultures the trickster is male, so I have invariably referred to the trickster as "he," especially given the prominence of the phallus in so many of his great doings and undoings, indeed of it being his "double and alter ego." However, true to his nature, the trickster's gender is sometimes more ambiguous. For instance, there are figurines of Eshu showing him/her back to back, not simply as two figures joined but as one, possessing the sexual organs of both sexes. As such, Eshu is not at all genderless, for that would defeat his sexual dynamic, but is, as Gates says, "dual-gendered," embracing in his/her person aspects of both sexes as complementary rather than as opposing forces (29–30). This capacity for inclusion and completion, the merging in the borderland of would-be opposites, those of man and woman among others, is in keeping with the trickster's nature in all matters.

7. And Poulakos goes further. Through an "analogical reading" of "Helen," he claims she "personifies rhetoric," that she and rhetoric in their common purpose share common characteristics: "both are attractive, both are unfaithful, and both have a bad reputation" ("Gorgias's" 5).

8. In this sense specifically, Helen is the consummate trickster (or the work of a trickster), surely a shape-shifter (or shape-shifted) by virtue of the power of language. Indeed, given the work that words make of her "reality," it is difficult to imagine a more fitting subject. See Bettany Hughes's rich and very engaging discussion of the changing and diverse images of Helen given over her long history and mythology, those of goddess, princess, and whore

among many others. Especially in her role as sex goddess, as Hughes develops it, these many images of Helen blend only to disperse once again, revealing and concealing, as Helen is then both mortalized and immortalized simultaneously in an allure of the flesh so intense as to transcend the flesh.

9. The reference is from *Being and Time*, where Heidegger's idea of "idle talk" is first developed, and where he says that as "Dasein maintains itself in idle talk, it is—as Being-in-the-world—cut off from its primary and primordially genuine relationships-of-Being towards the world, towards Dasein-with, and towards its very Being-in" (214). The passage not only marks the nature of "inauthenticity" developed in *Being and Time* but also forecasts Heidegger's later emphasis on Being, specifically on the "nonhuman" as constituting the critical ingredient of *dasein*.

10. The phrase is Lewis Hyde's apt variation of one from Hesiod's *Theogony*. The Muses accuse the poet of being "mere belly," a "wretched thing of shame" so burdened with the needs of the flesh that he is unable to grasp the truth, much less tell it. Hyde uses "belly free" and "belly free truth" regarding efforts and goals to overcome the trickster's influence (63–70). Though Hesiod himself is so much a part of the mythos of the trickster that he could scarcely identify it as such, he knows enough to claim the Muses "breathed a voice" into him "to sing the story" of past and future things. Of course, the purpose for raising the subject here is to say that in grasping "immortal truths" we must first be free of the wants and needs of the belly, of the flesh altogether, and achieving that, according to most readings of Plato, places us outside the province of language and rhetoric and into a transcendental realm of complete spirituality. And a realm, we can only surmise, also devoid of the breath of life.

11. See Diamond's "Introductory Essay" where he says in the primitive world of the trickster religion is never "connected with the hope of otherworldly reward" and "heaven" is conceived essentially as "the double of the earth." Nor, in this context, does religion function as a means of "evading moral contradiction" but embraces instead "the principle of ambiguity" (xii). According to Diamond these beliefs are then displaced by Christian beliefs in the immortal soul, eternal torment or reward in an afterlife and, above all, by the principle of the absolute distinction of good and evil in both this world and the next. Diamond traces this shift, in both its intellectual and ethical dimensions, back to Plato as much as any of the Church Fathers (xi—xxii). This transition, of course, is monumental, underscoring much of what I have been attempting to express in this book in terms of Being, language, and the Gorgian *kairos*.

12. We might take Heidegger at his word and see his phenomenology based as much in his purpose to retrieve Greek thought in its "authentic" nature as it is on the German Romantic tradition or even Continental philosophy as such. In *Truth and Method* Gadamer relates the difference between

Husserl and Heidegger in the way Heidegger himself attempted to draw the difference, claiming that Husserl's phenomenology was a methodological continuation of modern tendencies, whereas Heidegger differed from Husserl and all others in conceiving a phenomenology of retrieval that projects his work "not so much as the fulfillment of a long prepared development as, rather, a return to the beginnings of Western philosophy and the revival of the long-forgotten Greek argument about '[B]eing'" (227). We have seen this argument before from Gadamer (*Heidegger's* 142), but here it is pivotal to an adequate understanding of Heidegger's philosophy as it unfolds relative to *doxa*, language, and "the piety of thinking."

13. Once more, I choose Susanne Langer's translation of von Humboldt in Cassirer (9), as it possesses not only a greater felicity of expression but a far more precise meaning (insofar as there can be a difference) than I can possibly garner from the original German. See von Humboldt (53).

14. Again, I suspect the matter of *techne* is not so easily settled in Heidegger. The particular interpretation above follows Bruns's understanding (174–76) of Heidegger's statement in *Nietzsche I*, where Heidegger makes a distinction between *techne* in its original Greek sense, meaning the "knowledge that supports and conducts every human irruption into the midst of beings" and the later, "misuse" of the term to mean "technical," simply "making and manufacturing as such" (81). As will become apparent, Heidegger's treatment of *techne* can be problematic, being not only terribly complicated but, as many things Heideggerian, seeming to change and grow over time, as he returns to the subject again and again. Be that as it may, the proviso here—and an enormous one—is that *techne*, in its difference from *physis*, is ultimately the means or knowledge in the Heideggerian perspective through which we allow the power of *physis* to be expressed through us in our openness to Being, thus a power in which we participate by virtue of this *techne*.

15. See *On the Way to Language* (107). The importance of the "relation of all relations" cannot be overstated, being of great significance to everything Heidegger has to say about language. By means of this idea we can begin to appreciate the connection Heidegger draws between language and the ontological difference. William Richardson says "we must conceive the (ontological) difference as a scission (*-Shied*) between (*Unter-*) Being and beings [things] that refers them to each other *by the very fact that* it cleaves them in two" (579). Thus, the "scission" is at the same time "unifying," resulting in the paradoxical conclusion on Heidegger's part that in Saying what is hailed *and* what hails is the "scission" itself. Language, then, comes to presence *as* the unifying scission, and in this light we might understand Heidegger's cryptic statement that "the word itself is the relation which in each instance retains the thing within itself in such a manner that it 'is' a thing" (*Way* 66). We can think of the connection in a less technical sense, seeing the process as pure metaphor, as language connects things that come into being only

by virtue of their referring to each other, only by virtue of language as the metaphor creating their reality. Or we can think of the connection in simpler terms still, in terms of the trickster who personifies the ambiguity between what is within and what is without, making each derivative of their relation, as all things then are a matter of process and therefore not exactly "things" at all but events that transpire in the tension of the "betwixt-and-between."

16. Many years ago, as the result of an inner inflammation of the left eye, I underwent a vitrectomy and, subsequently, a lens implant. After the second surgery I remarked to my ophthalmologist how much clearer, yet more bluish, things appeared through the artificial lens. With great alacrity he informed me that that is how the world really is, what it really looks like instead of the distorted way I see it through my other eye with the original lens provided by nature. So now, I surmise, I see the world simultaneously half as it really is through my left eye and half as it is clouded and crusty because of nature's old, yellowed lens darkened by many years of hard use in my right eye. It fits my disposition. Maybe even accounts for it. Eshu would understand.

17. See Richardson's discussion of "technicity" (374). Its difference from Heidegger's sense of *techne* presents the same problematic touched upon earlier and to be touched upon again, as in some contexts and interpretations Heidegger seems to imply no difference between the two.

18. In another sense, Heidegger's use of Aristotle is not so remarkable. Gadamar says that the value of Heidegger's interpretations of Aristotle "lie in their ability to wipe away the scholastic overlay" and in that manner to serve as a model of his hermeneutical methods (*Heidegger's* 141). Perhaps of even greater importance is that these interpretations are pivotal to Heidegger's method of "destruction," as Gadamer claims they then served "as a springboard for his inquiry into the pre-Socratic beginnings" (144).

19. *The Question Concerning Technology* contains one of Heidegger's more detailed discussions of *techne*, though it is but an extension of that given in *Nietzsche*, Volume I (81). The crucial difference, however, is that in *The Question Concerning Technology* the distinction between *techne* and technicity is achieved through Heidegger's emphasis on language, and seemingly language alone in his explanation that "*techne* does not lie at all in making and manipulating nor in using of means, but rather in . . . revealing," specifically in revealing "whatever does not bring itself forth and does not yet lie before us." In this sense, as *techne* "belongs to bringing-forth," it belongs to *poiesis*, being "something poietic" (12–13).

20. On the difference between *techne* and *poiesis* see Veronique Foti's *Heidegger and the Poets*, where, it seems, there is no difference at all but a virtual congruence of the two; except as she very helpfully presents the essential "problematic" of *techne*, "with its double kinship to *poiesis*, on the one hand, and contemporary technicity on the other" (xvi). It is, of course, the former

meaning of *techne*, in its kinship to *poiesis*, that Foti exploits, defining it, as she does *poiesis*—and as Heidegger defines each—as a "bringing forth into unconcealment."

21. As many Greek words, the meaning of *dike* is imprecise in translation, and seems especially so after Heidegger has done his work of etymology. Apparently in his use of the word *dike* retains its meaning as "justice," not strictly in terms of divisions between right and wrong, good and evil, but in the more general sense as balance or harmony. For instance, in this sense killing one's father and marrying one's mother is a violation of *dike*, the "justice" of the natural order of things. This meaning also reflects what is perhaps the more Homeric sense of the term, as it has to do with the normal course of nature, certainly of the power of *physis* where ideals of right and wrong and true and false are dependent on the archaic sense of *doxa*, of matters of custom, usage, and tradition yet also of growth and becoming. In this sense *dike* is also associated with deception, not only "because evil is a necessary part of cosmic Becoming and of human life," as Untersteiner says, but "because the universe is irrational" (112).

22. It is an old, familiar theme. Poets and rhapsodes, Homer in particular, were often taken to task in antiquity for their "deceptions." Indeed, on the authority of Hieronymus (of Rhodes?), we have it that Pythagoras descended to Hades and saw Homer and Hesiod suffering unspeakable horrors in punishment for the lies they told of the gods (Diogenes 347). The more general implications of this coincidence of poetry and deception in the ancient world, in conjunction with a very detailed discussion of the place of *aletheia* in this connection, are explored in Louise Pratt's *Lying and Poetry from Homer to Pindar*.

23. In his discussion of Georg Trakl, Heidegger discusses the soul as it is presented in Trakl's poems as opposed to the sense of soul in Plato's works. In each, the soul is something "strange" on earth, yet in the former "the soul seeks the earth" and "does not flee from it," in this way fulfilling its nature "in her wandering to seek the earth so that she may poetically build and dwell upon it, and thus be able to save the earth as earth" (*Way* 163). I have cited the passage before and do so again because the distinction is vital also in the above context. In Plato's doctrine the soul is regarded as imperishable, supraterrestrial, while in the other the soul's (the *psyche*'s) destiny is earth, morality, language.

24. Plato says the "Field of *Lethe*" or "Plain of Oblivion" lies within the Underworld, a region crossed by the River of Forgetfulness whose waters, if we drink of them too deeply, cause the truth of Being or knowledge of reality to be concealed or forgotten (*Republic* 621a—d). Heidegger uses the paradigm to express the idea of "Being" without revelation or disclosure, of total "withdrawing and concealing" where "everything disappears." Appro-

priately, it is then the place where "the uncanny dwells in a peculiar exclusivity," thus being preeminently the "demonic" (*Parmenides* 118–19).

25. Werner Jaeger's work is most helpful in understanding the Greek *paideia*, especially as I am attempting to present it here in the context of the above interpretation of *doxa*. He discusses *doxa* generally in the terms of the philosophers, Parmenides in particular, and presents it in the conventional sense of "opinion," yet he shares with Heidegger the peculiar Romantic attachment to ancient Greece, and here the connection with Heidegger's sense of *doxa* is explicit in his discussion of the Homeric *paideia*, of the Greek purpose to keep alive the tradition of glory, the great deeds of men and gods. The first of Jaeger's volumes of significance is *Early Christianity and the Greek Paideia*; and the second, more pertinent to our purposes, is *Paideia: The Ideals of Greek Culture*, Vol. 1, *Archaic Greece: The Mind of Athens*. In terms of this theme of the Greek *paideia*, I doubt that few other works, at least none by Heidegger, give a better understanding of what Greek life and thought was in its "original beginning."

26. Except for its singular emphasis on the ontological aspect of our nature, the idea here is similar or very much the same as Arthur Koestler's of "cognitive dissonance." Janice Lauer was the first to apply the idea to rhetorical studies and only now, at the end of this book, I make reference to her work when, in fact, this whole project comes at her inspiration, both intellectually and professionally.

Works Cited

Arendt, Hannah. *The Life of the Mind*. Vol 1. *Thinking*. Vol 2. *Willing*. One-vol. edition. New York: Harcourt Brace Jovanovich, 1971.
Aristotle. *The Basic Works of Aristotle*. Ed. Richard McKeon. Trans. E. M. Edgehill, *et al*. New York: Harper & Row, 1941.
—. *Nicomachean Ethics*. Trans. L.H.G. Greenwood. New York: Arno, 1973.
—. "On Interpretation (*De Interpretatione*)." *The Basic Works of Aristotle*. Ed. Richard McKeon. Trans. E. M. Edgehill, *et al*. New York: Harper & Row, 1941. 40–61.
—. "On Melissus, Xenohanes, Gorgias." *Minor Works*. Trans. W. S. Hett. Cambridge: Harvard UP, 1963.
—. "On Sophistical Refutations (*De Sophisticis Elenchis*)." Trans. W.A. Pickard-Cambridge. *The Basic Works of Aristotle*. Ed. Richard McKeon. Trans. E. M. Edgehill, *et al*. New York: Harper & Row, 1941. 207–212.
—. *The Rhetoric and the Poetics of Aristotle*. Trans. W. Rhys Roberts. Ed. Friederich Solmsen. New York: Random House, 1954.
Armstrong, A. MacC. "The Fullness of Time." *Philosophic Quarterly* 6 (1956): 209–22.
Atwell, Janet M. "*Techne*." *Encyclopedia of Rhetoric and Composition: Communication from Ancient Times to the Information Age*. Ed. Theresa Enos. New York: Garland, 1996. 719.
Ballif, Michelle. *Seduction, Sophistry, and the Woman with the Rhetorical Figure*. Carbondale: Southern Illinois UP, 2001.
Bambach, Charles. *Heidegger's Roots: Nietzsche, National Socialism, and the Greeks*. Ithaca: Cornell UP, 2003.
Barnes, Jonathan. *Early Greek Philosophy*. New York: Penguin, 1987.
—. *The Presocratic Philosophers*. London: Routedge & Kegan Paul, 1982.
Barrett, William. *Irrational Man*. Garden City: Doubleday, 1958.
Baumlin, James S. "Decorum, Kairos, and the 'New' Rhetoric." *PRE/TEXT* 5 (1984): 171–83.
Beaufret, Jean. "Heraclitus and Parmenides." Maly and Emad 69–88.
Bernasconi, Robert. *The Question of Language in Heidegger's History of Being*. Atlantic Highlands, NJ: Humanities International, 1985.
Bersani, Leo. "Variations on a Paradigm." Rev. of *The Sense of an Ending* by Frank Kermode. *New York Times Book Review* 11 June 1967: (6, 45).

Bitzer, Lloyd. "The Rhetorical Situation." *Philosophy and Rhetoric* 1 (1968): 3–17.
Bizzell, Patricia, and Bruce Herzberg. *The Rhetorical Tradition: Readings from Classical Times to the Present.* Boston: St. Martin's, 1990.
Black, Edwin. "Plato's View of Rhetoric." *Plato: True and Sophistic Rhetoric.* Ed. Keith V. Erickson. Amsterdam: Rodopi, 1979. 171–91.
Black Elk. *Black Elk Speaks: Being the Life Story of a Holy Man of the Oglala Sioux.* As told to John Neihardt. Lincoln: U of Nebraska P, 2000.
Bourdieu, Pierre. *Outline of a Theory of Practice.* Trans. Richard Nice. New York: Cambridge UP, 1977.
Brock, Werner. "An Account of 'Being and Time.'" Martin Heidegger. *Existence and Being.* Ed. Brock. South Bend: Regency/Gateway, 1979. 11–116.
Bruns, Gerald L. *Heidegger's Estrangements: Language, Truth, and Poetry in the Later Writings.* New Haven: Yale UP, 1981.
Bruzina, Ronald. "Heidegger on the Metaphor and Philosophy." *Heidegger and Modern Philosophy.* Ed. Michael Murray. New Haven: Yale UP, 1978. 184–200.
Bryant, Donald C. "Rhetoric: Its Function and Its Scope." *Rhetoric: A Tradition in Transition.* Ed. Walter R. Fisher. East Lansing: Michigan UP, 1974. 195–230.
Buber, Martin. *The Knowledge of Man: A Philosophy of the Interhuman.* Trans. Maurice Friedman and Ronald Gregor Smith. New York: Harper & Row, 1965.
Burnet, John. "The Socratic Doctrine of the Soul." *Proceedings of the British Academy* 7 (1915–16): 235–59.
Byatt, A.S. "Connoisseur of Order." Rev. of *The Sense of an Ending* by Frank Kermode. *New Statesman* 4 Aug 1967: 146.
Campbell, Joseph. *The Masks of God: Primitive Mythology.* New York: Viking, 1959.
Camus, Albert. *The Myth of Sisyphus and Other Essays.* Trans. Justin O'Brien. New York: Vintage, 1955.
Caputo, John. *Demythologizing Heidegger.* Bloomington: Indiana UP, 1993.
Caputo, Philip. *A Rumor of War.* New York: Holt, Rinehardt and Winston, 1977.
Carter, Michael. *Art and the Genesis of Discourse.* Diss. Purdue U, 1986.
—. "*Stasis* and *Kairos*: Principles of Social Construction in Classical Rhetoric." *Rhetoric Review* 7.1 (1988): 97–112.
Cascardi, Anthony J. "The Place of Language; or, The Uses of Rhetoric." *Philosophy and Rhetoric* 16 (1983): 217–27.
Cassirer, Ernst. *Language and Myth.* Trans. Susanne K. Langer. New York: Harper & Row, 1946.
Chamberlain, Joshua Lawrence. *"Bayonet! Forward": My Civil War Reminiscences.* Gettyburg: Stan Clark Military, 1994.

Chatman, Seymour. *Story and Discourse: Narrative Structure in Fiction and Film*. Ithaca: Cornell UP, 1978.
Cicero. *On The Orator*. Book 3. On Fate. Stoic Paradoxes. The Divisions of Oratory. Trans. H. Rackham. Cambridge: Harvard UP, 1942.
—. *Orator*. Trans. H.M. Hubbell. Cambridge: Harvard UP, 1949.
The Civil War. Dir. Ken Burns. 1990. DVD. Florentine Films, 2002.
Claus, David. *Toward the Soul: An Inquiry into the Meaning of (Psyche) before Plato*. New Haven: Yale UP, 1981.
Cleve, Felix M. *The Giants of Pre-Socratic Greek Philosophy*. The Hague: Martinus Nijhoff, 1969.
Consigny, Scott. *Gorgias: Sophist and Artist*. U of South Carolina P, 2001.
—. "The Styles of Gorgias." *Rhetoric Society Quarterly* 22 (1992): 43–53.
Cornford, Francis Macdonald. *From Religion to Philosophy: A Study in the Origins of Western Speculation*. Princeton: Princeton UP, 1991.
Cook, Arthur Bernard. *Zeus: A Study in Ancient Religion*. Cambridge: Cambridge UP, 1925.
Coulter, James A. "The Relation of the *Apology of Socrates* to Gorgias's *Defense of Palamedes* and Plato's Critique of Gorgianic Rhetoric." Erickson 31–69.
Crowley, Sharon. "Of Gorgias and Grammatology." *College Composition and Communication* 30 (1979): 279–84.
Crusius, Timothy W. *A Teacher's Introduction to Philosophical Hermeneutics*. Urbana, IL: NCTE, 1991.
D'Angelo, Frank. *Composition in the Classical Tradition*. Boston: Allyn & Bacon, 2000.
Dastur, Francoise. *Heidegger and the Question of Time*. Trans. Francois Raffoul and David Pettigrew. Atlantic Highlands: Humanities, 1998.
—. "Language and *Ereignis*." *Reading Heidegger*. Ed. John Sallis. Bloomington: Indiana UP, 1993. 355–69.
Deloria, Vine, Jr. *God is Red*. New York: Dell, 1973.
DeMallie, Raymond J., ed. *The Sixth Grandfather: Black Elk's Teachings Given to John Neihardt*. Lincoln: U of Nebraska P, 1984.
de Man, Paul. "Semiology and Rhetoric." *Textual Strategies: Perspective in Post-Structuralist Criticism*. Ed. Josue V. Harari. Ithaca: Cornell UP, 1979. 121–40.
Derrida, Jacques. *Dissemination*. Trans. Barbara Johnson. Chicago: U of Chicago P, 1981.
—. "Heidegger's Ear: Philopolemology." Trans. John P. Leavey, Jr. *Reading Heidegger*. Ed. John Sallis. Bloomington: Indiana UP, 1993. 163–218.
—. *Margins of Philosophy*. Trans. Alan Bates. Chicago: U of Chicago P, 1982.
—. *Of Grammatology*. Trans. Gayatri Chakravorty Spivak. Baltimore: Johns Hopkins UP, 1998.

—. *Of Spirit: Heidegger and the Question.* Trans. Geoffrey Bennington and Rachel Bowley. Chicago: U of Chicago P, 1989.

—. *Speech and Phenomena, and Other Essays on Husserl's Theory of Signs.* Trans. David B. Allison. Evanston: Northwestern UP, 1973.

—. "Structure, Sign, and Play in Discourse of the Human Sciences." *Writing and Difference.* Trans. Alan Bates. Chicago: U of Chicago P, 1978. 278–94.

de Romilly, Jacqueline. *Magic and Rhetoric in Ancient Greece.* Cambridge: Harvard UP, 1975.

de Vogel, C. J. *Pythagoras and Early Pythagoreanism.* The Hague: Royal Van-Gorcum, 1959.

Diamond, Stanley. "Introductory Essay: Job and the Trickster." Paul Radin. *The Trickster: A Study in American Indian Mythology.* New York: Schocken Books, 1972. xi—xxii.

Diels, Hermann, and Walther Kranz, eds. *Die Fragmente der Vorsokratiker* 3 vols. Tubingen: Niemeyer, 1953.

Dieter, Otto. *"Stasis." Speech Monographs.* 17 (1950): 345–69.

Dillon, John and Tania Gergel, eds. and trans. *The Greek Sophists.* New York: Penguin, 2003.

Diogenes Laertius. *The Lives and Opinions of Eminent Philosophers.* Trans. C. D. Yonge. London: Henry G. Bohn, 1853.

Dodds, E.R. *The Greeks and the Irrational.* Berkeley: U of California P, 1968.

Dostal, Robert. "Time and Phenomenology in Husserl and Heidegger." Guignon 141–69.

Dreyfus, Herbert L. "Heidegger's Hermeneutic Realism." *The Interpretative Turn: Philosophy, Science, Culture.* Ed. David R. Hiley, James F. Bohman, and Richard Shusterman. Ithaca: Cornell UP, 1991. 25–41.

Durant, Will. *The Story of Philosophy.* New York: Washington Square, 1952.

Elias, Julius A. *Plato's Defence of Poetry.* Albany: SUNY P, 1984.

Ellenberger, Henri. "A Clinical Introduction to Psychiatric Phenomenology and Existential Analysis." *Existence: A New Dimension in Psychiatry and Psychology.* Ed. Rollo May, Ernest Angel, and Henri Ellenberger. New York: Basic, 1958. 92–126.

Engell, Richard A. "Implications for Communication of the Rhetorical Epistemology of Gorgias of Leontini." *Western Speech* 37 (1973): 175–81.

Enos, Richard Leo. "The Classical Period." *The Present State of Scholarship in Historical and Contemporary Rhetoric.* Ed. Winifred Bryan Horner. Columbia: U of Missouri P, 1983: 10–39.

Enos, Richard Leo. "The Epistemology of Gorgias's Rhetoric: A Re-examination." *The Southern Speech Communication Journal* 42 (1976): 35–51.

—. "Inventional Constraints on the Technographers of Ancient Athens: A Study of *Kairos*." *Rhetoric and Kairos: Essays in History, Theory, and Prax-*

is. Ed. Phillip Sipiora and James S. Baumlin. Albany: SUNY P, 2002. 77–88.
Enos, Theresa, ed. *Encyclopedia of Rhetoric and Composition: Communication from Ancient Times to the Information Age*. New York: Garland, 1996.
Erickson, Keith V., ed. *Plato: True and Sophistic Rhetoric*. Amsterdam: Rodopi, 1979.
"Fallen Heroes: A Rendezvous with History." *Archaeology*. 16 Feb. 2000. n. pag. Web. 25 May 2004.
Faulkner, William. *Light in August*. New York: Modern, 1968.
FitzGerald, Frances. *Fire in the Lake: The Vietnamese and the Americans in Vietnam*. New York: Little, Brown, & Co., 1972.
Foti, Veronique M. *Heidegger and the Poets: Poiesis/Sophia/Techne*. Atlantic Highlands: Humanities Press International, 1995.
Frankfort, Henri, *et. al. Before Philosophy: The Intellectual Adventure of Ancient Man*. Baltimore: Penguin, 1967.
Freeman, Kathleen, trans. *Ancilla to the Pre-Socratic Philosophers*. Cambridge: Harvard UP, 1948.
Furley, D. J. "The Early History of the Concept of the Soul." *Bulletin of the Institute of Classical Studies of the University of London* 3 (1956): 1–18.
Gadamar, Hans-Georg. *Heidegger's Ways*. Trans. John Stanley. Albany: SUNY P, 1994.
—. "Man and Language." *Philosophical Hermenuetics*. Trans. and ed. David E. Linge. Berkeley: U of California P, 1976. 59–68.
—. *Truth and Method*. Trans. Garrett Barden and John Cumming. New York: Crossroads, 1985.
—. "The Western View of the Inner Experience of Time and the Limits of Thought." *Time and the Philosophies*. Paris: UNESCO, 1977. 33–48.
Gage, John T. "A New Way into the Phaedrus and Composition: A Review." *Rhetoric Society Quarterly* 11 (1981): 29–34.
Gaonkar, Dilip. "Plato's Critique of Protagoras's Man-Measure Doctrine." *PRE/TEXT* 10.1–2 (1989): 71–80.
Gates, Henry Louis, Jr. *The Signifying Monkey: A Theory of African-American Literary Criticism*. New York: Oxford UP, 1988.
Gay, Peter. *Weimar Culture, the Outsider as Insider*. New York: Harper Torchbook, 1970.
Glover, Carl. "*Kairos* and Composition: Modern Perspectives on an Ancient Idea." Diss. U of Louisville, 1990.
Gomperz, Theodor. *Greek Thinkers: A History of Ancient Philosophy*. Trans. Laurie Magnus. London: Wiliam Cloves & Sons, 1964.
Gorgias. "82. Gorgias: A. *Leben und Lehre;* B. *Fragmente;* C. *Imitation*." Diels and Kranz 271–307.
---. "82. Gorgias: A. Life and Teachings; B. Fragments; C. Imitation." Trans. George Kennedy. Sprague 30-67.

---. "82. Gorgias of Leontini." Trans. Kathleen Freeman. Freeman 127–139.
Graeser, Andreas. "On Language, Thought, and Reality in Ancient Greek Philosophy." *Dialectica* 31 (1977): 360–88.
Gramm, Kent. *Gettysburg: A Meditation on War and Values*. Bloomington: Indiana UP, 1994.
Grassi, Ernesto. *Heidegger and the Question of Renaissance Humanism: Four Studies*. Binghamton: Medieval and Renaissance Texts, 1983.
—. "Italian Humanism and Heidegger's Thesis of the End of Philosophy." *Philosophy and Rhetoric* 13.2 (1980): 79–98.
—. *Rhetoric as Philosophy: The Humanistic Tradition*. Trans. John Kroise and Azizeh Azodi. Carbondale: Southern Illinois UP, 2001.
Gray, Glenn J. Introduction. *What is Called Thinking*. By Martin Heidegger. Trans. J. Glenn Gray. New York: Harper & Row, 1968. vi—xvi.
Gronbeck, Bruce E. "Gorgias on Rhetoric and Poetic: A Rehabilitation." *The Southern Speech Communication Journal* 38 (1972): 27–38.
Guignon, Charles B., ed. *The Cambridge Companion to Heidegger*. New York: Cambridge UP, 1993.
Guthrie, W. K. C. *Socrates*. Cambridge: U of Cambridge P, 1971.
—. *The Sophists*. Cambridge: U of Cambridge P, 1969.
Haar, Michel. "The Enigma of Everydayness." Trans. Michael B. Naas and Pascale-Anne Brault. *Reading Heidegger*. Ed. John Sallis. Bloomington: Indiana UP, 1993. 20–28.
—. *The Song of the Earth: Heidegger and the Grounds of the History of Being*. Trans. Reginald Lilly. Bloomington: Indiana UP, 1993.
Halloran, S.M. "On the End of Rhetoric, Classical and Modern." *College English* 36 (1975): 621–31.
Harries, Karsten. "Fundamental Ontology and the Search for Man's Place." *Heidegger and Modern Philosophy*. Ed. Michael Murray. New Haven: Yale UP, 1978. 65–79.
—. "Heidegger as a Political Thinker." *Heidegger and Modern Philosophy*. Ed. Michael Murray. New Haven: Yale UP, 1978. 304–28.
Harris, Joel Chandler. *The Complete Tales of Uncle Remus*. New York: Houghton Mifflin, 2002.
Havelock, Eric. *The Liberal Temper in Greek Politics*. New Haven: Yale UP, 1957.
—. *The Literate Revolution in Greece and Its Cultural Consequences*. Princeton: Princeton UP, 1982.
—. *Preface to Plato*. Cambridge: Harvard UP, 1963.
—. "The Socratic Self as It Is Parodied in Aristophanes' *Clouds*." *Yale Classical Studies* 22 (1972): 1–18.
Heidegger, Martin. *The Basic Problems of Phenomenology*. Trans. Albert Hofstadter. Bloomington: Indiana UP, 1982.
—. *Basic Writings*. Ed. David Farrell Krell. New York: Harper & Row, 1977.

—. *Being and Time*. Trans. John Macquarrie and Edward Robinson. New York: Harper & Row, 1962.
—. *The Concept of Time*. Trans. William McNeill. Cambridge: Blackwell, 1992.
—. *Discourse on Thinking*. Trans. John M. Anderson and E. Hans Freund. New York: Harper & Row, 1966.
—. *Early Greek Thinking*. Trans. David Farrell Krell and Frank Capuzzi. San Francisco: Harper & Row, 1975.
—. *Einfuhrung in die Metaphysik*. Tubingen: Max Niemeyer, 1966.
---. *The Essence of Reasons*. Trans. Terrence Malick. Evanston: Northwestern UP, 1969.
—. *Existence and Being*. Ed. Werner Brock. South Bend: Regnery/Gateway, 1949.
—. *Identity and Difference*. Trans. Joan Stambaugh. New York: Harper & Row, 1979.
—. *An Introduction to Metaphysics*. Trans. Ralph Manheim. New York: Doubleday, 1959.
—. "Letter on Humanism." Trans. Frank A. Capuzzi and J. Glenn Gray. David. *Basic Writings*. Ed. David Farrell Krell. New York: Harper & Row, 1977. 193–242.
—. *Nietzsche*. Trans. David Farrell Krell. 4 vols. San Francisco: Harper and Row, 1979.
—. "On the Essence of Truth." Trans. R. F. C. Hull and Alan Crick. Heidegger. *Existence and Being*. Ed. Werner Brock. South Bend: Regnery/Gateway, 1949. 292–324.
—. *On the Way to Language*. Trans. Joan Stambaugh. New York: Harper & Row, 1969.
—. *On Time and Being*. Trans. Joan Stambaugh. New York: Harper & Row, 1972.
—. *Parmenides*. Trans. Andre Schuwer and Richard Rojcewicz. Bloomington: Indiana UP, 1992.
—. *The Piety of Thinking*. Trans. James G. Hart and John C. Maraldo. Bloomington: Indiana UP, 1976.
—. "Plato's Doctrine of Truth." Trans. John Barlow. *Philosophy in the Twentieth Century*. Vol 3. Ed. William Barrett and Henry D. Aiken. New York: Random House, 1962.
—. *Poetry, Language, and Thought*. Trans. Albert Hofstadter. New York: Harper & Row, 1972.
—. Preface. Trans. William Richardson. *Heidegger: Through Phenomenology to Thought*. By William Richardson. The Hague: Martinus Nijhoff, 1974. viii—xxiii.
—. *The Principle of Reason*. Trans. Reginald Lilly. Bloomington: Indiana UP, 1996.

—. *The Question of Being*. Trans. Jean Wilde and William Kluback. New Haven: Twayne, 1958.
—. *The Question Concerning Technology and Other Essays*. Trans. William Lovitt. New York: Harper & Row, 1977.
—. "The Way Back into the Ground of Metaphysics." Trans. Walter Kaufmann. *Existentialism from Dostoevsky to Sartre*. Ed. Walter Kaufmann. Cleveland: World, 1956. 206–21.
—. *What is Called Thinking*. Trans. J. Glenn Gray. New York: Harper & Row, 1968.
—. "What is Metaphysics." Trans. R. F. C. Hull and Alan Crick. *Existence and Being*. Ed. Werner Brock. South Bend: Regnery/Gateway, 1949. 325–61.
—. *What is Philosophy*. Trans. William Kluback and Jean T. Wilde. New York: Twayne, 1958.
Heidegger, Martin and Eugen Fink. *Heraclitus Seminar 1966/67*. Trans. Charles H. Seibert. Tuscaloosa: U of Alabama P, 1979.
Heine, Steven. *Existential and Ontological Dimensions of Time in Heidegger and Dogen*. Albany: SUNY P, 1985.
Heim, Michael R. "Philosophy as Ultimate Rhetoric." *The Southern Journal of Philosophy* 19 (1981): 181–95.
Hesiod. *The Work and Days. Theogony. The Shields of Herakles*. Trans. Richmond Lattimore. Ann Arbor: U of Michigan P, 1959.
Hill, Carolyn Erickson. "Changing Times in Composition Classes: *Kairos*, Resonance, and the Pythagorean Connection." Sipiora and Baumlin 211–25.
Hill, Douglas. "Trickster." *Man, Myth and Magic: An Illustrated History of the Supernatural*. Vol 21. Ed. Richard Cavendish. New York: Marshall Cavendish, 1970. 2881–85.
Holy Bible: Authorized (King James) Version. Philadelphia: National Bible P, 1970.
Homer. *The Homeric Hymns*. Trans. Apostolos N. Athanassakis. Baltimore: Johns Hopkins UP, 1976.
—. *The Iliad*. Trans. Robert Fitzgerald. New York: Anchor, 1989.
—. *The Iliad*. Trans. Richard Lattimore. Chicago: U of Chicago P, 1951.
—. *The Iliad*. Trans. W.H.D. Rouse. New York: Mentor Books, 1938.
Hughes, Bettany. *Helen of Troy: Goddess, Princess, Whore*. New York: Alfred A. Knopf, 2005.
Hunt, Everett Lee. "Plato on Rhetoric and Rhetoricians." *Quarterly Journal of Speech* 6 (1920): 33–53.
—. "On the Sophists." *The Province of Rhetoric*. Ed. Joseph Schwartz and John A. Rycenga. New York: Ronald, 1965. 69–84.
Hyde, Lewis. *Trickster Makes This World: Mischief, Myth, and Art*. New York: Farrar, Straus and Giroux, 1998.

Hyde, Michael J. "The Hermeneutic Phenomenon and the Authenticity of Discourse." *Visible Language* 2 (1983): 146–62.
Hyde, Michael J., and Craig R. Smith. "Hermenuetics and Rhetoric: A Seen But Unobserved Relationship." *Quarterly Journal of Speech* 65 (1979): 347–63.
—. "Rhetorically, Man Dwells: On the Making-Known Function of Discourse." *Communication* 7 (1983): 201–20.
Ijssling, Samuel. *Rhetoric and Philosophy in Conflict*. Trans. Paul Dunphy. The Hague: Martinus Nijhoff, 1976.
Jaeger, Werner. *Early Christianity and Greek Paideia*. Cambridge: Harvard UP, 1961.
—. *Paideia: The Ideals of Greek Culture. Archaic Greece: The Mind of Greece*. Vol. 1. Trans. Gilbert Highet. New York: Oxford UP, 1939.
Jaques, Elliot. *The Form of Time*. New York: Crane, Russak & Co., 1982.
Jarratt, Susan C. *Rereading the Sophists: Classical Rhetoric Refigured*. Carbondale, Southern Illinois UP, 1991.
—. "The Role of the Sophists in Histories of Consciousness." *Philosophy and Rhetoric* 23 (1990): 85–95.
—. "The First Sophists and the Uses of History." *Rhetoric Review* 6 (1987): 67–77.
—. "The Return of the Sophists." Rhetoric Society of America. Arlington TX. 15 June 1986. Reading.
Jaynes, Julian. *The Origin of Consciousness in the Breakdown of the Bicarmerl Mind*. Boston: Houghton Mifflin, 1976.
Johnson, Barbara. Introduction. *Dissementions*. By Jaques Derrida. Trans. Barbara Johnson. Chicago: U of Chicago P, 1981. vii—xxxiii.
Johnstone, Henry W. *Validity and Rhetoric in Philosophical Argument*. University Park: Dialogue, 1978.
Kant, Immanuel. *Critique of Pure Reason*. Trans. Norman Kemp Smith. New York: St. Martin's, 1929.
Kaufmann, Walter, ed. *Existentialism from Dostoevsky to Sartre*. Cleveland: World, 1956.
—. *Tragedy and Philosophy*. New York: Anchor, 1969.
Kazantzakis, Nikos. *The Last Temptation of Christ*. Trans. P. A. Bien. New York: Simon & Schuster, 1960.
Kearney, Richard. "Heidegger and the Possible." *Philosophical Studies* 27 (1980): 176–95.
Kelman, Harold. "The Auspicious Moment." *American Journal of Psychoanalysis* 29 (1969): 59–83.
Kennedy, George. *The Art of Persuasion in Greece*. Princeton: Princeton UP, 1963.
—. *Classical Rhetoric and Its Christian and Secular Tradition from Ancient to Modern Times*. Chapel Hill: U of North Carolina P, 1980.

Kerenyi, Karl. "The Trickster in Relation to Greek Mythology." Trans. R. F. C. Hull. *The Trickster: A Study in American Indian Mythology.* Ed. Paul Radin. New York: Schocken Books, 1972. 173–191.

Kerford, G. B. *The Sophistic Movement.* Cambridge: Cambridge UP, 1981.

Kermode, Frank. *The Sense of an Ending: Studies in the Theory of Fiction.* New York: Oxford UP, 1966.

Kinneavy, James. "Contemporary Rhetoric." *The Present State of Scholarship in Historical and Contemporary Rhetoric.* Ed. Winifred Bryan Horner. Columbia: U of Missouri P, 1983: 167–213.

—. "*Kairos* in Aristotle's Rhetoric." College Composition and Communication Conference. Nashville. 18 March 1994. Reading.

—. "*Kairos* in Classical and Modern Rhetorical Theory." *Rhetoric and Kairos: Essays in History, Theory, and Praxis.* Ed. Phillip Sipiora and James S. Baumlin. Albany: SUNY P, 2002. 58–76.

—. "*Kairos*: A Neglected Concept in Classical Rhetoric." Purdue University Rhetoric Seminar Paper. TS. 6 June 1984. Collection of Bernard Miller, Ypsilanti.

—. "*Kairos*: A Neglected Concept in Classical Rhetoric." *Rhetoric and Praxis: The Contributions of Classical Rhetoric to Practical Reasoning.* Ed. Jean Dietz Moss. Washington: Catholic U of America, 1986. 79–105.

—. "The Relation of the Whole to the Part in Interpretation Theory in the Composing Process." *Rhetoric and Composition: An Inquiry.* Ed. Donald McQuade. New York: McMillian, 1980. 1–23.

Kluback, William, and Jean T. Wilde. Introduction. *What is Philosophy.* By Martin Heidegger. Trans. William Kluback and Jean T. Wilde. New York: Twayne, 1958. 7–17.

Kisiel, Theodore. "The Language of the Event: The Event of Language." *Heidegger and the Path of Thinking.* Ed. John Sallis. Pittsburgh: Duquesene UP, 1970. 85–104.

—. Translator's Introduction. *Heidegger and the Tradition.* By Werner Marx. Trans. Theodore Kisiel and Murray Greene. Evanston: Northwestern UP, 1971. xvii—xxxiii.

Kittel, Gerhard, ed. *Theological Dictionary of the New Testament.* Grand Rapids: WM B. Eerdmans, 1965.

Kitto, H.D.F. *The Greeks.* Baltimore: Penguin, 1951.

Kockelmans, Joseph J., ed. and trans. *On Heidegger and Language.* Evanston: Northwestern UP, 1972.

—. "Language, Meaning, and Ek-sistence." *On Heidegger and Language.* Trans. and ed. Kockelmans. 3–32.

—. "Theoretical Problems in Phenomenological Psychology." *Phenomenology and the Social Science.* Vol. 1. Ed. Maurice Natanson. Evanston: Northwestern UP, 1973. 225–80.

Kohanski, Alexander S. *The Greek Mode of Thought in Western Philosophy.* Cranbury: Fairleigh Dickinson UP, 1984.
Krell, David Farrell. "Where Deathless Horses Weep." *Reading Heidegger.* Ed. John Sallis. Bloomington: Indiana UP, 1990. 95–106.
Krupat, Arnold. *The Voice in the Margin: Native American Literature and the Canon.* Berkeley: U of California P, 1989.
Kung, Hans. *Does God Exist? An Answer for Today.* Trans. Edward Quinn. Garden City: Doubleday, 1980.
Langan, Thomas. *The Meaning of Heidegger.* New York: Columbia UP, 1959.
Lauer, Janice, and Kelly Pender. *Invention in Rhetoric and Composition.* West Lafayette: Parlor, 2003.
Leff, Michael C. "The Forms of Reality in Plato's Phaedrus." *Rhetoric Society Quarterly* 11 (1981): 21–23.
Leitch, Vincent B. *Deconstructive Criticism: An Advanced Introduction.* New York: Columbia UP, 1983.
Lilly, Reginald. Translator's Preface. *Song of the Earth: Heidegger and the Grounds of the History of Being.* By Michel Haar. Trans. Reginald Lilly. Bloomington: Indiana UP, 1993. xv—xvi.
Loraux, Nicole. *Born of the Earth: Myth and Politics in Athens.* Trans. Selina Stewart. Ithaca: Cornell UP, 2000.
—. *The Invention of Athens: The Funeral Orations in the Classical City.* Trans. Alan Sheridan. New York: Cambridge UP, 1986.
Lunsford, Andrea A., ed. *Reclaiming Rhetorica: Woman in the Rhetorical Tradition.* Pittsburgh: U of Pittsburgh P, 1995.
Magliola, Robert R. *Derrida on the Mend.* West Lafayette: Purdue UP, 1984.
—. *Phenomenology and Literature.* West Lafayette: Purdue UP, 1977.
Mailloux, Steven, ed. *Rhetoric, Sophistry, Pragmatism.* New York: Cambridge UP, 1995.
Maly, Kenneth, and Parvis Emad, eds. *Heidegger on Heraclitus: A New Reading.* Lewiston, New York: Edwin Mellen, 1987.
Marcel, Gabriel. *Mystery of Being.* Vol. 1. *Reflection and Mystery.* Trans. G. S. Fraser. Vol. 2. *Faith and Reality.* Trans. Rene Hague. Chicago: Henry Regnery, 1960.
Mansfeld, J. "Heraclitus on the Psychology and Physiology of Sleep and on Rivers." *Mnemosyne* 20 (1967): 1–29.
Marx, Werner. *Heidegger and the Tradition.* Trans. Theodore Kisiel and Murray Greene. Evanston: Northwestern UP, 1971.
May, Rollo. "The Origins and Significance of the Existential Movement in Psychology." *Existence: A New Dimension in Psychiatry and Psychology.* May, Ernest Angel, and Henri Ellenberger, eds. New York: Basic, 1958. 3–36.

—. "Contributions of Existential Psychotherapy." *Existence: A New Dimension in Psychiatry and Psychology*. May, Ernest Angel, and Henri Ellenberger, eds. New York: Basic, 1958. 37–91.

May, Rollo, Ernest Angel, and Henri Ellenberger, eds. *Existence: A New Dimension in Psychiatry and Psychology*. New York: Basic, 1958.

McComiskey, Bruce. *Gorgias and the New Sophistic Rhetoric*. Carbondale: Southern Illinois UP, 2002.

McNeley, James Kale. *Holy Wind in Navajo Philosophy*. Tucson: U of Arizona P, 1997.

Mehta, J. L. *The Philosophy of Martin Heidegger*. New York: Harper & Row, 1971.

Merlan, Philip. "Time Consciousness in Husserl and Heidegger" *Philosophy and Phenomenological Research* 8 (1947): 23–53.

Merton, Thomas. "War and the Crisis of Language." *The Critique of War: Contemporary Philosophical Explorations*. Ed. Robert Ginsberg. Chicago: Henry Regency, 1969. 99–119.

Minar, Edwin L. "The Logos of Heraclitus." *Classical Philology* 34 (1939): 323–41.

Moline, Jon. *Plato's Theory of Understanding*. Madison: U of Wisconsin P, 1981.

Momaday, N. Scott. *House Made of Dawn*. New York: Harper, 1999.

—. *The Man Made of Words: Essay, Stories, Passages*. New York: St. Martin's, 1997.

—. "The Man Made of Words." *Literature of the American Indians: Views and Interpretations*. Ed. Abraham Chapman. New York: New American, 1975. 96–110.

—. "The Morality of Indian Hating." *The Man Made of Words: Essay, Stories, Passages*. New York: St. Martin's, 1997. 57–79.

—. Personal Interview. 22 March 1979.

—. *The Way to Rainy Mountain*. Albuquerque: U of Mexico P, 1977.

Moon, Sheila. *A Magic Dwells: A Poetic and Psychological Study of the Navaho Emergence Myth*. Middletown: Wesleyan UP, 1970.

Morrow, Glenn R. "Plato's Conception of Persuasion." *Plato: True and Sophistic Rhetoric*. Ed. Keith V. Erickson. Amsterdam: Rodopi, 1979. 339–54.

Mugerauer, Robert. *Heidegger's Language and Thinking*. Atlantic Highlands, NJ: Humanities P International, 1988.

Murphy, James J. *A Synoptic History of Classical Rhetoric*. New York: Random House, 1972.

Murray, Michael, ed. *Heidegger and Modern Philosophy*. New Haven: Yale UP, 1978.

Nietzsche, Friedrich. *Beyond Good and Evil: Prelude to a Philosophy of the Future*. Trans. Walter Kaufmann. New York: Random House, 1966.

—. *The Birth of Tragedy and the Case of Wagner.* Trans. Walter Kaufmann. New York: Random House, 1967.
—. *Thus Spoke Zarathustra.* Trans. R. J. Hollingdale. Baltimore: Penquin, 1961.
—. *Twilight of the Idols and the Anti-Christ.* Trans. R. J. Hollingdale. Baltimore: Penguin, 1968.
O'Brien, Tim. *Going After Cacciato.* New York: Dell, 1975.
—. *The Things They Carried.* New York: Penguin, 1991.
O'Connor, Flannery. *Mystery and Manners: Occasional Prose.* Ed. Sally and Robert Fitzgerald. New York: Farrar, Straus, and Giroux, 1969.
Okrent, Mark. *Heidegger's Pragmatism: Understanding, Being, and the Critique of Metaphysics.* Ithaca: Cornell UP, 1988.
Olafson, Frederick A. *Heidegger and the Philosophy of Mind.* New Haven: Yale UP, 1987.
Ong, Walter J. *Interfaces of the Word: Studies in the Evolution of Consciousness and Culture.* Ithaca: Cornell UP, 1977.
—. *Orality and Literacy.* New York: Methuen, 1982.
Osborne, Catherine. *Rethinking Early Greek Philosophy: Hippolytus of Rome and the Presocratics.* London: Duckworth, 1987.
Ott, Hugo. *Martin Heidegger: A Political Life.* Trans. Allan Blunden. London: HarperCollins, 1993.
Parain, Brice. *A Metaphysics of Language.* Trans. Mary Mayer. New York: Anchor, 1971.
Parkes, Graham, Ed. *Heidegger and Asian Thought.* Honolulu: U of Hawaii P, 1987.
Partee, Morris Henry. "Plato on the Rhetoric of Poetry." *Plato: True and Sophistic Rhetoric.* Ed. Keith V. Erickson. Amsterdam: Rodopi, 1979. 385–98.
—. "Plato's Theory of Language." *Foundations of Language* 8 (1972): 113–32.
Pelton, Robert D., *The Trickster in West Africa: A Study of Mythic Irony and Sacred Delight.* Berkeley: U of California P, 1980.
Pirsig, Robert. *Zen and the Art of Motorcycle Maintenance.* New York: Bantam, 1975.
Plato. *The Dialogues of Plato.* Trans. B. Jowett. Glasgow: Oxford UP, 1953.
—. *Phaedrus.* Trans. W.C. Helmbold and W.G. Rabinowitz. Indianapolis: Bobbs-Merrill Educational, 1956.
—. *Plato: The Collected Dialogues.* Ed. Edith Hamilton and Huntington Cairns. Trans. Benjamin Jowett, et al. Bollingen Series. Princeton: Princeton UP, 1971.
Pöggeler, Otto. "Being as Appropriation." Trans. Rüdiger H. Grimm. Murray 84–115.
—. *Martin Heidegger's Path of Thinking.* Trans. Daniel Magurshak and Sigmund Barber. Atlantic Highlands: Humanities Press International, 1989.

Poulakos, John. "Gorgias's *Encomium to Helen* and the Defence of Rhetoric." *Rhetorica* 1 (1983): 1–16.

—. "*Kairos* in Gorgias's Rhetorical Compositions." *Rhetoric and Kairos: Essays in History, Theory, and Praxis*. Ed. Phillip Sipiora and James S. Baumlin. Albany: SUNY P, 2002. 89–96.

—. "Rhetoric, the Sophists, and the Possible." *Communication Monographs* 51 (1984): 215–26.

—. *Sophistical Rhetoric in Classical Greece*. U of South Carolina P, 1995.

—. "Toward a Sophist Definition of Rhetoric." *Philosophy and Rhetoric* 16 (1983): 35–48.

Poulakos, Takis. "The Historical Intervention of Gorgias's Epitaphios: The Genre of Funeral Oration and the Athenian Institution of Public Burials." *PRE/TEXT* 10.1–2 (1989): 90–99.

Poulet, Georges. *Studies in Human Time*. Trans. Elliot Coleman. New York: Harper & Row, 1956.

Pratt, Louise H. *Lying and Poetry from Homer to Pindar: Falsehood and Deception in Archaic Greek Poetics*. Ann Arbor: U of Michigan P, 1993.

Proust, Marcel. *Remembrance of Things Past*. Trans. C.K. Scott-Moncrieff. 2 vols. New York: Random House, 1932–34.

Puech, Henri-Charles. "Gnosis and Time." *Man and Time: Papers from the Eranos Yearbooks*. Vol. 3. Ed. Joseph Campbell. Trans. Ralph Manheim. New York: Bollingen Foundation; Princeton UP, 1957. 38–84.

Quimby, Rollin W. "The Growth of Plato's Perception of Rhetoric."*Plato: True and Sophistic Rhetoric*. Ed. Keith V. Erickson. Amsterdam: Rodopi, 1979. 21–30.

Quintilian. *Institutio Oratoria*. Trans. H. E. Butler. Cambridge: Harvard UP, 1920.

Radin, Paul. *The Trickster: A Study in American Indian Mythology*. New York: Schocken Books, 1972.

Raffoul, Francois, and David Pettigrew. "Introduction to the English Edition." *Heidegger and the Question of Time*. By Francoise Dastur. Trans. Francois Raffoul and David Pettigrew. Atlantic Highlands: Humanities, 1998. xxi—xxviii.

Richards, I.A. *The Philosophy of Rhetoric*. New York: Oxford UP, 1936.

Remarque, Erich Maria. *All Quiet on the Western Front*. Trans. A. W. Wheen. New York: Ballantine, 1982.

Richardson, William J. *Heidegger: Through Phenomenology to Thought*. The Hague: Martinus Nijhoff, 1974.

Ricoeur, Paul. *The Conflict of Interpretations*. Trans. Kathleen McLaughlin. Evanston: Northwestern UP, 1974.

—. *The Rule of Metaphor. Multi-Disciplinary Studies of the Creation of Meaning in Language*. Trans. Robert Czerny. Toronto: U of Toronto P, 1977.

—. "The Task of Hermeneutics." Trans. David Pellauer. *Heidegger and Modern Philosophy.* Ed. Michael Murray. New Haven: Yale UP, 1978. 141–160.
Robinson, John M. "On Gorgias." *Exegesis and Argument.* Ed. E. N. Lee. The Netherlands: Koninklijke Van Gorcum, 1973. 49–61.
Rodhe, Erwin. *Psyche: The Cult of Souls and Belief in Immortality among the Greeks.* Trans. W. B. Hillis. Freeport: Book for Libraries, 1970.
Rorty, Richard. *Essays on Heidegger and Others.* New York: Cambridge UP, 1991.
—. *Philosophy and the Mirror of Nature.* Princeton: Princeton UP, 1970.
Rosen, Stanley H. "Heidegger's Interpretation of Plato." *Essays in Metaphysics.* Ed. Carl G. Vaught. University Park: Penn State UP, 1970: 51–77.
Rosenmeyer, Thomas G. "Gorgias, Aeschylus, and Apate." *American Journal of Philosophy* 76 (1955): 225–60.
Sallis, John. *Echoes After Heidegger.* Bloomington: Indiana UP, 1990.
—. Foreword. Haar. *The Song of the Earth: Heidegger and the Grounds of the History of Being.* By Michel Haar. Trans. Reginald Lilly. Bloomington: Indiana UP, 1993. xi—xiii.
—, ed. *Reading Heidegger.* Bloomington: Indiana UP, 1993.
—. "Toward the Showing of Language." *Southwestern Journal of Philosophy* 4 (1973): 75–84.
Santayana, George. *Realms of Being.* New York: Cooper Square, 1972.
—. *Scepticism and Animal Faith.* New York: Dover, 1955.
Sartre, Jean Paul. "Existentialism is a Humanism." *Existentialism from Dostoevsky to Sartre.* Ed. Walter Kaufmann. Cleveland: World, 1956. 287–311.
Schiappa, Edward. *Protagoras and Logos: A Study in Greek Philosophy and Rhetoric.* Columbia: U of South Carolina P, 1991.
—. Rhetorike: What's in a Name? Toward a Revised History of Early Greek Rhetorical Theory." *Quarterly Journal of Speech* 78 (1992): 1–15.
—. "Sophistic Rhetoric: Oasis or Mirage?" *Rhetoric Review* 10 (1991): 5–18.
Scholes, Robert. *Textual Power: Literary Theory and the Teaching of English.* New Haven: Yale UP, 1985.
Schrag, Calvin O. *Existence and Freedom.* Evanston: Northwestern UP, 1961.
—. *Experience and Being.* Evanston: Northwestern UP, 1969.
—. *Radical Reflection and the Origin of the Human Sciences.* West Lafayette: Purdue UP, 1980.
—. "Rhetoric Resituated at the End of Philosophy." *Quarterly Journal of Speech* 71 (1985): 164–74.
—. *Communicative Praxis and the Space of Subjectivity.* Bloomington: Indiana UP, 1986.
Segal, Charles P. "Gorgias and the Psychology of Logos." *Harvard Studies in Classical Philology* 66 (1962): 99–155.

Seidel, George Joseph. *Martin Heidegger and the Pre-Socrates*. Lincoln: U of Nebraska P, 1964.
Sherover, Charles M. *Heidegger, Kant and Time*. Bloomington: Indiana UP, 1971.
Sipiora, Phillip. "*Kairos*: The Rhetoric of Time and Timing in the New Testament." *Rhetoric and Kairos: Essays in History, Theory, and Praxis*. Ed. Phillip Sipiora and James S. Baumlin. Albany: SUNY P, 2002. 114–27.
Sipiora, Phillip, and James S. Baumlin, ed. *Rhetoric and Kairos: Essays in History, Theory, and Praxis*. Albany: SUNY P, 2002.
Smith, John E. "Time and Qualitative Time." *Rhetoric and Kairos: Essays in History, Theory, and Praxis*. Ed. Phillip Sipiora and James S. Baumlin. Albany: SUNY P, 2002. 46–57.
Snell, Bruno. "*Die Sprache* Heraklits." *Hermes* 61 (1926): 353–81.
Song of the South. Dirs. Harve Foster and Wilfred Jackson. Perf. James Bassett. Disney, 1946. Film.
Sonnemann, Ulrich. *Existence and Therapy: An Introduction to Phenomenological Psychology and Existential Analysis*. New York: Grune & Stratton, 1954.
Sprague, Rosamond Kent, ed. *The Older Sophists: A Complete Translation by Many Hands of the Fragments in* Die Fragmente Der Vorsokratiker. Columbia: U of South Carolina P, 1972.
Stambaugh, Joan. Introduction. *Identity and Difference*. By Martin Heidegger. Trans. Joan Stambaugh. New York: Harper & Row, 1979. 7–18.
Steiner, George. *After Babel: Aspects of Language and Translation*. Oxford: Oxford UP, 1975.
—. *Extraterritorial: Papers on Literature and the Language Revolution*. New York: Atheneum, 1976.
—. *Martin Heidegger*. New York: Penguin, 1978.
Steinhoff, Virginia N. "The *Phaedrus* Idyll as Ethical Plato: The Platonic Stance." *The Rhetorical Tradition and Modern Writing*. Ed. James J. Murphy. New York: MLA, 1982.
Sutton, Jane. "The Death of Rhetoric and Its Rebirth in Philosophy." *Rhetorica* 4 (1986): 203–26.
Swearingen, Jan C. "The Rhetor as Eiron: Plato's Defense of Dialogue." *PRE/TEXT* 3 (1982): 289–335.
Thompson, George. "From Religion to Philosophy." *The Journal of Hellenic Studies* 73 (1953): 77–83.
Tiles, J. E. "*Techne* and Moral Expertise." *Philosophy* 59 (1984): 49–66.
Tillich, Paul. *A History of Christian Thought*. Ed. Carl Braaten. New York: Simon & Schuster, 1967.
—. *The Interpretation of History*. Trans. N. A. Rasetzki and Elsa L. Talmey. New York: Schribners, 1936.
—. *Systematic Theology*. Vols. 2 & 3. Chicago: U of Chicago P, 1963.

Tint, H. "Heidegger and the 'Irrational.'" *Proceedings of the Aristolelian Society* 57 (1957): 253–65.
Untersteiner, Mario. *The Sophists*. Trans. Kathleen Freeman. Oxford: Alden, 1954.
Vail, L.M. *Heidegger and Ontological Difference*. University Park: Penn State UP, 1972.
Van Hook, LaRue. "The Encomium on Helen, by Gorgias." *The Classical Weekly* 6 (1913): 122–23.
Vatz, Richard E. "The Myth of the Rhetorical Situation." *Philosophy and Rhetoric* 6 (1973): 148–61.
Velie, Alan. "The Trickster Novel." *Narrative Chance: Postmodern Discourse on Native American Indian Literatures*. Ed. Gerald Vizenor. Norman: U of Oklahoma P, 1993. 121—139.
Versenyi, Laszlo. *Socratic Humanism*. New Haven: Yale UP, 1963.
Vitanza, Victor J. *Negation, Subjectivity, and the History of Rhetoric*. Albany: SUNY P, 1997.
Vivante, Paolo. *Homer*. New Haven: Yale UP, 1985.
Vizenor, Gerald, ed. *Narrative Chance: Postmodern Discourse on Native American Indian Literatures*. Norman: U of Oklahoma P, 1993.
—. "Trickster Discourse: Comic Holotropes and Language Games." *Narrative Chance: Postmodern Discourse on Native American Indian Literatures*. Ed. Gerald Vizenor. Norman: U of Oklahoma P, 1993. 187–211.
Von Humboldt, Wilhelm. *Schriften zur Sprache*. Stuttgart: Philipp Reclam, 1973.
Ward, Geoffrey C. *The Civil War: An Illustrated History*. New York: Alfred A Knopf, 1990.
Wartofsky, Marx. *Conceptual Foundations of Scientific Thought: An Introduction to the Philosophy of Science*. New York: MacMillan, 1968.
Weaver, Richard. *The Ethics of Rhetoric*. South Bend: Regnery/Gateway, 1953.
Wedberg, A. "The Theory of Ideas." *Plato: A Collection of Critical Essays*. Vol. 1. *Metaphysics and Epistemology*. Ed. Gregory Vlastos. Garden City, NY: Anchor, 1971. 28–52.
Weiss, Helene. "The Greek Conceptions of Time and Being in the Light of Heidegger's Philosophy." *Philosophy and Phenomenological Research* 2 (1941): 173–87.
Welsh, Kathleen E. "The Platonic Paradox: Plato's Rhetoric in Contemporary Rhetoric and Composition Studies." Rhetoric Society of America. Arlington, TX. 15 June 1986. Reading.
Wheeler, Everett L. *Mnemosyne: Stratagem and the Vocabulary of Military Trickery*. New York: E. J. Brill, 1988.
White, Eric Charles. *Kaironomia: On the Will-to-Invent*. Ithaca: Cornell UP, 1987.

Wills, Garry. *Lincoln at Gettysburg: The Words that Remade America*. New York: Simon and Schuster, 1992

Wold, Henry G. *Plato and Heidegger: In Search of Selfhood*. East Brunswick: Buckle UP, 1981.

Wolz, Henry G. *Plato and Heidegger: In Search of Selfhood*. Lewisburg, PA: Bucknell UP, 1981.

Worsham, Lynn. "The Question Concerning Invention: Hermeneutics and the Genesis of Writing." *PRE/TEXT* 8.3 (1987): 197–243.

Wyschogrod, Michael. *Kierkegaard and Heidegger*. London: Routledge & Kegan Paul, 1954.

Zimmerman, Michael E. *Heidegger's Confrontation with Modernity: Technology, Politics, and Art*. Bloomington: Indiana UP, 1990.

—. *Eclipse of the Self: The Development of Heidegger's Concept of Authenticity*. Athens: Ohio UP, 1981.

Index

Achilles, 184, 244, 339
aho, 90, 123, 156, 210-211
aletheia, 58, 116, 143, 192, 218, 270, 280, 294-295, 311-312, 345-346, 358
anamnesis, 99, 106, 110
ananke, 96
Anaxagoras, 89
Andecken (memory), 202
Andromache, 127, 131
aneu logou, 8, 15, 17, 20, 44-45, 52, 55-56, 63-64, 104, 106, 151, 323, 330, 341
Angel, Ernst, 329
animal rationale, 153
Anthanasius, 95
antithesis, 242
apate (deception), 16-17, 58, 93, 99, 120, 170, 184, 189, 276-278, 296-297, 300
Aquinas, 21, 42
Arendt, Hannah, 106, 323, 330, 343
arete, 129, 131, 133
Aristophanes, 47
Armstrong, A. MacC., 36-37
Atwell, Janet, 268, 270
Augenblick (blink of an eye), 26-27, 119-120, 167, 221-223, 233, 277, 290-292, 319-320
autochthony, 10, 32-33, 286, 304, 315

Bach, Johann Sebastian, 338
Ballif, Michelle, 16
Bambach, Charles, 33, 304
Barnes, Jonathan, 334
Baumlin, James, 70-71
Beaufret, Jean, 243
Befindlichkeit, 139, 154, 156
Ben Tre, 199, 351
Bernasconi, Robert, 325-326
Bizzell, Patricia, 324
Black Elk, 209, 328, 343-344
Black, Edwin, 32, 64, 209, 328, 331, 343-344
Blut und Boden (blood and earth), 302, 304
Bourdieu, Pierre, 55, 149, 330
Brock, Werner, 146
Bruns, Gerald, 117-118, 226, 238, 261, 346, 351, 356
Bruzina, Ronald, 315, 350
Bryant, Donald C., 142
Buber, Martin, 353
Burnet, John, 53, 87-88
Burns, Ken, 130-131, 133-134, 254; *The Civil War*, 130, 343
Butcher, S. H., 328
Butler, H. E., 333
Byatt, A. S., 330

Callicles, 75
Cambrey (France), 41, 204
Campbell, Joseph, 216, 305
Camus, Albert, 145, 282, 346
Caputo, Philip, 337

379

Carter, Michael, 71-72, 333
Cascardi, Anthony, 8, 19, 112, 325
Cassirer, Ernst, 50-52, 136, 356
Chamberlain, Joshua Lawrence, 187-189, 233; Little Round Top, 187-188, 233, 304
Chatman, Seymour, 344
chronos, 38, 40, 42, 59, 290, 328, 333
Cicero, 14, 35, 46, 66, 68-71, 333
Claus, David, 89, 107
Cleve, Felix, 53, 112, 332
Consigny, Scott, 16, 94, 102, 324, 327, 334, 337
Cook, Arthur Bernard, 36
Copernican Revolution, 5, 11, 321
Cornford, Francis Macdonald, 53, 87, 101
Coulter, James, 335
Crusius, Timothy, 178
Crassus, 68
Crowley, Sharon, 60, 141, 345
Custer, George Armstrong, 180, 206, 217

Dastur, Francoise, 171, 196, 230, 263, 289, 292, 298
de Man, Paul, 170
de Romilly, Jacqueline, 18, 45, 96, 109-110, 338-340, 350
de Vogel, C. J., 18
de-cision, 25, 191, 223-227, 230, 233, 273-275, 280, 285, 290, 292, 300-302, 307-308, 312; *Ent-scheidung* (de-cision), 25, 223
decorum, 14, 17, 35, 44, 67-71, 212, 232, 319; *to prepon* (decorum), 35, 67-68, 71
deep remembrance, 41, 204-205
Deloria, Vine, 208; *God is Red*, 208
DeMallie, Raymond, 353
Demosthenes, 10

Derrida, Jacques, 35, 53, 60, 102, 113, 229, 280-281, 291, 322-323, 341, 350
Descartes, Rene, 54, 142, 322, 346
Diamond, Stanley, 355
Diels, Hermann Alexander, 50, 75, 77, 84, 222, 239, 332-335, 337-339
Dieter, Otto, 71, 333
dike, 272-274, 294, 358
Dillon, John, 337-338, 343
Diodorus Siculus, 75, 95
Diogenes Laertius, 358
Diomedes, 88
Disraeli, Benjamin, 72
dissoi logoi, 65, 78-79
Durant, Will, 353

eidos (Platonic forms), 35, 52, 208, 332
Eigentlichkeit (authenticity), 163
ekplexis (turmoil), 115
Eleatics, 77
Elias, Julius, 331
Ellenberger, Henri, 39, 329
Empedocles, 77-78, 97, 335
Engell, Richard, 19, 324, 337
Enos, Richard, 45, 66, 79, 287-288, 337, 349
Entruckung (rapture), 290
episteme (true knowledge), 56, 63, 93, 100, 121, 125, 138, 148, 351
Ereignis (appropriation), 140, 224, 228
Eshu (Elegbara), 30, 237, 245-246, 249-250, 259-261, 264-265, 279, 286, 290, 296-298, 300, 306, 317, 353-354, 357
existential psychotherapy, 39-40, 329

FitzGerald, Frances, 150
Foti, Veronique, 357-358

Frankfort, Henri, 51
Freeman, Kathleen, 306, 332-333, 335-336, 343, 345
Funaioli, Gino, 46, 66-67
fundamental ontology, 176, 191
funeral oration, 10-11, 28, 82-83, 94-95, 114, 181-183, 187, 189, 210, 302, 304, 338; *epitaphios logos*, 10
Furley, D. J., 87-89, 336

Gadamer, Hans-Georg, 24, 95, 98, 118, 123, 158, 191, 204, 238, 257-258, 279-280, 314, 325, 335, 356-357
Gage, John, 103
Gaonkar, Dilip, 345
Gates, Henry Louis, 245, 279, 306, 354
Gay, Peter, 327
gegenuber, 146
George, Stefan, 15, 21, 46, 51, 57, 94, 121, 172, 179, 184, 239, 289, 315, 324, 333, 338
Gerede (idle talk), 174, 198, 205, 225, 250
Gettysburg, 133, 180, 183-187, 346-347, 351
Gettysburg Address, 133, 183, 186, 347
Geworfenheit (thrown), 139
Glaukos, 130
Glover, Carl, 328
Gomperz, Theodor, 53, 334-335
Gospel of Judas, 329
Graeser, Andreas, 58
Gramm, Kent, 186, 351
Grassi, Ernesto, 23, 53
Gray, J. Glenn, 257
great *dynastes*, 91-93, 96, 98, 164, 168, 178, 218, 224, 229, 261, 265-266, 271, 275-276, 278, 296, 316, 319

Greek Revival, 184
Gronbeck, Bruce, 94, 259, 337
Guignon, Charles, 157, 258
Guthrie, W. K. C., 34, 72-73, 107, 349

Haar, Michael, 26, 293, 300, 350
Halloran, S. M., 134
Hamlet, 118, 182, 185, 211
Harries, Karsten, 155-156, 293
Harris, Joel Chandler, 354
Havelock, Eric, 9, 48-50, 54, 59, 95, 100, 107, 151, 331, 332
hearkening, 26, 227
Hector, 88, 128-131, 135, 138-139, 184, 233, 339
Heilsgeschichte (salvation history), 39
Helen of Troy, 18, 32, 77, 91, 93, 115, 135, 164, 248-249, 272, 334, 354-355
Hellman, Lillian, 134
Heraclitus, 1, 50-51, 53, 57, 81, 89, 117, 123, 129, 143, 160, 164, 166, 169, 198, 208, 230, 242-243, 246, 291, 325, 338, 343, 345
Hercules, 244, 354
Hermagoras, 71
hermeneutic circle, 307, 310, 312, 321, 327
Hermes, 244, 260, 354
Herzberg, Bruce, 324
Hesiod, 7, 47, 66, 327, 355, 358
heuristic *angst*, 162
heyokas, 328
Hill, Douglas, 37, 328
Hippasus, 241, 250
Hölderlin, Friedrich, 327
homo mensura docrtrine, 26, 77, 140, 141-146, 345
homoioteleuton, 94, 338

382 *Index*

house of Being, 9-10, 196, 250, 258, 267
Hughes, Bettany, 327, 354-355
Hühnerfeld, Paul, 327
Hunt, Everett Lee, 324, 331
Husserl, Edmund, 21, 330, 341, 356
Huxley, Aldous, 147
Hyperenor, 88
Hyde, Lewis, 243, 245, 297-298, 354-355
Hyde, Michael, 115, 192, 325

Ijselling, Samuel, 56
Iliad, 88, 90, 92, 116, 127-129, 133, 155, 179, 186, 233, 336, 339
immanence, 6, 9-10, 50, 59, 81, 85, 116, 121-122, 127, 137-139, 143, 156, 160, 176, 178, 190, 196, 201, 234, 248, 260, 274, 276, 281, 284, 289, 293-294, 298, 305, 308, 315, 319, 322
irruption, 190-191, 225, 229, 288, 300-302, 356

Jarratt, Susan, 16
Jaeger, Werner, 359
Jaynes, Julian, 228, 339
Johnson, Barbara, 5, 321
Johnstone, Henry, 23, 247, 325

Kairos (mythological figure), 12-13, 26, 29-30, 34, 36-38, 43, 46, 60, 65-67, 74, 109, 166, 232, 233, 237, 279, 290, 319-320, 328, 333
Kant, Immanuel, 4-5, 11, 21, 310, 321-322, 338
Kaufmann, Walter, 331, 339
Kazantzakis, Nikos, 329
Kelman, Harold, 39-40, 43

Kennedy, George, 15, 46, 66-68, 73, 94-96, 324, 331, 333, 338-339, 343
Kerenyi, Karl, 354
Kerford, G. B., 58, 335
Kermode, Frank, 42-43, 330
kinesis, 71
King, Martin Luther, 128, 134, 303
Kinneavy, James, 20, 44, 46, 63, 65-67, 73, 324-325, 328, 330, 333
Kisiel, Theodore, 114, 161, 201, 208, 325
Kittel, Gerhard, 328
Kitto, H. D. F., 128
Kluback, William, 21, 24
Kockelmans, Joseph, 329-330, 350
Koestler, Arthur, 359
koine, 328
koyemshi, 37, 328
Kranz, Walther, 222, 332
Krell, David Farrell, 179
Krupat, Arnold, 347-348
Kung, Hans, 38
Kurosawa, Akira, 300; *Rashomon* 300

Lakota, 217, 328, 343, 352
Lattimore, Richard, 336
Lauer, Janice, 359
Lebenswelt (life-world), 6
Leff, Michael, 61, 108-110, 113, 340
legein, 194-195, 198-199, 262, 270
Legia, 306
Lichtmetaphysik, 53
Lilly, Reginald, 239
Lincoln, Abraham, 133-134, 183-189, 233, 303, 343, 346-347, 351
lumin naturale, 53

Macquarrie, John, 342
das Man ("they"), 174, 220
Manheim, Ralph, 321, 343
Marcel, Gabriel, 40, 171, 226
Marx, Werner, 152, 158, 172, 335, 354
May, Rollo, 329
McComiskey, Bruce, 16
McNeley, James, 90
Meade, George, 183
Menelaus, 88
Merleau-Ponty, Maurice, 179
Merton, Thomas, 351
metaphysics of presence, 8, 28, 35, 60, 68, 157, 176, 211, 223, 323
Miller, Cassandra (Cassie), 180-181
Minar, Edwin, 50
Moline, John, 105-106
Momaday, N. Scott, 1, 150-152, 205-207, 210, 212-215, 217-218, 220, 231, 307, 337, 347-348, 352-353; Abel, 211-212, 214-215, 217-218, 224, 227, 231-233, 307; Francisco, 212, 214-215, 227, 231, 233, 307; *House Made of Dawn*, 207, 211, 214, 218, 220, 224, 231, 353; John "Big Bluff" Tosamah, 212-214, 352
Moon, Sheila, 239
Morrow, Glenn, 64, 331
Mugerauer, Robert, 341-342
Murphy, James, 331

native soil, 9-10, 13, 28, 32, 83, 114, 140, 151, 184, 187, 204, 303-305, 310, 315
Navaho, 90, 156, 211
Neihardt, John, 209
Nietzsche, Friedrich Wilhelm, 3, 9, 32, 95, 136, 142, 160, 185, 261, 282, 289, 313-314, 323, 338, 339, 345-357

nonhuman, 144, 158, 162-164, 189, 235, 242, 264-265, 271-272, 280, 283-289, 294, 298-301, 305, 307, 309, 312, 316, 318-319, 335, 355
nunc stans, 59, 72, 291

O'Brien, Tim, 344, 347, 352
O'Connor, Flannery, 294
Odysseus, 78, 89, 130, 138, 244
Oedipus, 9, 137-138
Okrent, Mark, 321
Olafson, Frederick, 139, 154, 156
Ong, Walter, 48-50, 58, 151, 157, 332
ontological difference, 120-122, 124, 126, 156, 165, 185, 200, 223-224, 230-231, 251-253, 257, 263-264, 272, 275-276, 286-287, 291, 299, 307, 342, 356
ontotheological, 59
Oriki Esu, 237, 353
Orisha, 353
Orozco, Jose Clemente, 308-309, 311-313, 315
Osborne, Catherine, 323
ousia, 50, 56-57, 116, 159, 219, 349

paideia, 47, 76, 266, 302, 359
palinode, 135
Pandaros, 88
para-*doxa*, 30, 37, 83, 115, 235, 237-238, 276, 292, 298-299, 308, 313, 316, 319, 351
Parain, Bruce, 111, 113
parison, 94
Parkes, Graham, 352
Parmenides, 34, 62, 71, 77-78, 81, 111-112, 143, 239, 285, 287-288, 295, 301-302, 316, 335, 344, 359

Partee, Morris, 45, 57
Passio, 253
peitho, 16-17
Pelton, Robert, 244, 253, 260, 306-307, 317, 354
perceptio, 136
peri hekaston ("in any given case"), 330
Pericles, 10, 184, 186, 338
phainesthai, 122, 125, 160, 345
pharmakon, 97, 102
piety of thinking, 258-259, 262, 356
Pindar, 14, 66, 123, 358
Pirsig, Robert, 45-46, 127-129, 144, 147-148, 330-331
Platonic Stance, 103, 109
Plutarch, 75, 339
Pöggeler, Otto, 123, 218, 222
poiesis, 136, 270-272, 357-358
Poulakos, John, 16, 21, 44, 87, 110, 141, 247-249, 266, 325, 350, 354
Poulet, Georges, 41
Pratt, Louise, 100, 358
pre-Socratics, 21, 34, 53, 78, 87, 89-90, 117, 123, 138, 142, 146, 176-177, 205, 269, 332-333, 338, 345, 348-349
Priam, 128
primary rhetoric, 19-20
prison house of language, 92, 112, 170, 175, 267
Professor Kantorek, 327
propriety, 14, 35, 44, 62-63, 67-70, 233, 319
Protagoras, 77, 140, 142, 168-169, 331, 335, 344-345
Proust, Marcel, 40-42, 204, 220, 329
psychagogue, 60, 86, 107
Ptolemy, 5
Puech, Henri-Charles, 329

Pythagoreans, 53, 66, 87, 240-242, 250, 354

quiet power of the possible, 161, 167, 248-250, 256-257, 259, 261, 267, 270, 274, 284, 308, 318
Quimby Rollin, 331
Quintilian, 14, 35, 46, 68-71, 333

racial memory, 150-151, 188, 206-207, 210, 212, 214-215, 217-218, 231, 347-348
Radin, Paul, 328, 354
Rede (authentic speech), 51, 197, 198
Remarque, Erich, 327
res, 3, 5, 17-19, 25, 32, 42, 45-46, 50-51, 66, 68-69, 71, 73, 82, 92, 105, 111, 124, 129, 172, 206, 219, 237
rhapsode, 52
Richards, I. A., 135
Richardson, William, 21, 121, 143, 161, 191, 197, 224-225, 227, 289, 326, 342, 348, 353, 356-357
Ricoeur, Paul, 153, 197, 326, 331
Roberts, W. Rhys, 45
Rodhe, Erwin, 87-88, 102, 336, 337, 339
Rorty, Richard, 12, 23, 32
Rosen, Stanley, 110-111, 121
Rosenmeyer, Thomas, 18, 81, 115, 337
Rostagni, Augusto, 66-67
Rouse, W. H. D., 336
Russell, Bertrand, 73

sacrifice, 11, 82, 128, 133-134, 163, 184, 187, 189, 204, 206, 210, 216, 231, 232-233, 303-306, 327, 329

Sagen (Saying), 151, 197, 219
Sallis, John, 293, 304
Santayana, George, 46, 54, 102, 339
Sarpedon, 130, 138, 184, 233, 304
Schiappa, Edward, 168, 327, 344-345
Scholes, Robert, 321
Schrag, Calvin, 23-24, 93, 112, 121, 163, 176, 325
secondary rhetoric, 15, 20, 23
Segal, Charles, 18, 25, 86, 91, 96-97, 115, 138, 200, 201, 334, 337
Seidel, George Joseph, 21, 57, 146, 342, 348
Seienden (things), 120-121
Sein (Being), 120-121, 143, 161, 342
Seinsfrage (question of Being), 345
Seven Sages, 14
Sipiora, Phillip, 328
Smith, John, 73, 325
Snell, Bruno, 51, 89
Socrates, 34, 48, 52-53, 55-57, 66, 76, 95, 99, 101, 107, 109-111, 113, 135, 140, 167-168, 275, 324, 333, 335, 338, 340
Solmsen, Friedrich, 45
Solon, 14, 55
Spartoi, 310
St. John of the Cross, 201
Stambaugh, Joan, 200
Standing Rock (Indian Reservation), 217
stasis, 35, 70-72, 110, 208, 291, 308, 333
Steiner, George, 100, 121-122, 162, 179, 180, 184-185, 239, 289, 296, 305, 315, 338, 353
Steinhoff, Virginia, 103, 107, 109
Stesichorus, 135, 138, 276, 277
Sutton, Jane, 15
Swearingen, Jan, 110

techne, 25, 28, 30, 57, 93-94, 96-98, 103, 107, 115, 167-171, 238, 250, 254, 259, 261-266, 268-279, 294, 296, 310-316, 319, 324, 338, 350, 356-358
techne of *apate*, 94, 96, 259, 266, 271
thanc, 202, 205, 209, 211, 215-216, 234, 257, 269, 271
thaumazein, 280, 322, 343
things-for-us, 4-6
things-in-themselves, 4-6, 192
Thompson, George, 51, 94, 338
Thrasymachus, 75
Tiles, J. E., 268, 338
Tillich, Paul, 38-39, 43-45, 223, 328-330
Tint, H., 348
Tisias, 96, 124
Tolstoy, Leo, 40
tragedy of knowledge, 80-81, 83, 85, 102, 162, 300-301, 308, 334
Trakl, Georg, 358
trickster, 29-30, 37, 212-213, 236, 239-240, 243-250, 252-256, 258-268, 271, 273, 277-279, 282, 284-286, 290, 292, 294, 296-300, 302, 304, 306-308, 312-313, 316-319, 324, 328-329, 347, 352-355, 357

uncanny, 30, 36-37, 126, 160, 162, 188, 190, 230, 235, 238, 242, 265-268, 271, 275, 280, 282, 283-290, 293-299, 302, 306-307, 317-318, 359; *Unheimliche* (uncanny), 282
Unter-Scheid (difference), 223
Untersteiner, Mario, 12, 16, 18-19, 46, 66, 67, 71, 73-74, 78-79, 81, 83, 85-86, 99, 113, 139, 159, 162, 223, 272, 276-278, 297, 306, 324, 334-337, 345, 349, 358

Vail, L. M., 197, 222
Van Gogh, Vincent, 309
Van Hook, LaRue, 337-338
verba, 17-19, 23, 25, 45, 46, 50, 58-59, 68-69, 73, 82, 92, 111, 219, 233
veritas, 58, 113, 116, 192, 295, 309, 311, 346
Versenyi, Laszlo, 78, 96-97, 334, 349
Vitanza, Victor, 16, 324, 327, 334
Vivante, Paolo, 155
von Humboldt, Wilhelm, 51, 258, 260, 356

Walking Buffalo, 208
Wartofsky, Marx, 354
Wasichu, 328, 352-253
Weaver, Richard, 61
Wedberg, A., 340
Weiss, Helene, 59
Welsh, Kathleen, 324

Wheeler, Everett, 100
White, Eric, 16, 333-334
Whitehead, Alfred North, 34
Whitman, Walt, 351
Wilde, Jean, 21, 24
Wills, Garry, 183-184, 187, 189, 346-347
Winnebago Cycle, 243
Wittgenstein, Ludwig, 338
Wolz, Henry, 332
Worsham, Lynn, 325, 350

xa, 150-151, 188, 193-194, 196, 198-199, 202-204, 210, 249, 284, 347

Yoruba, 30, 245, 260, 296, 298, 317, 353-354

Zeno, 77-79
Zimmerman, Michael, 160, 195, 326

About the Author

Bernard Miller was born and raised in North Dakota and attended universities there for undergraduate and graduate degrees. Between degrees he was drafted into the US Army, spent a year at the Defense Language Institute at Fort Bliss, Texas, studying the Vietnamese language. He was sent to Vietnam and upon his return taught for three years at a community college on an Indian reservation in North Dakota and for three years at a state university in Minnesota. He obtained his PhD from Purdue University in 1987 and since that time has taught courses in writing, American Indian literature, and freshman composition at Eastern Michigan University. His research and publications have dealt primarily with rhetorical theory, with an emphasis on cross-cultural studies and the various connections between pre-Platonic and postmodern thought. His current interest concerns the relationship of words and war, and he is currently working on a book-length manuscript titled *The Rhetoric of War: Words as Power and Betrayal*.

www.ingramcontent.com/pod-product-compliance
Lightning Source LLC
Chambersburg PA
CBHW031702230426

43668CB00006B/85